Low Voltage Electrician Certification Training Course

低压电工考证培训教程

杨清德　余明飞　孙红霞　主编
Qingde Yang　Mingfei Yu　Hongxia Sun

黄国平　王丽红　俞　娜　译
Guoping Huang　Lihong Wang　Na Yu

化学工业出版社

· Beijing ·

This book is the English version of the *Low Voltage Electrician Certification Training Course (Video version)* (ISBN: 978-7-122-35692-5). It includes 12 chapters: safety production instructions for electricians safety laws and regulations, basic knowledge of electrician, electrical tools and safety signs, low voltage electrical apparatus, asynchronous motor, electrical line, electrical lighting, power capacitor, use of safety tools, safe operation technology, elimination of potential safety hazards at the operation site, and emergency response at operation site. Furthermore, the examination question bank is also attached. It has all the characteristics of the new form of teaching materials.

图书在版编目（CIP）数据

低压电工考证培训教程＝Low Voltage Electrician Certification Training Course：英文/杨清德，余明飞，孙红霞主编；黄国平，王丽红，俞娜译. —北京：化学工业出版社，2023.8
ISBN 978-7-122-43291-9

Ⅰ.①低… Ⅱ.①杨… ②余… ③孙… ④黄… ⑤王… ⑥俞… Ⅲ.①低电压-电工-岗位培训-教材-英文 Ⅳ.①TM08

中国国家版本馆CIP数据核字（2023）第065096号

责任编辑：高墨荣　刘丽宏　李佳伶　　　装帧设计：刘丽华
责任校对：宋　夏

出版发行：化学工业出版社（北京市东城区青年湖南街13号　邮政编码100011）
印　　刷：三河市航远印刷有限公司
装　　订：三河市宇新装订厂
710mm×1000mm　1/16　印张26　字数524千字　2023年9月北京第1版第1次印刷

购书咨询：010-64518888　　　　　　　　　　售后服务：010-64518899
网　　址：http://www.cip.com.cn
凡购买本书，如有缺损质量问题，本社销售中心负责调换。

定　　价：298.00元　　　　　　　　　　　　　　　　版权所有　违者必究

Preface

Working with certificates for special operations is a mandatory requirement of *The Production Safety Law of the People's Republic of China, the Regulations on the Administration of Safety Technology Training and Assessment for Special Operation Personnel* (Order No. 30 of the State Administration of Work Safety) and other local laws and regulations, which is related to safe production and the safety of the national economy and people's lives and property.

The Low-voltage Electrician Certificate is one of the most widely used and representative one, but it currently has the following problems:

① Although there are hundreds of thousands of foreign employees in China, there are many deficiencies and blanks in relevant training and certification, one of which is the lack of textbooks;

② With the new *Vocational Education Law of the People's Republic of China* as the prelude, the reform of vocational education in the new era would go overseas: the recruitment and training of foreign students in vocational education has been in full swing, but the teaching materials are almost blank;

③ The anti-globalization originated by the west would accelerate the "going overseas" of China's industry and production capacity, especially for BRI countries, which would inevitably be accompanied by the "going overseas" of technology and standards, but the relevant carrier medias are extremely limited or even blank.

Therefore, the translating team hopes to help solve the above problems through the publishing of this book.

The translation of this book has received strong support from all parties: Prof. Qingde Yang, the chief editor of the Chinese version, has made a decision to support this job from stranger to full support through a 5-minute phone call, and has provided countless direct and comprehensive help throughout the translation process. After knowing that the funds were limited, Prof. Yang and his team immediately decided to waive the copyright fee. Their noble character and integrity contributed to the publication of the book; Mr. Gang Li, secretary of the Automobile Engineering School of Beijing Polytechnic, and Mr. Weijie Gou, director of the Automobile Manufacturing Technology Department of the same college, provided support in the whole process from project approval, funds, and until solving other difficulties in the process, making the publication of the book possible; Ms. Na Yu, the senior manager of the training department of Beijing Benz Automobile Co., Ltd., immediately provided various supports and volunteered to become a member of the translating team after knowing this job, sharing a lot of translation and proofreading work. I would like to express my sincere thanks to the numerous leaders, colleagues and friends who have given us help.

The translation and publishing time of the book is extremely tight, and the translating members have their own main jobs, all the jobs are completed in spare time, which is very heavy and urgent even if they are full-time. Here, I also pay high tribute to the members of the translating team.

Due to the limited capability of translation and the limited time, it is inevitable that there will be some mistakes or omissions. We hope our potential readers could give us your advice (qinhe848@163.com), with a view to revising it when it is republished.

Guoping Huang (Mr.)

Contents

Chapter 1 Safety Production Instructions for Electricians 001

1.1　Laws and regulations on safety production 002
 1.1.1　*Regulations on Management of Safety Technology Training and Assessment for Special Operation Personnel* 002
 1.1.2　*The Production Safety Law of the People's Republic of China* 003
 1.1.3　Other laws and regulations 006
1.2　Safety protection and electric shock prevention measures 008
 1.2.1　Organizational measures to ensure the safety of electricians 008
 1.2.2　Technical measures to prevent electric shock 011
1.3　Electrical accident and first aid for electric shock 019
 1.3.1　Harm of electric current to human body 019
 1.3.2　Causes and types of electric shock 021
 1.3.3　Electric shock mode 022
 1.3.4　Electrical accident 024
 1.3.5　First aid operation for electric shock 025
1.4　Electrical fire protection, explosion-proof, lightning protection and anti-static 029
 1.4.1　Electrical fire and explosion prevention 029
 1.4.2　Electrical lightning protection and anti-static 032

Chapter 2 Basic knowledge of electrical safety technology 035

2.1　Basic knowledge of direct current (DC) circuit 036
 2.1.1　Circuit and its basic physical quantities 036
 2.1.2　Ohm's Law 039
 2.1.3　Series and parallel connection of resistance 040
 2.1.4　Kirchhoff's Law and its application 042
 2.1.5　Series and parallel connection of capacitors 043

2.2 Basic electromagnetic knowledge 044
- 2.2.1 Magnetic field of current.. 044
- 2.2.2 Basic physical quantity of magnetic field 046
- 2.2.3 Left-hand rule.. 047
- 2.2.4 Electromagnetic induction phenomenon and Lenz's Law 048
- 2.2.5 Self-inductance and mutual inductance 050

2.3 Fundamentals of alternating current (AC).. 051
- 2.3.1 Single phase sinusoidal AC.. 051
- 2.3.2 Three phase AC.. 054

2.4 Fundamentals of electronic technology 061
- 2.4.1 Crystal diode and rectifier circuit.. 061
- 2.4.2 Crystal triode and its amplification circuit.. 066
- 2.4.3 Voltage stabilizing circuit.. 070

Chapter 3 Electrical tools and safety signs 073

3.1 Use of electrical tools and safety tools 074
- 3.1.1 Use of common electrical tools.. 074
- 3.1.2 Use of other electrical tools 077
- 3.1.3 Use of manual electric tools... 080
- 3.1.4 Use of electrical safety appliances... 082

3.2 Use of common electrical instruments.. 083
- 3.2.1 Use of multimeter 083
- 3.2.2 Use of clamp ammeter 094
- 3.2.3 Use of insulation resistance meter 097
- 3.2.4 Use of electric energy meter 100
- 3.2.5 Use of ammeter 102
- 3.2.6 Use of voltmeter.. 104

3.3 Electrical safety signs 105
- 3.3.1 Electrical safety color 105
- 3.3.2 Electrical safety sign board.. 106

Chapter 4 Low-voltage electrical apparatus...................... 109

4.1 Basic knowledge of low-voltage apparatus 110

4.1.1	Classification of low-voltage apparatus	110
4.1.2	Common terms of low-voltage apparatus	113
4.1.3	Selection and installation of low-voltage electrical apparatus	114
4.2	**Low-voltage circuit breaker and leakage protector**	**116**
4.2.1	Low-voltage circuit breaker	116
4.2.2	Leakage protector	118
4.3	**Contactor and Relay**	**119**
4.3.1	Contactor	119
4.3.2	Relay	122
4.4	**Fuse and master switch**	**128**
4.4.1	Fuse	128
4.4.2	Master switch	130

Chapter 5 Asynchronous motor ... 133

5.1	**Basic knowledge of asynchronous motor**	**134**
5.1.1	Brief introduction to motor	134
5.1.2	Structure and principle of asynchronous motor	138
5.1.3	Characteristics of asynchronous motor	142
5.2	**Starting and running of asynchronous motor**	**145**
5.2.1	Starting of asynchronous motor	145
5.2.2	Running of asynchronous motor	150
5.3	**Fault inspection and maintenance of asynchronous motor**	**152**
5.3.1	Fault inspection and maintenance of three phase cage asynchronous motor	153
5.3.2	Fault inspection and maintenance of three phase wound asynchronous motor	157

Chapter 6 Electrical line ... 159

6.1	**Types and characteristics of electrical line**	**160**
6.1.1	Overhead line	160
6.1.2	Cable line	163
6.1.3	Indoor line	165
6.2	**Safety of electrical lines**	**165**
6.2.1	Conductivity	165
6.2.2	Mechanical strength	166

6.2.3	Spacing	167
6.2.4	Identification and connection of conducting wires	168
6.2.5	Line protection	175
6.2.6	Line management	176

Chapter 7 Electrical lighting 179

7.1	**Electrical lighting mode and type**	**180**
7.1.1	Modes of electric lighting	180
7.1.2	Types of electrical lighting	182
7.2	**Installation and maintenance of lighting equipment**	**183**
7.2.1	Installation of lighting equipment	183
7.2.2	Lighting circuit trouble shooting	189

Chapter 8 Power capacitor 195

8.1	**Brief introduction to power capacitor**	**196**
8.1.1	Classification and application of power capacitors	196
8.1.2	Structure and compensation principles of power capacitor	197
8.2	**Installation requirements and wiring of power capacitor**	**198**
8.2.1	Installation requirements of power capacitor	198
8.2.2	Wiring of power capacitor	200
8.3	**Safe operation of power capacitor**	**202**
8.3.1	Operation and maintenance of capacitor	202
8.3.2	Trouble shooting of capacitors	203

Chapter 9 Use of safety tools (K1) 205

9.1	**Safe use of electrical instruments (K11)**	**206**
9.2	**Use of electrical safety appliances (K12)**	**219**
9.3	**Identify electrical safety signs (K13)**	**223**

Chapter 10 Safe operating technology (K2) 227

10.1	**Practical examination platform and examination instructions**	**228**

10.1.1 Cognition of practical examination platform 228
10.1.2 Instruction to practical examination 232
10.2 Simulation equipment examination training project 234
10.2.1 Wiring and safe operation of electric drive circuit 234
10.2.2 Wiring and safe operation of lighting circuit 244
10.3 Equipment examination training project 255
10.3.1 Wiring and safe operation of electric drive circuit 255
10.3.2 Wiring and safe operation of lighting circuit 267

Chapter 11 Elimination of potential safety hazards at the operation site (K3) 275

Chapter 12 Emergency response at operation site (K4) .. 287

12.1 Emergency response and first aid for electric shock 288
12.2 Fire extinguishing simulation with fire extinguisher 294

Appendix 1 Single choice question bank 299

Appendix 2 Judgment question bank 349

References 405

Safety Production Instructions for Electricians

1.1 Laws and regulations on safety production

1.1.1 *Regulations on Management of Safety Technology Training and Assessment for Special Operation Personnel*

On May 24, 2010, the State Administration of Work Safety formulated and issued the *Regulations on the Administration of Safety Technology Training and Assessment for Special Operation Personnel*. The second amendment was made on May 29, 2015, with the main purpose of implementing the system for special operation personnel to work with certificates, improving the safety technology level of special operation personnel, and preventing and reducing casualties.

(1) What is a special operation

Special operation refers to the operation that is prone to casualties and has significant harm to the safety of operators, others and surrounding facilities. Those who are directly engaged in special operations are called special operators. The regulations clearly list high-voltage electrician operations, low-voltage electrician operations, and explosion-proof electrical operations as special operations. High-voltage electrician operation refers to the operation, maintenance, installation, overhaul, transformation, construction, commissioning of high-voltage electrical equipment 1kV and above, and the test with insulation tools and apparatus of them. Low-voltage electrician operation refers to the operation of the installation, commissioning, operation, maintenance, overhaul, reconstruction and test of low-voltage electrical equipment below 1kV. Explosion-proof electrical operation refers to the installation, overhaul and maintenance of various explosion-proof electrical equipment, which is applicable to explosion-proof electrical operations other than underground coal mines.

(2) Regulations on special operation personnel

① Special operation personnel must take special theoretical study and practical operation training of safety technology valid for the type of operation, and could work only after passing the examination and obtaining the special operation certificate. The certificate is shown in Fig. 1-1.

② The principle of unified supervision, hierarchical implementation and separation of teaching and examination should be applied to the safety technology training, assessment, certificate issuance and review of special operation personnel.

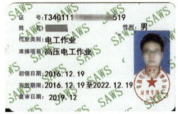

Fig. 1-1　Certificate format

③ The assessment of special operation personnel includes examination and approval. The examination should be in the charge of the examination issuing authority or the organization entrusted by it; The examination and approval should be in the charge of the examination and certification authority.

④ The period of validity of the special operation certificate is 6 years, which is valid nationwide.

⑤ The special operation certificate should be reviewed once every 3 years. During the validity period of the special operation certificate, if the employee has been engaged in this type of work for more than 10 years, the review time could be extended to once every 6 years. Once the special operation certificate needs to be reviewed, the applicant or the applicant's employer should submit an application to the original examination and certificate issuing authority, or the local examination and certificate issuing authority where the employee is engaged, within 60 days before the expiration of the time limit.

【Special reminder】

Before training, special operation personnel must go to the local hospital above the county level for a physical examination. Only those who pass the physical examination could participate in the safety technology theoretical training and practical operation training corresponding to the special operation they are engaged in. When registering for training, it is required to fill in the *Application Form for Training of Special Operators*, submit 3 pieces of 1-inch photos taken recently (white background or blue background), one copy of an ID card and one copy of the academic certificate, which should be reviewed and sealed.

1.1.2　The Production Safety Law of the People's Republic of China

(1) Safety production policy

Safety production refers to the relevant activities in production and business that take corresponding accident prevention and control measures to avoid accidents that cause personal injury and property loss, so that the production process could be carried out under the specified conditions, and the smooth operation of production and business activities could be ensured. It ensures the personal safety and health of employees, and avoids damage to equipment, facilities and the environment. *The Production Safety Law of the People's Republic of China* which came into force on December 1, 2014, contains 114 articles in 7 chapters. The law points out that production safety should be people-oriented, adhere to safety development, and adhere to the policy of "safety first, prevention first, and comprehensive governance".

① Safety first refers to putting safety, especially the personal safety of employees and other personnel, in the first place in production, to effectively protect the life safety and health of operators. Adhering to safety first is an inevitable requirement for building a harmonious socialist society.

② Prevention first is to move the safety production work forward, take precautions in advance, and establish a progressive and three-dimensional accident prevention system, to improve the safety situation and prevent safety accidents. Prevention first is to build a solid safety defense line by building a safety culture, improving the safety legal system and the level of safety science and technology, implementing safety responsibilities and increasing safety investment. Although safety accidents could not be completely avoided in production activities, accidents could be prevented and reduced as long as people pay attention to them and take proper preventive measures.

③ Comprehensive governance refers to adapting to the requirements of the situation of production safety in China, consciously complying with the law of production safety. Everyone should face up to the long-term, arduous and complex nature of production safety, and grasp the main contradictions and key links in production safety. And comprehensively use economic, legal, administrative and other means, manage people, rule by law, implement preventing technology in multiple ways, give full play to the supervision role of society, operators, and public opinion, and solve problems in the field of safety production effectively.

【Special reminder】

The policy of safety production is an organic and unified whole. "Safety first" is the commander and soul of prevention first and comprehensive governance. Without the idea of safety first, prevention first would lose ideological support, and comprehensive governance would lose the basis for remediation. "Prevention first" is the fundamental way to achieve safety first. "Comprehensive governance" is a means and method to implement safety first and prevention first.

(2) Relevant regulations on employees

① Rights of employees

a. The right to know and make suggestions. Employees have the right to know the risky factors, preventive measures and accident emergency measures existing in their workplaces and posts, and have the right to put forward suggestions on their own operation safety.

b. The right to criticize, report and accuse. The employees should have the right to criticize, report and accuse the problems existing in the production safety work in their own entities.

c. Right of refusal. Employees have the right to refuse illegal commands and forced risky operations.

d. Withdrawal right. When employees find an emergency that directly endangers their personal safety, they have the right to stop operations or leave the workplace after taking emergency measures. The production and business operation entity should not reduce the wages, welfare and other benefits of the employees or terminate the labor contracts concluded with them because the employees stop operation in an emergency or take emergency evacuation measures.

e. The right to claim beyond insurance. In addition to enjoying social insurance for work-related injuries according to law, the employees who have suffered injuries and (or) losses due to production safety accidents, if they still have the right to be compensated according to the relevant civil laws, have the right to claim compensation from their own employing entity.

② Obligations of employees

a. Comply with the rules and regulations. In the operation process, it is required to strictly abide by the rules and regulations on safe production and operating procedures of the employing entity, obey the management, and correctly wear and use labor protection articles.

b. Education and training. They should receive education and training on production safety, master the knowledge of production safety required for their own work, improve their production safety skills, and enhance their ability to prevent accidents and deal with emergencies.

c. Report the potential risks. If any potential risk or other unsafe factors are found, they should report to the on-site safety production management personnel or the person in charge of the employing entity immediately, as shown in Fig. 1-2; the person receiving the report should deal with it in a timely manner.

Fig. 1-2　Employees have the obligation to report the potential risks

【Special reminder】

The labor union should have the right to request correction of the acts of the business operation entity that violate ***The Production Safety Law of the People's Republic of China***; Have the right to propose solutions to illegal command, forced risky operation or potential accidents have been found; Have the right to participate in accident investigation according to the law, put forward handling suggestions to relevant departments, and require relevant personnel to be held accountable.

1.1.3　Other laws and regulations

(1) The Fire Control Law of the People's Republic of China

The Fire Control Law of the People's Republic of China was adopted on April 29, 1998, revised at the fifth meeting of the Standing Committee of the 11th National People's Congress on October 28, 2008, and revised at the 10th meeting of the Standing Committee of the 13th National People's Congress on April 23, 2019, which came into force on April 23, 2019. The regulations of the new fire control law on the rights and obligations of employees in fire protection work mainly include:

① Everyone has the obligation to maintain fire safety, protect fire-fighting facilities, prevent fires and report fire alarms. Any adult has the obligation to participate in organized fire fighting.

② No one is allowed to damage, misappropriate, dismantle or stop firefighting facilities and equipment. No one is allowed to bury, occupy or block fire hydrants or occupy fire separation spaces, and occupy, block or close evacuation passages, emergency exits and fire truck passages.

③ Anyone who finds a fire should call the fire agency immediately. Any person should provide convenience for reporting to the fire agency free of charge and should not obstruct reporting to the fire agency. The false fire alarm is strictly prohibited.

④ After the fire is extinguished, the relevant personnel should protect the site according to the requirements of the fire agency and the public security agency, accept the accident investigation, and truthfully provide the information related to the fire.

【Special reminder】

Electricians should enhance their awareness of fire prevention, pay attention to fire safety at all times, and be good propagandists of enterprise fire prevention.

(2) The Labor Law of the People's Republic of China & The Labor Contract Law of the People's Republic of China

① *The Labor Law of the People's Republic of China* was adopted at the 8th meeting of the Standing Committee of the 8th National People's Congress on July 5, 1994, and came into force as of January 1, 1995. The first amendment was made at the 10th meeting of the Standing Committee of the 11th National People's Congress on August 27, 2009, and the second amendment was made at the 7th meeting of the Standing Committee of the 13th National People's Congress on December 29, 2018. *The Labor Law of the People's Republic of China* stipulates that workers have the right to equal employment and choice of occupations, the right to obtain labor remuneration, the right to rest and take vacations, the right to obtain protection of labor safety and health, the right to receive vocational skills training, the right to enjoy social insurance and benefits, the right to submit for settlement of labor disputes, and other labor rights stipulated by law. Labors should complete their labor tasks, improve their vocational skills, implement the rules on labor safety and health, and observe labor discipline and professional ethics.

② *The Labor Contract Law of the People's Republic of China* was adopted on June 29, 2007, and came into force on January 1, 2008. It was amended at the 30th meeting of the Standing Committee of the 11th National People's Congress on December 28, 2012. *The Labor Contract Law of the People's Republic of China* stipulates that an employer should establish a labor relationship with an employee from the date of employment. The conclusion of a labor contract should follow the principles of legality, fairness, equality, voluntariness, consensus through consultation, honesty and credibility. The labor contract concluded according to law is binding, and the employer and the employee should perform the obligations agreed in the labor contract.

(3) The Regulation on Industrial Injury Insurance

The Regulation on Industrial Injury Insurance is formulated to ensure that employees who suffer from accidents or occupational diseases at work could obtain medical treatment and economic compensation, which promote occupational injury prevention and occupational rehabilitation, and disperse the risk of occupational injury of employers. It was issued by the State Council on April 27, 2003, and effective as of January 1, 2004. It was revised by Decree No. 586 of the State Council in 2010 and implemented on January 1, 2011.

(4) Regulations on Emergency Handling and Investigation of Electric Power Safety Accidents

The Regulations on Emergency Handling and Investigation of Electric Power Safety Accidents was adopted at the 159th executive meeting of the State Council on June 15, 2011, and came into force on September 1, 2011. The regulations consist of 6 chapters and 37 articles. The classification of power safety accidents involves the adoption of corresponding emergency response measures, the application of different investigation and handling procedures, and the determination of corresponding accident responsibilities, which are very necessary to be clarified in the regulations. According to the extent to which the accident affects the safe and stable operation of the power system or the normal supply of power, the regulations classify power safety accidents into four levels: extraordinarily serious accidents, serious accidents, major accidents and ordinary accidents.

【Special reminder】

There are many laws and regulations that electricians should know and master. The readers could read the original text of the law through books or the Internet further, so that you can understand and abide by the law in your work better.

1.2 Safety protection and electric shock prevention measures

1.2.1 Organizational measures to ensure the safety of electricians

(1) Work card system

The work card is a written order for permission to work on electrical equipment, as well as a written basis for specifying safety responsibilities, conducting safety disclosure to staff, performing work permit procedures, work interruption, transfer and termination procedures, and implementing technical measures to ensure safety. There are two types of work cards: type I and the type II.

1) The type I of electrical work card

Scope of application: The work on high-voltage equipment requires complete or partial power cut-off; The work on the secondary wiring and lighting circuits in the high-voltage room needs to cut off the power of high-voltage equipment or take safety measures; other work that needs to cut off the high-voltage equipment or take safety measures; The maintenance or test of 400V low-voltage equipment that requires safety measures.

2) The type II of electrical work card

Scope of application: Live working and working on the housing of live equipment; The work on control panel, low-voltage distribution panel, distribution box and power main line; The work on the secondary wiring circuit that does not need to cut off the high-voltage equipment; The work on the rotating generator, excitation circuit or high-voltage motor rotor resistance circuit; The off duty personnel should use insulating rod and voltage mutual inductor to decide the phases or clamp ammeter to measure the current of high-voltage circuit; Replacement of lighting bulbs in production areas and production-related areas.

【Special reminder】

The filling scope of various work cards is clearly specified in the **Electric Power Safety Work Regulations.**

Verbal or telephone orders are only applicable to the completion of work other than the type I and the type II of work cards.

(2) Work permit system

The work permit system is that the work approver (electrician on duty) sends the order of work permit to the person in charge after taking safety technical measures for equipment power cut-off according to the contents of the low-voltage work card or low-voltage safety measure card. Then the person in charge of the work could start the work. In the maintenance process, work interruption, transfer and work termination must be approved by the work approver. All these organizational procedures are called the work permit system.

The work permit system is a system in which the work approver reviews the safety measures listed in the work card and decides whether to permit work. The person in charge of the work could start the work only after the work approver issues the order; During the maintenance process, the work interruption, transfer and work termination must be approved by the work approver.

The work approver should be the person on duty, but the work cards should not be issued. In other words, the issuer of the work cards should not concurrently serve as the person in charge of the work (the work approver).

(3) Work monitoring system

After completing the work permit process, the person in charge of the work (supervisor) should explain the on-site safety measures, live parts and other precautions to the work shift personnel.

The person in charge of the work (supervisor) must always be at the operation site to check and timely correct the violation of safety work procedures and safety measures by the staff of the work team during the work process. Especially when a worker moves a part of body close to the live part or the work shift personnel transfers the workplace, position, posture and angle during work, the monitoring should be strengthened to avoid danger.

In case of total power cut-off, the person in charge of the work (supervisor) could participate in the work. In case of partial power cut-off, only when the safety measures are reliable and the personnel is concentrated in one working place, where they could participate in the work without accidentally touching the live parts. During the work, if the person in charge of the work must leave the work place for some reason, a competent person should be appointed to temporarily replace him, and the staff of the work shift should be informed. The original person in charge of the work should also go through the same procedures after returning.

(4) Work interruption, transfer and termination system

All safety measures should remain in place when work is interrupted. For work on power lines, if the work team has to leave the operation site temporarily, safety measures must be taken and personnel must be assigned to guard against people and animals getting close to the excavated pit, the unstable tower, the loaded lifting and mechanical drive devices. Before resuming work, the integrity of safety measures such as grounding wires should be checked.

For the work that could not be completed on the day, the operation site should be cleaned, the closed road should be opened, and the work cards should be returned to the person on duty. When the work resumes on the next day, the work card should be obtained with the permission of the person on duty.

The person in charge of the work must check carefully whether the safety measures meet the requirements of the work card before work. When the same electrical connection part is transferred to several operation sites with the same work order, all safety measures should be completed by the person on duty at one time before the commencement of work, and no transfer formalities are required. However, the person in charge of the work should explain the electrification scope, safety measures and precautions to the staff when transferring the operation site. After all the work is completed, the work team should clean and tidy up the site.

The person in charge of the work should check carefully first, and lead the staff to leave the site after confirming there is no problem. Then explain the problems found in the repaired project, test results and existing problems to the person on duty, and jointly check the equipment condition with the person on duty to see if there are any leftovers, whether it is clean, etc., and then fill in the work termination time on the work cards. The work cards would be terminated only after both parties sign.

【Special reminder】

The terminated work cards should be kept for at least 3 months.

1.2.2 Technical measures to prevent electric shock

(1) Technical measures to prevent direct electric shock

1) Insulation

Insulation is closing or isolating the electrified body with insulating materials so that the equipment could work safely and normally for a long time. At the same time, it could prevent the human body from touching the electrified part to avoid electric shock accidents.

Good insulation is a necessary condition for the normal operation of equipment and lines, and an important measure to prevent electric shock accidents. In electrical appliances, electrical equipment, devices and electrical engineering, the commonly used insulation materials are bakelite, plastic, rubber, mica and mineral oil.

The insulation material would be damaged after being used for a period of time, so the insulation needs to be tested regularly to ensure safety and reliability of electrical insulation. Insulation inspection includes insulation test and appearance inspection. Only insulation resistance tests could be conducted on site, including insulation resistance and absorption ratio measurement.

Common insulation safety appliances include insulating gloves, insulating boots, insulating shoes, insulating pads, insulating platforms, etc. Insulation safety appliances could be divided into basic safety appliances and auxiliary safety appliances.

【Special reminder】

When working on low-voltage electrified equipment, insulating gloves, insulating shoes (boots) and insulating pads could be used as basic safety appliances, and could only be used as auxiliary safety appliances under high-voltage conditions.

2) Shielding

Shielding refers to the use of barriers, fences, shields, covers or isolation plates, box gates, etc., to isolate the electrified body from the outside world. The shielding protection has the functions of preventing electric shock (the human body touches or gets too close to the electrified body), preventing electric arc from injuring people, preventing arc short circuit fire, and facilitating maintenance and production safety.

Generally, the movable part of switching appliances could not be insulated, but needs shielding. For high-voltage equipment, it is often difficult to fully insulate, and when people get close to a certain extent, serious electric shock accidents would occur. Therefore, no matter whether the high-voltage equipment has been insulated, shielding or other measures to prevent access should be taken.

The shielding device does not contact with the electrified body directly and has no strict requirements on the electrical properties of the materials used. The materials used for the shielding device should have sufficient mechanical strength and good fire resistance. However, in order to prevent electric shock caused by accidental electrification, the shielding device made of metal materials must be grounded or connected to neutral.

The shielding devices include permanent shielding devices (such as barriers of power distribution devices and covers of switches), temporary shielding devices (such as temporary barriers used in maintenance work), and fixed shielding devices (such as protective nets of buses).

 【Special reminder】

Shielding such as barriers and fences should be provided with obvious signs, hung with signboards, and locked when necessary.

3) Safety spacing

Safety spacing refers to a certain safety distance between the electrified body and the ground, between the electrified body and other facilities and equipment, and between the electrified bodies.

The purpose of setting safety spacing is to prevent electric shock accidents caused by human touching or to approach electrified bodies; Prevent vehicles or other objects from colliding with or getting too close to the electrified bodies to cause accidents; Prevent electrical short circuit accidents, over-voltage discharge and fire accidents; And easy to operate.

The main electrical safety spacing includes line spacing, equipment spacing, maintenance spacing, etc.

【Special reminder】

The safety spacing depends on the voltage, equipment type and installation method.

4) Safety voltage

The safety voltage standards determined in China are 42V, 36V, 24V, 12V and 6V.

【Special reminder】

Equipment with safe voltage belongs to equipment category Ⅲ.

5) Safety colors and safety signs

The national safety colors include red, blue, yellow and green. The meanings of each color are shown in Table 1-1.

Table 1-1 Meaning of safety colors

No.	Type	Meaning
1	Red	It indicates prohibition, stop and danger, and is used for prohibition signs, stop signals and places where people are prohibited from touching
2	Blue	It indicates instructions and regulations that must be observed, such as the signs that must wear certain protective tools and the signs that guide vehicles to drive, which are painted in blue
3	Yellow	It indicates warning and caution, and all kinds of warnings such as "Be careful, beware of electric shock" are shown in yellow
4	Green	It indicates prompt, safety status and passage, such as the safe passage and fire-fighting equipment inside the workshop, which are shown in green

In production, safety colors often use other colors (i.e. contrast colors) as the background color to make them more eye-catching, so as to improve the discrimination of safety colors. For example, red, blue and green use white as the contrast color, and yellow use black as the contrast color. The yellow and black stripes alternate, with good visibility, and are generally used to indicate warning danger. The red and white intervals are often used to indicate "No Crossing".

According to *Safety Signs and Guidelines for Their Use* (GB 2894—2008), safety signs are used to express specific safety information and are composed of graphical symbols, safety colors, geometric shapes (frames) or words.

Safety signs are classified into four categories: prohibition signs, warning signs, instruction signs and prompt signs, as well as supplementary signs. See Table 1-2 for safety signs related to electricity.

Table 1-2 Safety signs

Type	Definition	Geometry	Safety signs related to electricity
Prohibition signs	Prohibit or stop certain actions	A ring with a slash, in which the ring is connected with the slash in red; Black for graphic symbols and white for background	No combustibles, No smoking, No entry, No fire, No fire-fighting with water, No kindling, No start, No rotating during repair, No refueling during operation, No crossing, No riding, No climbing, etc.
Warning Signs	Warning of possible dangers	Black equilateral triangle, black symbol and yellow background	Caution, Danger! High Voltage, Beware of explosion, Beware of fire, Beware of corrosion, Warning poisoning, Warning mechanical injury, Warning hand injuries, Beware of lifting objects, Warning splinter, Warning falling objects, Beware of falling, Beware of vehicles, Warning arc, Warning roof fall, Warning gas, Warning collapse, Warning hole, Warning ionizing radiation, Warning fission matter, Warning laser, Warning microwaves, Watch Your Step, etc.
Instruction signs	Must comply	Round, blue background, white graphic symbol	Must wear safety helmet, Must wear protective shoes, Safety belt buckled, Must wear protective glasses, Must wear gas mask, Take care of noise, Must wear protective gloves, Must wear protective clothing, etc.
Prompt signs	Indicate the direction of the target	Square, green and red background, white graphic symbols and text	There are 6 general warning signs (green background): safety passage, emergency exit, etc.; There are 7 warning signs for fire-fighting equipment (red background): fire alarm bell, fire alarm telephone, underground fire hydrant, ground fire hydrant, fire hose, fire extinguisher, fire pump connector
Supplementary signs	Supplementary description of the above four signs to prevent misunderstanding	Supplementary signs are divided into horizontal and vertical. The horizontal writing is rectangle, which is written below the sign, and can be connected with the sign or separated; For vertical ones, the text is written vertically on the top of the sign. Color of supplementary signs: black characters on white background for vertical signs, white characters on red background for horizontal signs, black characters on white background for warning signs, and white characters on blue background for command signs	

(2) Protective neutral connection and protective grounding

Indirect contact electric shock is the electric shock in a fault system. Protective neutral connection and protective grounding are the basic technologies to prevent indirect contact electric shock.

1) Protective neutral connection

Protective neutral connection is to closely connect the uncharged metal part of electrical equipment under normal conditions with the zero line (or neutral line) of the power grid. The application scope of protective neutral connection is mainly used in three-phase-four-wire power supply system (TN-C) and three-phase-five-wire power supply system (TN-S). In the factory, it is used for 380/220V low-voltage electrical equipment.

In the three-phase-four-wire power supply system (TN-C) with neutral point grounding, the protective neutral line (PE) and the working neutral line (N) are combined into one, that is the working neutral line also acts as the protective neutral line, as shown in Fig. 1-3. In the three-phase-five-wire power supply system (TN-S), the special protective neutral line (PE) and working neutral line (N) are not connected and are strictly separated, except that they are grounded together at the transformer neutral point, as shown in Fig. 1-4.

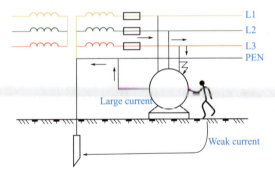

Fig. 1-3　Schematic diagram of TN-C system protective neutral connection

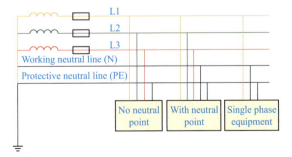

Fig. 1-4　Schematic diagram of TN-S system protective neutral connection

2) Protective grounding

Protective grounding generally refers to the electrical device for the purpose of safety, uses the grounding system including a grounding electrode, grounding bus, and grounding wire to make an electrical connection with the ground; Or the electrical device is electrically connected to a reference potential point, normally to the reference ground. See Table 1-3 for types of protective grounding.

Table 1-3 Types of protective grounding

Grounding mode	Details	Schematic diagram
Working grounding	In a three-phase AC power system, the low-voltage neutral point grounding of power transformer for power supply is called working grounding. It could reduce the risk of high voltage jumping into low voltage, and reduce the risk of electric shock when a certain phase of low voltage is grounded. Working grounding is the main safety facility for low-voltage power grid, and the working grounding resistance must be less than 4Ω.	Neutral point — L1 / L2 / L3; Working grounding; Grounding body
Protective grounding	To prevent the danger caused by accidental electrification of the exposed uncharged conductor of electrical equipment, the method that connects the electrical equipment through the protective grounding wire with the buried grounding body is called protective grounding. Protective grounding is the main safety measure of neutral point ungrounded low-voltage system. In a normal low-voltage system, the protective grounding resistance should be less than 4Ω.	L1 / L2 / L3; Grounding body; Protective grounding

Grounding mode	Details	Schematic diagram
Lightning protection and grounding	To prevent electrical equipment and buildings from being damaged by lightning, lightning rods, lightning conductors, lightning arresters and other lightning protection equipment are grounded, which is called lightning protection grounding	
Anti static grounding	The grounding set to eliminate static electricity generated in the production process is called anti-static grounding	
Shield grounding	To prevent electromagnetic induction, the metal housing, shield cover, shield wire sheath of power equipment, or metal shield of buildings are grounded, which is called shield grounding	
Repeated grounding	One or more neutral lines (or neutral points) of a three-phase-four-wire system are reliably connected with the ground through grounding devices, which is called repeated grounding	
Common grounding	In the grounding protection system, connecting the grounding trunk line or branch line at multiple points with the grounding device is called common grounding	

How does protective grounding protect personal safety? If there is a motor without a protective grounding device, when its internal insulation is damaged and the housing is charged, once the human body contacts it, the current path between the charged metal housing and the earth is connected through the human body, and the current on the metal housing flows into the earth through the human body to cause electric shock, as shown in Fig. 1-5 (a).

After the metal housing of the motor is reliably electrically connected with the ground with a conductor, as shown in Fig. 1-5 (b), if the metal housing is charged due to the insulation damage of the motor, when the human body contacts it, two parallel current paths would be formed between the metal housing and the ground: one is to discharge the current to the ground through the protective grounding wire, and the other is to discharge the current to the ground through the human body. In these two parallel circuits, the resistance of the protective grounding wire is very small, usually only about 4Ω, and the minimum resistance of the human body is above 500Ω. According to the principle that the current in the parallel circuits is inversely proportional to the resistance, the current that the human body passes through is much less than the current that passes through the protective grounding wire. Meanwhile, the human body has no sense of electric shock. Moreover, because the resistance of the protective grounding wire is too small, which is close to the short circuit between the motor and the ground, there would be a large current passing through the protective grounding wire. This large current would make the protective equipment in the circuit act, automatically cut off the circuit, and protect the safety of people and equipment.

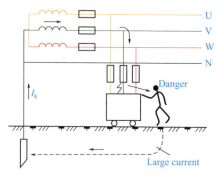
(a) Electric shock without protective grounding

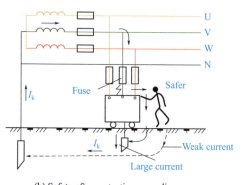
(b) Safety after protective grounding

Fig. 1-5　Principle diagram of protective grounding

【Special reminder】

Protective grounding and protective neutral connection are two technical measures for safe use of electricity with similarities and differences. See Table 1-4 for comparison.

Table 1-4　Comparison between protective grounding and protective neutral connection

Comparison	Protective grounding	Protective neutral connection
Similarities	Both are protective measures taken to prevent the metal housing of electrical equipment from being charged	
	Applicable electrical equipment is basically the same	
	Both are required for good grounding or neutral connection device	

Comparison		Protective grounding	Protective neutral connection
Differences	Valid for different system	Suitable for high and low-voltage power supply systems with the ungrounded neutral point	Suitable for low-voltage power supply system with grounding neutral point
	Different line connections	The grounding wire is directly connected to the grounding system	The protective neutral line is directly connected with the neutral line of the power grid, and then grounded through the neutral line
	Different requirements	Each electrical appliance is required to be grounded	Only the neutral point of the three-phase-four-wire system is required to be grounded

(3) Electric leakage protection

An electric leakage protector is a device that could automatically disconnect the circuit or give an alarm when the leakage (electric shock) current (mA) value in the circuit reaches or exceeds its specified value under specified conditions. The electric leakage protector acts sensitively and cuts off the power in a short time. It is mainly used to prevent indirect and direct contact with electric shocks. Besides protecting personal safety, it could also prevent leakage fire and monitor one phase grounding faults.

There are many kinds of electric leakage protectors, which could be divided into voltage type and current type according to the action principle; According to the number of poles, it could be divided into types of two-poles, three-poles and four-poles; According to the action time, it could be divided into quick action type, delay type and inverse time type electric leakage protection devices. The selection of electric leakage protection devices should be based on the different requirements of the protection objects, which should be both technically effective and economically reasonable. Unreasonable type selection not only fails to achieve the protection purpose, but also causes the electric leakage protection device to refuse to operate or misoperate. A proper and reasonable selection of leakage protection device is the key to implementing electric leakage protection measures.

1.3 Electrical accident and first aid for electric shock

1.3.1 Harm of electric current to human body

(1) Electric shock current

The greater the current passing through the human body, the more obvious the physiological response of the human body, and the shorter the time required to cause ventricular fibrillation, the greater the risk of death. For power frequency alternating current (PFAC), the human body presents different states according to the magnitude of the current passing through the human body. The current could be divided into three levels: sensing current, escape current and lethal current, as shown in Table 1-5.

Table 1-5 Three levels of human body electric shock current   mA

Type	Description	Adult males		Adult females
Sensing current	The minimum current that causes people to feel. Meanwhile, people feel slight numbness and tingling	PFAC	1.1	0.7
		DC	5.2	3.5
Escape current	The maximum current that people could get escaped from an electric shock. Meanwhile, the feeling of heating and tingling increases. When the current reaches a certain level, the person who gets an electric shock would grasp the electrified body tightly due to muscle contraction and spasm, and cannot get rid of the electricity by himself	PFAC	16	10.5
		DC	76	51
Lethal current	Life-threatening current in a short time	PFAC	\multicolumn{2}{c}{30~50}
		DC	\multicolumn{2}{c}{1300(0.3s), 50(3s)}

(2) Injury to human body by current magnitude

Whether AC or DC, the higher the voltage and the greater the current intensity, the greater the danger to people. In general, the magnitude of current and the degree of human injury are shown in Table 1-6.

Table 1-6 Current magnitude and human injury degree

Current(mA)	Human perception
1	People would feel "numb electricity"
5	Painful feeling
8~10	Feel unbearable pain
20	Muscles contract violently and lose freedom of movement
50	Life threatening
100	Death

(3) Current path and human injury

When an electric shock occurs, the greater the current intensity through the heart, lungs and central nervous system, the more serious the consequences would be. See Table 1-7 for the percentage of cardiac current passing through different paths.

Table 1-7 Percentage of cardiac current passing through different paths

Current path	Left hand→Feet	Right hand→Feet	Right hand→Left hand	Left foot→right foot
Percentage%	6.7	3.7	3.3	0.4

1.3.2 Causes and types of electric shock

(1) Causes of electric shock accidents

① Lack of electrical safety knowledge, such as operating high-voltage isolating switch with load on; After the low-voltage overhead line is broken, the power is not cut off, and the live wire is touched by hand by mistake; Live connection under weak light and touch live body by mistake; Touch the switch with broken rubber cover.

② Violation of safety operation rules, such as operating high-voltage isolating switch with load on; Live temporary lighting line setting; The installation and wiring are not standardized.

③ The equipment is unqualified, such as high and low voltage cross lines, and low voltage lines are mistakenly set over the high voltage lines. The incoming and outgoing lines of electrical equipment are not wrapped up and exposed; People touch unqualified temporary lines, etc.

④ Poor equipment management, such as not to deal with the situation in time when the low-voltage line is broken and the pole is blown down by strong wind; The housing is charged for a long time due to the damaged wiring of the water pump motor.

⑤ Other accidental factors, such as strong wind breaking the power line and touching the human body, and human body being struck by lightning.

(2) Regular pattern of electric shock accidents

The general pattern of electric shock accidents are as follows.

① There is obvious seasonality. There are many accidents in the second and third quarters of the year, with the most concentrated accidents from June to September.

② Low voltage electric shock is more than high voltage electric shock.

③ More electric shock accidents occurred in rural areas than in urban areas.

④ Electric shock accidents are easy to occur at electrical connection parts. For example, terminals, crimps, welding heads, lamp holders, plugs, sockets, controllers, contactors, fuses, etc.

⑤ More electric shock accidents occur to portable and mobile equipment.

⑥ Many electric shock accidents are caused by illegal operation and misoperation.

(3) Type of electric shock accident

Generally speaking, there are two types of injuries caused by electric current: electric stroke and electric injury. Through the analysis of many electric shock accidents, the two types of electric shock injuries would occur at the same time.

① Electric shock. Electric stroke is the damage caused by the electric current passing through the human body and damaging the normal work of the human heart, nervous system and lungs. Electric stroke might be caused by the human body touching the live wire, the housing of the leakage equipment or other charged bodies, as well as by lightning stroke or capacitor discharge.

② Electric injury. Electric injury refers to the partial injury caused by the thermal effect, chemical effect or mechanical effect of current to human body, including electric arc burn, scald, electric branding, skin metallization, electrical mechanical injury, electrooptic eye and other different forms of injury.

【Special reminder】

Whether it is electric stroke or electric injury, it would endanger people's health and even life.

1.3.3 Electric shock mode

According to the way the human body touches the electrified body and the way the current flows through the human body, electric shock could be divided into single-phase electric shock, two-phase electric shock and step voltage electric shock.

(1) Single-phase electric shock

When one part of the human body come into contact with the earth, and the other part come into contact with a single phase charged body, a single-phase electric shock occurs, which is shown in Fig. 1-6.

(2) Two-phase electric shock

When electric shock occurs, different parts of the human body touch two-phase charged bodies at the same time, which is called two-phase electric shock, that the human body plays the role of load between the phases to form a circuit current, as shown in Fig. 1-7. Meanwhile, the current flowing through the human body depends entirely on the current path and the voltage of the power supply network.

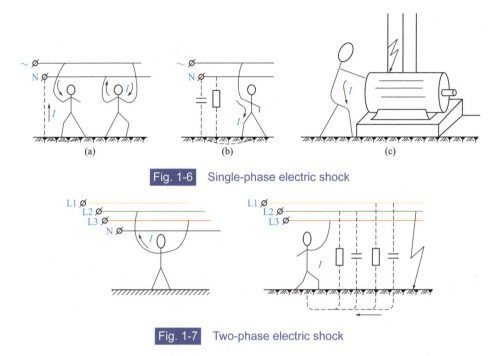

Fig. 1-6 Single-phase electric shock

Fig. 1-7 Two-phase electric shock

(3) Step voltage electric shock

When the electrical equipment contacts with the housing or one phase of the power system is short circuited to the ground, the current flows out from the grounding electrode in all directions, forming different potential distributions on the ground. When people approach the short-circuit point, the potential difference between two feet is called step voltage. The electric shock caused by the human body touching the step voltage is called step voltage electric shock, as shown in Fig. 1-8.

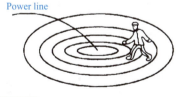

Fig. 1-8 Step voltage electric shock

When a step voltage threat is detected, people should quickly put their feet together, or jump with one or two legs as soon as possible to leave the danger area 20m away.

Furthermore, the electric shock modes include contact voltage electric shock, induced voltage electric shock and residual charge electric shock as well.

【Special reminder】

Electric shock in any way is dangerous, very dangerous! In daily work, if there are no necessary safety measures, we should not touch the low-voltage electrified body, nor close to the high-voltage electrified body.

1.3.4 Electrical accident

(1) Types of electrical accident

According to the different action forms of electric energy, the types of electrical accidents are shown in Table 1-8.

Table 1-8 Types of electrical accidents

No.	Type	Description
1	Current injury accident	It refers to the electric shock casualty accident caused by the human body touching the electrified body and the current passing through the human body
2	Electromagnetic field injury accident	It refers to the harm to human body caused by absorbing radiation energy under the action of electromagnetic field
3	Lightning accident	It refers to the damage to building facilities, casualties of people and animals, and fire and explosion accidents caused by lightning
4	Electrostatic accident	Electrostatic discharge could occur in the place of explosive mixture and cause explosion
5	Electrical equipment accident	Circuit fault belongs to electrical equipment accident, but equipment accident is often associated with personal accident

(2) Causes of electrical accidents

① Operators do not master the electrical principles well; ② Equipment is not well installed; ③ Electrical failures accumulate, causing major accidents; ④ The original design has certain defects; ⑤ The operator did not do well on safety protection.

(3) Preventive measure

① Strengthen the safety awareness of operators and strictly implement *The Power Safety Work Regulations of State Grid Corporation of China*.

② The employees could master the electrical principles and have the necessary business knowledge and technical skills via training.

1.3.5 First aid operation for electric shock

Electric shock could occur in any place where there are wires, electrical appliances and electrical equipment. Once an electric shock accident occurs, the person should be disconnected from the power supply as soon as possible, and rescue the person with electric shock against time. Principles for on-site rescue of electric shock are: quick, on-site, accurate and persistent.

(1) Operation of disconnecting the person from the power supply

1) Operation of disconnecting from low-voltage power supply

In case of 220V electric shock, the power supply should be cut off immediately.

◆ **Operating tips** ◆

If people get electric shock, cut off the power source first. Cut the wires one by one.
In case of emergency, remove the wire with a bar. The bar should be dry wood or bamboo.
In severe emergency, put on gloves and stand on plank to drag the person.
Proper measures should be taken to ensure safety. One-handed dragging is the safest.
Do not drag people directly to prevent them from being connected together by electricity.

【Special reminder】

It is better to use one hand to disconnect the person who gets an electric shock from the power supply.

2) Operation of cutting off high-voltage power supply

If someone gets an electric shock due to high voltage, other people should not touch or get close to the high-voltage electrical equipment or lines. For high-voltage electric shock accidents, the method introduced in the operating tips could be used to make the person who gets an electric shock get away from the power supply quickly. The person who gets an electric shock should also be taken away from the live wire to a distance of 8~10m, and first aid should be started immediately.

◆ **Operating tips** ◆

Call the power supply department cut off the power in case of high voltage electric shock.
If the safety measures are not well taken, it should be 8~10 meters away.

> Wear insulating boots and gloves, and pull off the switch with insulating tools.
> Throw a short circuit line near the overhead high-voltage line.
> If the line is short-circuited and grounded, the protection devices act to power off.
> Take the injured to a safe place immediately, after disconnection from the power supply.
> Do first aid until the medical staff arrives.

(2) First aid operation after disconnection from the power supply

After the electric shock person is disconnected from the power supply, first aid should be carried out according to the specific situation, as shown in Table 1-9. The timelier first aid is carried out, the greater the probability of being saved.

Table 1-9 First aid measures for the person with electric shock after leaving the power supply

No.	Symptom	First-aid measures
1	Conscious, able to breathe and heartbeat autonomously	Lie the injured person flat and observe closely. Do not stand up or walk temporarily to prevent secondary shock or heart failure
2	Apnea, but with heat beating	Lie the injured person flat, loosen the buttons, open the respiratory tract, and conduct the mouth-to-mouth artificial respiration immediately
3	Heartbeat stops, breathing exists	Take external cardiac compression immediately to give first aid
4	No breathing and heartbeat	In terms of on-site rescue, it is better for two people to perform mouth-to-mouth artificial respiration and external cardiac compression respectively at a ratio of 2∶15, that is, 2 times of artificial respiration and 15 times of cardiac compression. In terms of single manual CPR operation, during the actual operation of the exam, it is required to use mouth-to-mouth artificial respiration and external cardiac compression alternately at 10∶150, and the total time from judging the carotid artery to the last blowing should not exceed 130 seconds.

【Special reminder】

If the injured person suffers from electric shock loses his/her consciousness, the rescuer should quickly judge the breathing and heartbeat of the injured person within 10s by watching, listening and testing. The doctor should make a conclusion on whether the person is dead.

Do not move the injured at on-site rescue. When moving the injured or sending them to the hospital, make sure to make the injured lie flat on the stretcher and put the back with a flat hard board, and continue to rescue. If the heartbeat and respiration stop, continue artificial respiration and external cardiac compression. The treatment could not be stopped until the hospital medical staff take over. If the person gets electric shock has skin burns, it could be washed and wiped with clean water, and then wrapped with gauze or handkerchief to prevent infection.

(3) Mouth-to-mouth artificial respiration

For the unconscious person who gets an electric shock, if his breathing is not complete or weak, or if his breathing stops and his heart beats, the mouth-to-mouth artificial respiration method should be used for rescue. As shown in Fig. 1-9, firstly, make the head of the person who gets an electric shock to one side, removes the blood clot, sputum or foam in the mouth, and take out the false teeth and other dirt in the mouth to make the respiratory tract smooth; The first aider inhales deeply, pinches the person's nose, blows deeply into the person's mouth, and then relaxes his nose to exhale himself. Repeat every 5s, without interruption until the person wakes up.

Operating tips

The injured lies on his/her back and unbuttons the collar to loosen the clothes.
Open the mouth, pinch the nose, lift the jaw, and blow close to the mouth to see the chest.
If it is difficult to open the mouth, blow the nostrils. Blowing once every 5 seconds is normal.
The amount of blowing depends on the injured, and adults and children should be moderate.
Artificial respiration should be unstopped. Contact the doctor as soon as possible.

Remove dirt from mouth

Tilt the head back as far as possible

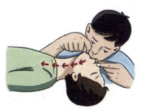
Blowing with mouth

Release and breathe

Fig. 1-9 Operation steps of mouth-to-mouth artificial respiration method

(4) External cardiac compression

The person who gets an electric shock with breathing but has a weak or irregular heartbeat, or whose heartbeat has stopped should be rescued by external cardiac compression method. As shown in Fig. 1-10, firstly, make the head of the person with electric shock tilt back, the first aider kneels at the hip of the person, place the right palm on the chest, put the left palm on the right one, squeeze it downward for 3~4cm, and then suddenly relax. The squeezing and relaxing actions should be rhythmic, once a second (three times in two seconds for children). The squeezing should be accurate, and the force should be appropriate. If the force is too strong, it would cause internal injury to the person who gets an electric shock. If the force is too small, it would be invalid. When rescuing children, the pressing force should be appropriately reduced, and should not be interrupted until the person who gets an electric shock wakes up.

Operating tips

The injured lies on his/her back in a hard bed with smooth respiratory tract guaranteed.
Press down the root of the palm without impact, and suddenly relax the hand.
The wrist is slightly bent and pressed one inch, once a second is more appropriate.
The neck pulse could be touched, which means the compression effect is enough.

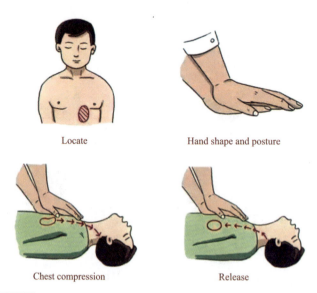

Fig. 1-10　Operation steps of external cardiac compression method

1.4 Electrical fire protection, explosion-proof, lightning protection and anti-static

1.4.1 Electrical fire and explosion prevention

(1) Causes of electrical fire and explosion

There are many reasons for electrical fire and explosion, and equipment defects and improper installation are important reasons. The heat generated by the current and the spark or arc generated by the circuit is the direct causes.

① Overheating of the electrical equipment. Overheating of electrical equipment is mainly caused by heat generated by current, including short circuit, overload, poor contact, iron core heating, poor heat dissipation, etc.

② Electric spark or arc. Electric spark is a breakdown discharge phenomenon between electrodes. A large number of electric sparks gather to form an arc, which causes the combustion of combustibles and results in a dangerous ignition source. It is easy to cause fire or explosion accidents in flammable and explosive places. Electric sparks mainly include working sparks and accident sparks.

Working sparks refer to sparks generated during normal operation of electrical equipment, such as small sparks at the sliding contact between DC motor brush and rectifier, tiny spark behind the brush at the sliding contact between AC motor brush and slip ring, sparks when switches or contactors are opened and closed, sparks when pins are pulled out or inserted, etc. The accident spark is the spark that occurs when the circuit or equipment fails. Such as sparks in case of short circuit or grounding, flashes in case of insulation damage, sparks in case of loose wire connection, sparks in case of fusant fusing, over-voltage discharge sparks, electrostatic sparks, induced electric sparks and sparks caused by incorrect operation in repair work.

(2) Electrical fire and explosion prevention measures

The occurrence of electrical fire and explosion requires three conditions: Firstly, flammable and explosive substances and environment; Secondly, ignition conditions; And thirdly, oxygen (air). Electric arcs, sparks and dangerous high temperature are often generated in the power, lighting, control, protection, measurement and other systems in the production site, and various electrical equipment and lines in the living place during normal work or accidents, which provides conditions for ignition or detonation. Objectively, many industrial sites meet the explosion conditions. When the mixed concentration of the explosive substances and oxygen is within the explosion limit, explosion would occur. Therefore, it is necessary to adopt explosion-proof measures. Fire and explosion prevention measures must be comprehensive measures, mainly including the following:

a. Select the right electrical equipment;

b. Keep necessary fire spacing;

c. To ensure the normal operation of electrical equipment, good ventilation conditions, fire-resistant facilities, and perfect relay protection devices are necessary;

d. Qualified grounding (neutral connection) measures.

(3) Identification of explosion-proof places and explosion-proof electrical equipment

① See Table 1-10 for the classification of hazardous areas.

Table 1-10　The classification of hazardous areas

Explosive substances	Area definition	Chinese Standards	North American Standards
Gas (CLASS I)	Places where explosive gas mixture exists continuously or for a long time under normal conditions	Zone 0	Div.1
	Places where explosive gas mixture may occur under normal conditions	Zone 1	-
	Places where explosive gas mixture is unlikely to occur under normal conditions, but only occasionally or for a short time under abnormal conditions	Zone 2	Div.2
Dust or fiber (CLASS II / III)	Under normal conditions, the mixture of explosive dust or combustible fiber and air may occur continuously, frequently for a short time or exist for a long time	Zone 10	Div.1
	Under normal conditions, the mixture of explosive dust or combustible fiber and air cannot appear, but only under abnormal conditions, occasionally or for a short time	Zone 11	Div.2

② Explosive gas hazardous area. Explosive gas hazardous areas are divided into Zone 0, Zone 1 and Zone 2 according to the degree of danger.

Zone 0──Environment with continuous or long-term explosive gas mixture.

Zone 1──Environment where explosive gas mixture may occur during normal operation.

Zone 2──Environment in which explosive gas mixtures are unlikely to occur during normal operation, or only exist for a short time even if they do occur.

Zone 0 generally only exists in sealed containers, storage tanks and other internal gas spaces. In the actual design process, Zone 1 also rarely exists, most of which belongs to Zone 2.

(4) Classification of explosion-proof electrical equipment

Class I ── Electrical equipment for underground coal mine.

Class II ── Electrical equipment used in places other than mines.

Class II electrical equipment is divided into Class IIA, IIB and IIC according to the maximum test safety clearance or minimum ignition current ratio applicable to explosive gas mixture; According to the highest surface temperature, they were divided into six groups: T1~T6.

(5) Firefighting of electrical fire

Prevention must be the top priority of electrical fire fighting measures. To put out electrical fire, appropriate methods must be taken according to the fire situation on-site to ensure the safety of fire fighters.

1) Cut off the power

In case of electrical fire, if the power supply has not been cut off at the site, it should be cut off first, which is an important measure to prevent expanding the scope of fire and avoiding electric shock accidents. Pay attention to the following points when cutting off the power supply:

① Reliable insulating tools must be used when cutting off the power supply to prevent electric shock during operation;

② The place where the power is cut off should be properly selected so as not to affect the firefighting work;

③ When cutting off conducting wires, conducting wires of different phases should be cut off at different positions to avoid short circuit;

④ If the conducting wire is loaded, remove the load as much as possible before cutting off the power supply.

2) Prevention of electric shock

In order to prevent electric shock accidents during firefighting, pay attention to keeping a necessary safety distance from the electrified parts when putting out fire with electricity. Water and foam extinguishers should not be used to extinguish the fire. Dry sand, carbon dioxide extinguisher and dry powder extinguisher should be used to extinguish the fire. Prevent directly contacting between the body, hands, feet, or the fire extinguisher to the electric part, or being too close to the electric part to cause an electric shock accident. Insulating rubber gloves should also be worn when putting out fire with electricity.

3) Firefighting of oil-filled equipment

When extinguishing the fire of oil-filled equipment, the following points should be noted:

① In case of fire on oil-filled equipment, carbon dioxide, 1211, dry powder and other fire extinguishers could be used to extinguish the fire; If the fire is serious, cut off the power immediately and use water to extinguish the fire.

② In case of fire inside the oil-filled equipment, the power supply should be cut off immediately, and the spray water gun should be used to extinguish the fire. If necessary, sand and soil could be used to extinguish the fire. The leaked oil fire could be extinguished with a foam extinguisher.

③ In case of fire on running generators and motors, in order to prevent deformation of the shafts and bearings, try to make them rotate slowly, use spray water guns to extinguish the fire and help them cool down. Carbon dioxide extinguisher, 1211 extinguisher and steam could also be used to extinguish fire.

【Special reminder】

In case of emergency or limited conditions, it should be noted to use carbon dioxide, 1211, dry powder and other fire extinguishers, because they are non-conductive and could be used for live extinguishing. When using the carbon dioxide extinguisher, ensure good ventilation, keep it away from the fire area properly, and pay attention to prevent the sprayed carbon dioxide from contaminating the skin. When putting out a live fire, keep a certain distance between the extinguisher body, nozzle and human body and the live body.

1.4.2　Electrical lightning protection and anti-static

(1) Lightning hazard

Lightning is a violent discharge phenomenon in the atmosphere, which is characterized by large current, high voltage, strong electromagnetic radiation, etc. Charged cumulus is the condition for lightning. The average current of lightning is 30kA, and the maximum current could reach 300kA. The voltage of lightning is very high, about 100 million to 1 billion volts. The temperature during lightning is as high as 20k ℃ .

There are four kinds of lightning: direct lightning, spherical lightning, electromagnetic pulse and cloud flash. Direct lightning and spherical lightning would cause harm to people and buildings, while electromagnetic pulse mainly affects electronic equipment, mainly due to induction; Cloud flash occurs between two clouds or between two sides of a cloud, so it is the least harmful to human beings. Lightning could be divided into intra cloud lightning, inter cloud lightning and cloud ground lightning according to its location.

The main hazards of lightning are as follows:

① Lightning generates a strong current, which generates high temperature when passing objects instantaneously, causing burning and melting; When reaching people and animals, it would cause casualties.

② Destroy the power supply and distribution system, causing a large-scale power failure. Lightning explosion and static electricity could cause objects such as trees and electric poles to be split and collapsed.

③ All kinds of power lines, telephone lines and communication lines have high voltage due to lightning strike, causing damage to electrical equipment.

(2) Lightning protection measures

① Improve the insulation level of the line itself.
② Strengthen the protection of weak points of insulation.
③ Insulator root grounding.
④ Install lightning protection devices to protect buildings from direct lightning. A complete set of lightning protection devices includes lightning rod, lightning network, lightning strip, lightning arrester, downlead and grounding device.
⑤ In case of thunderstorm, it is required to stay outdoors as little as possible. If it is necessary to conduct patrol inspection, it is required to wear waterproof raincoats such as plastic, and it is not allowed to work at heights. Do not stand on the top of a mountain, on the roof of a building or close to objects with high conductivity. Close all doors and windows in the station to prevent spherical lightning from entering the room. Keep away from lightning rod tower, chimney, solitary tree, street lamp pole, flag pole and other building facilities in the station as far as possible. Minimize the use of telephones and mobile phones.

(3) Electrostatic hazard and protection

Static electricity is a kind of charge in a static state or a charge that does not flow (the flowing charge forms a current). Static electricity is formed when electric charges collect on an object or surface. When static objects contact zero potential objects (grounded objects) or objects with potential differences between them, charge transfer occurs, which is a spark discharge phenomenon that we see in our daily life.

1) Hazards of static electricity

Firstly, it may cause explosion and fire. Although the energy of static electricity is not large, static sparks appear because of its high voltage and easy discharge; Secondly, electric shock might occur. Although the electric shock generated by static electricity would not cause death, it would often lead to secondary accidents, so we should also take precautions; Thirdly, it might affect production. In production, static electricity might affect the normal operation of instruments and equipment or reduce the quality of products. In addition, static electricity could also cause mis-operation of the electronic automatic components.

2) Eliminate electrostatic protection

① Create conditions to accelerate electrostatic leakage or neutralization. It mainly includes two methods: leakage method and neutralization method. Grounding, humidifying and adding antistatic agent belong to leakage methods; The use of induction static eliminator, radiation static eliminator and lon current electrostatic eliminator belongs to neutralization method. Generally, the enterprises adopt the grounding measures.

② Control the process, including measures taken in material selection, process design, equipment structure and operation management.

Basic knowledge of electrical safety technology

2.1 Basic knowledge of direct current (DC) circuit

2.1.1 Circuit and its basic physical quantities

(1) Classification and composition of circuits

The circuits could be divided into DC circuit and AC circuit according to the frequency of transmission voltage and current; According to their function, they could be divided into power circuit and electronic circuit.

(2) Composition and function of circuit

A simple circuit consists of four major parts, namely power supply, load, control device and connecting wire. The functions of each component of the circuit are shown in Table 2-1.

Table 2-1 Functions of each component of simple circuit

Component	Function	Examples
Power supply	It is the equipment that supplies electric energy, and its function is to provide electric energy for the load in the circuit	Dry battery, storage battery, generator, etc
Load	All kinds of electrical equipment are called loads. They are devices to retrieve electrical energy, and their role is to convert electrical energy into the required form of energy	The bulb converts electric energy into light energy, the motor converts electric energy into mechanical energy, and the electric furnace converts electric energy into heat energy, etc.
Control device	Distribute electric energy in the circuit and control the whole circuit according to the needs of the load	Switches, fuses and other devices or equipment that control the working state (on/off) of the circuit
Connecting wire	It is the intermediate link between the power supply and load to form a path, and plays the role of transmitting and distributing electric energy	Various connecting wires

【Special reminder】

Because there are many types of circuits, the actual circuits are more complex, so that different functions could be realized.

(3) Three working states of the circuit

Three working states of the circuit are shown in Fig.2-1:

① On load state (closed circuit). In the case of on load working state, the power supply is connected to the load, there is current in the circuit, and the load could obtain a certain voltage and electrical power.

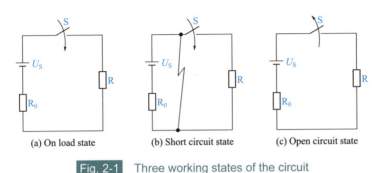

Fig. 2-1　Three working states of the circuit

The manufacturer has specified a value for the working current, voltage, power, etc., of electrical equipment, which is called the rated value of electrical equipment. The state when electrical equipment works at rated value is called rated working state. The on-load state of the circuit is divided into three situations. The rated working state of the circuit is called full load; less than the rated value is called under load; exceeding the rated value is called overload.

② Open circuit state. In the open circuit state, no current flows through the circuit. There are many reasons for open circuit in the circuit, such as switch disconnection, fusant melt, disconnection of electrical equipment and connecting wire, etc.

③ Short circuit state. Short circuit refers to the phenomenon that the places that should not be connected in the circuit are shorted together. In the case of short circuit, the output current is very large. In case of no protection measures, the power supply or load would be burned or even a fire would occur. Therefore, it is usually necessary to install fuse and other safety devices in the circuit to avoid adverse consequences in case of short circuit.

【Special reminder】

The three working states of the circuit have their own uses, for example, some regulation or control loops are often short-circuited.

(4) Basic physical quantity of circuit

The basic physical quantities of the circuit are shown in Table 2-2.

Table 2-2 Basic physical quantities of circuit

Physical quantities	Description	Formula	Symbol	Unit and symbol
Electromotive force (EMF)	The EMF is equal to the work done by the power supply force inside the power supply to move the unit positive charge from low potential (negative pole) to high potential (positive pole). EMF is a physical quantity that reflects the power source's ability to convert other forms of energy into electrical energy. The direction of the EMF is specified as from the negative pole to the positive pole inside the power supply.	$E = \dfrac{W}{q}$	E	Volt, V
Current	The electric quantity passing through any cross-section of conductor in unit time is called current intensity, which is called current for short. The direction of current stipulates that the flow direction of positive charge is the current direction. In a metal conductor, the direction of the current is opposite to that of the directional movement of free electrons	$I = \dfrac{q}{t}$	I	Ampere, A
Voltage	In a circuit, the potential difference between any two points is called the voltage of these two points. Voltage direction regulation: from high potential (positive) to low potential (negative)	$U_{ab} = \dfrac{W_{ab}}{q}$	U_{ab}	Volt, V
Potential	The potential is the voltage between a point in the electric field and a reference point. The reference point could be selected randomly. If the reference point changes, the potential would change.	$V = U_{ab}$	V	Volt, V
Resistance	Resistance is a physical quantity indicating the blocking degree of conductor to current. Resistance usually plays the role of voltage division or shunt in the circuit; For signals, both AC and DC signals could pass through resistance. The magnitude of resistance is related to conductor length, material, cross-sectional area, temperature and other factors	$R = \rho \dfrac{L}{S}$	R	Ohms, Ω

Physical quantities	Description	Formula	Symbol	Unit and symbol
Electric energy	In a period of time, the work done by the electric field force is called electric energy. For the unit of electric energy, people often do not use the Joule, but still use the illegal measurement unit "degree". The conversion relationship between Joule and "degree" is 1 degree (electricity) = 1kW·h = 3.6×10⁶J	$W = qU = IUt = I^2Rt = \dfrac{U^2}{R}$	W	Joule, J
Electric power	The work done by current in unit time is called electric power, which is a physical quantity describing the speed of work done by current. The electric power is equal to the voltage at both ends of the conductor and the current passing through the conductor. The power that the electrical appliance works normally under the rated voltage is called the rated power, and the power that the electrical appliance works under the actual voltage is called the actual power	$P = \dfrac{W}{t} = IU = I^2R = \dfrac{U^2}{R}$	P	Watt, W

【Special reminder】

Correctly understand the definitions of basic physical quantities, memorize their units and symbols, and unit conversion.

2.1.2 Ohm's Law

(1) Ohm's Law of partial circuits

The current I flowing through the conductor is proportional to the voltage U at both ends of the conductor and inversely proportional to the resistance of this conductor. I.e.

$$I = \dfrac{U}{R} \text{ or } U = IR$$

Ohm's Law reveals the relationship among current, voltage and resistance in a circuit. It is one of the basic laws of circuits and is widely used.

(2) Ohm's Law of full circuit

The current of the closed circuit is proportional to the power supply electromotive force and inversely proportional to the resistance of the whole circuit (the sum of internal resistance and external resistance). I.e.

$$I = \frac{E}{R+r} \text{ or } E = IR + Ir$$

The change rule of terminal voltage with the resistance of external circuit is as follows.

① When R increases, because $I = \frac{E}{R+r}$, I would decrease, Ir would decrease, and $U = E/r$, U would increase.

Special example: in case of open circuit, $R = \infty$, $I = 0$, $U = E$。Therefore, connect the voltmeter to both ends of the power supply, and the measured voltage is approximately equal to the power supply EMF (because the internal resistance of the voltmeter is very large). In other words, in open circuit, the terminal voltage is equal to the power supply EMF and the current is zero.

② When R decreases, because $I = \frac{E}{R+r}$, I would increase, Ir would increase, and $U = E - Ir$, U would decrease.

Special case: in the case of short circuit, $R = 0$, $I = E/r$, the current would be large, $U = 0$. In other words, in the case of short circuit, the terminal voltage is zero and the current in the circuit is the maximum. Therefore, it is not allowed to directly connect the ammeter to both ends of the power supply to measure the current (because the internal resistance of the ammeter is very small, it is easy to burn the ammeter and the power supply); At the same time, it is required that protective devices must be set in the circuit to avoid burning out the power supply and causing fire accidents.

【Special reminder】

Ohm's Law is only applicable to pure resistance circuit, while it is not valid to metal conduction, electrolyte conduction, gas conduction and semiconductor components.

2.1.3 Series and parallel connection of resistance

The characteristics and application of resistance series and parallel connection circuits are shown in Table 2-3.

Table 2-3 Characteristics and application of resistance series and parallel connection circuits

Item	Series connection	Parallel connection
Current	The current is equal everywhere, i.e., $I_1 = I_2 = I_3 = \cdots = I$	The total current is equal to the sum of all branch currents, i.e., $I = I_1 + I_2 + \cdots + I_n$
Voltage	The total voltage at both ends is equal to the sum of the voltages at both ends of each resistor, i.e. $U = U_1 + U_2 + U_3 + \cdots + U_n$	The total voltage is equal to the partial voltage, i.e., $U_1 = U_2 = \cdots = U = U_a U_b$
Resistance	The total resistance is equal to the sum of the resistances, i.e. $R = R_1 + R_2 + R_3 + \cdots + R_n$	The reciprocal of the total resistance is equal to the sum of the reciprocal of each parallel resistance, i.e., $\frac{1}{R} = \frac{1}{R_1} + \frac{1}{R_2} + \cdots + \frac{1}{R_n}$
Resistance and voltage division	The voltage distributed on both ends of each resistance is proportional to its resistance value, i.e $U_1 : U_2 : U_3 : \cdots : U_n = R_1 : R_2 : R_3 : \cdots : R_n$	The voltage on each branch resistance is equal
Resistance and shunt	No shunt	The current of each branch is inversely proportional to the resistance value, i.e., $I_1 : I_2 : \cdots : I_n = \frac{1}{R_1} : \frac{1}{R_2} : \cdots : \frac{1}{R_n}$
Power distribution	The power distributed by each resistor is proportional to its resistance value, that is, the current of each branch is inversely proportional to the resistance value, i.e $P_1 : P_2 : P_3 : \cdots : P_n = R_1 : R_2 : R_3 : \cdots : R_n$	The power distributed by each resistor is inversely proportional to the resistance value, i.e., $R_1 P_1 = R_2 P_2 = \ldots = R_n P_n = RP$
Examples	(1) Voltage division: In order to obtain the required voltage, the voltage divider is usually made by using the voltage divider principle of resistance series circuit; (2) For current limiting: a resistor is connected in series in the circuit to limit the current flowing through the load; (3) For voltmeter range expansion: the voltmeter could be modified by using the voltage division function of the series circuit, that is, the ammeter is connected in series with a voltage division resistor, so the ammeter is modified into a voltmeter	(1) Form multi-branch power supply network with equal voltage, such as 220V lighting circuit; (2) Shunt and ammeter range expansion. The shunt function of parallel circuit could be used to refit the ammeter to expand the range, that is, the ammeter is connected in parallel with a shunt resistor, and then the ammeter is refitted into a larger range ammeter

① If there are two resistors in series, the voltage division formula is

$$U_1 = \frac{R_1}{R_1 + R_2} U, \quad U_2 = \frac{R_2}{R_1 + R_2} U$$

② If there are two resistors in series, the power division formula is

$$\frac{P_1}{P_2} = \frac{R_1}{R_2}, \quad \frac{1}{R} = \frac{1}{R_1} + \frac{1}{R_2} + \cdots + \frac{1}{R_n}$$

③ If only two resistors are connected in parallel

$$R = \frac{R_1 R_2}{R_1 + R_2}$$

If three resistors are connected in parallel

$$R = \frac{R_1 R_2 R_3}{R_1 R_2 + R_1 R_3 + R_2 R_3}$$

④ If only two resistors are connected in parallel, the shunt formula is

$$I_1 = \frac{R_2}{R_1 + R_2} I$$

$$I_2 = \frac{R_1}{R_1 + R_2} I$$

The steps to solve the equivalent resistance of the mixed circuit are as follows:

① First, sort the circuit to make the connection relationship clear. Sort out the equivalent circuit diagrams of the series and parallel connection of each resistor.

② Simplify the branch, that is, calculate the equivalent resistance of each branch according to the characteristics of the resistance in series.

③ Merge the branches, that is, further simplify the circuit according to the characteristics of parallel resistance, as shown in Figure 2-2.

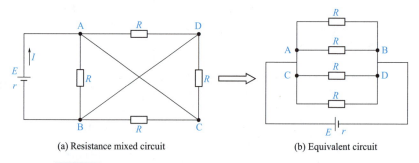

(a) Resistance mixed circuit   (b) Equivalent circuit

Fig. 2-2   Resistance mixed circuit and its equivalent circuit

【Special reminder】

Readers should correctly understand the characteristics and differences between series circuit and parallel circuit in order to solve practical problems in the circuit.

2.1.4　Kirchhoff's Law and its application

(1) Kirchhoff's First Law (KCL)

Kirchhoff's First Law is also called nodal current law, or KCL for short. It means that the sum of current flowing into any node at any time is equal to the sum of current flowing out of the node, i.e.,

$$\Sigma I_{in} = \Sigma I_{out}$$

If the current flowing into the node is positive and the current flowing out of the node is negative, the algebraic sum of the current flowing through any node at any time is equal to zero, which is another expression of Kirchhoff's Law, i.e.,

$$\Sigma I = 0$$

【Special reminder】

The node current law is not only applicable to nodes, but also could be extended to any assumed closed plane. It is still valid.

The reference direction is any assumed direction. If the calculation result is positive, it indicates that the actual direction of the vector is the same as the reference direction; if the calculation result is negative, indicating that the actual direction of the vector is opposite to the reference direction.

(2) Kirchhoff's Second Law (KVL)

Kirchhoff's Second Law is also called loop voltage law, or KVL for short, which determines the relationship between the voltages of each part of a closed loop. At any time, along any circuit bypass direction in the circuit, the algebraic sum of voltages in each section of the circuit is equal to zero, i.e.,

$$\Sigma U = 0$$

【Special reminder】

When Kirchhoff's Law is applied to circuit analysis, it is only related to the connection mode of the circuit, but not to the nature of the components constituting the circuit.

2.1.5 Series and parallel connection of capacitors

The characteristics of capacitor series and parallel circuits are shown in Table 2-4.

Table 2-4 Characteristics of capacitor series and parallel circuits

Physical quantity	Series circuit	Parallel circuit
Quantity of electric charge	$Q = Q_1 = Q_2 = \cdots = Q_n$	$Q = Q_1 + Q_2 + \cdots + Q_n$
Voltage	$U = U_1 + U_2 + \cdots + U_n$ Voltage distribution is inversely proportional to capacitance $\dfrac{U_1}{U_2} = \dfrac{C_1}{C_2}$	$U = U_1 = U_2 = \cdots = U_n$
Capacitance	$\dfrac{1}{C} = \dfrac{1}{C_1} = \dfrac{1}{C_2} = \cdots = \dfrac{1}{C_n}$ When n capacitors with capacitance of C_0 are connected in series $C = \dfrac{C_0}{n}$	$C = C_1 + C_2 + \cdots + C_n$ When n capacitors with capacitance of C_0 are connected in parallel $C = nC_0$

(1) Application of capacitors in series

When the actual working voltage is higher than the rated working voltage of the capacitor, the capacitor could be used in series. It should be noticed that the total capacitance would decrease.

(2) Application of capacitors in parallel

In the actual circuit, when the capacitance of one capacitor is insufficient, several capacitors could be used in parallel to increase its capacitance. The rated working voltage should be the lowest withstand voltage of the capacitors connected in parallel.

【Special reminder】

Although the characteristics of capacitors in series and parallel are similar to those of resistors in series and parallel, they are quite different, and should be compared.

2.2 Basic electromagnetic knowledge

2.2.1 Magnetic field of current

(1) Magnetic induction line

① The tangent direction of a point on the magnetic induction line is the same as the magnetic field direction of the point, as shown in Fig. 2-3 (a).

② The density of the magnetic induction line indicates the strength of the magnetic field, as shown in Fig. 2-3 (b).

③ The magnetic induction lines of uniform magnetic field are parallel lines with uniform distribution.

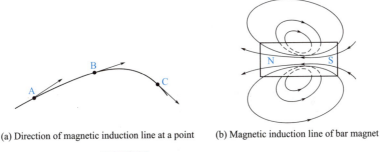

(a) Direction of magnetic induction line at a point   (b) Magnetic induction line of bar magnet

Fig. 2-3　Magnetic induction line

(2) Magnetic field of current

The direction of the linear current and the current around the energized solenoid and the direction of the magnetic field could be determined by the right-hand spiral rule, as shown in Fig. 2-4.

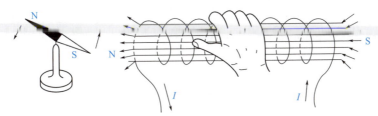

Fig. 2-4　Right-hand spiral rule determines magnetic field of energized coil

The magnetic induction line has the following characteristics.

① The direction of the magnetic induction line outside the magnet is from pole N to pole S, while inside the magnet is from pole S to pole N, forming a closed loop.

② The magnetic induction lines do not intersect each other, that is, the magnetic field direction at any point in the magnetic field is unique.

③ The stronger the magnetic field, the denser the magnetic induction line.

④ When there is magnetic material, the magnetic induction line mainly tends to pass through the magnetic material. The magnetic fields of the energized straight conductor and the energized solenoid are judged by the right-hand spiral rule.

⑤ In the electrified straight conductor, the thumb points to the current direction and the four fingers bend in the direction of the magnetic induction line.

⑥ In the energized solenoid, the four fingers point to the current direction, and the thumb refers to the N-pole direction of the magnetic induction line in the solenoid.

2.2.2 Basic physical quantity of magnetic field

Magnetic induction intensity, magnetic flux, magnetic permeability and magnetic field intensity are the four basic physical quantities describing the magnetic field, as shown in Table 2-5.

Table 2-5 Basic physical quantities of magnetic field

Physical quantity	Symbol	Formula	Description
Magnetic induction intensity	B	$B = \dfrac{F}{IL}$	It is a physical quantity describing the force effect of magnetic field and represents the strength and direction of magnetic field at any point in the magnetic field
Magnetic flux	ϕ	$\phi = BS$	The product of the magnetic induction intensity and a section area perpendicular to it is called the magnetic flux passing through the area
Magnetic permeability	μ	$\mu = \mu_r \mu_0$	Used to measure the magnetic conductivity of materials
Magnetic field intensity	H	$H = \dfrac{B}{\mu}$	It is the ratio of the magnetic induction intensity of a point in the magnetic field to the permeability of the magnetic medium

(1) Magnetic induction intensity

The magnetic induction intensity is a vector, and its direction (namely the magnetic induction direction) is the direction that the N pole points to when the small magnetic needle is placed at the point.

(2) Relative magnetic permeability

Relative permeability is the ratio between the permeability of special medium μ and vacuum permeability μ_0, $\mu_r = \mu / \mu_0$. According to relative permeability μ_r, materials could be divided into 3 categories.

① Materials with $\mu_r < 1$ are called diamagnetic materials, such as hydrogen, copper, graphite, silver, zinc, etc.

② Materials with $\mu_r > 1$ are called paramagnetic substances, such as air, tin, aluminum, lead, etc.

③ Materials with $\mu_r = 1$ are called ferromagnetic materials, such as iron, steel, nickel, cobalt, etc.

(3) Magnetic field intensity

The magnetic field intensity is a vector whose direction is consistent with the direction of magnetic induction intensity.

【Special reminder】

Each of the 4 basic physical quantities has different emphasis on reflecting the properties of magnetic field. The magnetic induction intensity mainly reflects the strength and direction of the magnetic field at a certain point in the magnetic field, and its magnitude is related to the medium in the magnetic field, namely, the magnetic permeability; Magnetic flux reflects the magnetic field of a certain section in the magnetic field, which is also related to the magnetic medium; The magnetic field strength reflects the magnetic field at a certain point in the magnetic field, and is related to the excitation current and conductor shape, but it has nothing to do with the magnetic medium in the magnetic field, that is, the magnetic permeability. It is just a physical quantity introduced to simplify the calculation.

2.2.3 Left-hand rule

(1) Effect of magnetic field on current carrying conductor

① In a uniform magnetic field, the magnitude of the magnetic force on a section of energized conductor perpendicular to the magnetic field direction could be deduced from the definition formula of magnetic induction intensity $B=\dfrac{F}{IL}$, that is $F=BIL$. The direction of the force could be determined by the left-hand rule.

② If in a uniform magnetic field, the current direction is at an angle θ with the magnetic field direction. The force exerted by the magnetic field on the energized conductor
$$F=BIL\sin\theta$$
③ The force couple moment of the magnetic field on the rectangular coil is
$$M=NBIS\cos\theta$$

(2) Application of the left-hand rule

Extend the left hand, make the thumb perpendicular to the other four fingers and in a plane with the palm, let the magnetic induction line penetrate the palm, and the four fingers point to the current direction. The direction pointed by the thumb is the direction of the force exerted on the energized straight wire in the magnetic field, as shown in Fig. 2-5.

Memorizing tips of left-hand rule

When electric current passes through a straight wire, electromagnetic force could be generated. The left hand is used for force judgment, the four fingers and the thumb are perpendicular. Put flat left hand into magnetic field, and N pole heads to the palm center. The four fingers point to the current direction and the thumb to the electromagnetic force.

---

(3) Acting torque of magnetic field on energized coil

It could be found from the left-hand rule that: $F \perp B$, $F \perp I$, that is, F is perpendicular to the plane determined by B and I.

As shown in Fig. 2-6, when the coil plane is parallel to the magnetic induction line, the force arm is the largest and the magnetic torque of the coil is the largest; When the coil plane is perpendicular to the magnetic induction line, the force arm is zero and the magnetic moment of the coil is zero.

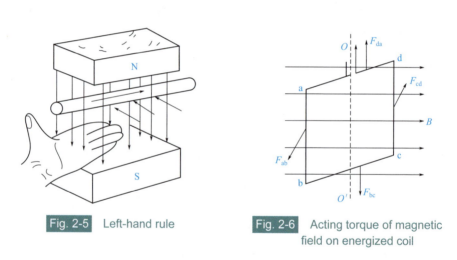

Fig. 2-5 Left-hand rule

Fig. 2-6 Acting torque of magnetic field on energized coil

2.2.4 Electromagnetic induction phenomenon and Lenz's Law

(1) Conditions of induced EMF and induced current

① As long as the magnetic flux passing through the closed loop changes, induced EMF and induced current would be generated in the loop. If the loop is not closed, there is only induced EMF and no induced current.

② As long as some conductors in the closed circuit cut the magnetic induction line in the magnetic field, induced current would be generated in the circuit. This is only a special case of electromagnetic induction. Because when a part of the conductor in the closed circuit cuts the magnetic induction line in the magnetic field, the magnetic flux in the circuit changes.

(2) Lenz's Law

① Lenz's Law: The direction of the induced EMF in the coil always tries to make the magnetic field of the induced current generated by it to block the change of the original magnetic flux.

② The specific steps to determine the direction of induced current by Lenz's Law are as follows:

a. Determine the direction of the original magnetic flux;

b. Determine whether the change of the original magnetic flux passing through the circuit is increasing or decreasing;

c. Determine the magnetic field direction of the induced current according to Lenz's Law;

d. According to the right-hand spiral rule, the direction of the induced current is determined by the direction of the magnetic field of the induced current.

③ Scope of application of Lenz's Law: Lenz's Law is a general law for judging the direction of induced current. It is not only applicable to the determination of the direction of the induced current generated by the movement of partial conductor in the closed circuit cutting the magnetic induction line in the magnetic field, which is the same as the result determined by the right-hand rule, but also applicable to the determination of the direction of the induced current generated when the magnetic flux passing through the closed circuit changes.

(3) Determination of induced EMF and induced current direction

① The content of the right-hand rule: extend the right hand flat, make the thumb perpendicular to the other four fingers, the palm facing the incoming direction of the magnetic induction line, and the thumb pointing to the moving direction of the conductor cutting the magnetic induction line, then the four fingers point to the direction of the induced current, as shown in Fig. 2-7.

② Scope of application of the right-hand rule: The right-hand rule is applicable to the determination of the direction of the induced current generated by the movement of partial conductor in the closed circuit cutting the magnetic induction line in the magnetic field.

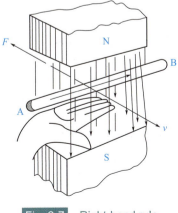

Fig. 2-7　Right-hand rule

Memorizing tips of right-hand rule

The wire cuts the magnetic induction line, and the induced potential is generated.
The conductor is externally connected with a closed circuit, and the induced current is judged by the right-hand rule.
Put the flat right hand into magnetic field, and the palm faces to the N pole.
The thumb points to the direction of wire movement, and the four finger point to the direction of current.

2.2.5 Self-inductance and mutual inductance

(1) Self-induction phenomenon

① The phenomenon that the coil itself causes induced EMF due to the change of current in the coil is called self induction.

② The inductance L of the hollow coil is a constant, that is $L = \dfrac{\psi}{I}$, its magnitude is related to the coil size, geometry, turns and other factors, and has nothing to do with the magnitude of the current. The self-inductance coefficient L of the coil is the property of the coil itself, which reflects the ability of the coil to generate self induced EMF and store magnetic field energy.

③ Application of self-induction phenomenon: self-induction phenomenon is a special electromagnetic induction phenomenon, which is widely used in electrical equipment and radio technology. The ballast of fluorescent lamp is an example of using coil self-induction phenomenon.

 【Special reminder】

Self-induction also has a negative side. In the circuit with large self-inductance coefficient and strong current (such as the stator winding of large motor), at the moment of cutting off the power, due to the great change of current in a very short time, high self-inductance EMF would be generated, which would ionize the air between the switch knife and the fixed clip into a conductor and form an arc. Therefore, a special safety switch must be used to cut off this circuit.

(2) Mutual inductance

① For two adjacent coils, when the current in one coil changes, the magnetic flux of the other coil changes. This phenomenon is called mutual inductance.

② Mutual inductance coefficient: the mutual inductance coefficient of two coils completely depends on the structure, size, turns and relative position of the two coils, independent of the current of the coil. It reflects the ability of the two coils to generate mutual inductance magnetic flux and mutual inductance EMF.

2.3 Fundamentals of alternating current (AC)

2.3.1 Single phase sinusoidal AC

(1) Characteristics of sinusoidal AC

Sinusoidal AC has three characteristics.
① Instantaneous: In a cycle, the instantaneous values at different times are different.
② Periodicity: every other same time interval, the change rule is the same.
③ Regularity: change according to the law of sine function.

(2) Physical quantities and analytic expressions of sinusoidal AC

The current, voltage and EMF, whose magnitude and direction change periodically with time according to the sinusoidal law are called sinusoidal AC, voltage and EMF. The instantaneous value of at a certain time t could be expressed by an analytical formula (trigonometric function formula), i.e.

$$i(t) = I_m \sin(\omega t + \varphi_{i0})$$
$$u(t) = U_m \sin(\omega t + \varphi_{u0})$$
$$e(t) = E_m \sin(\omega t + \varphi_{e0})$$

Where, I_m, U_m and E_m are respectively called the amplitude of AC current, voltage and EMF (also called peak value or maximum value), the unit of current is ampere (A), and the unit of voltage and EMF is volt (V). ω is the angular frequency of AC, in radians per second (rad/s), which represents the electrical angle of sinusoidal AC changes per second; φ_{i0}、φ_{u0}、φ_{e0} are respectively the initial phase of current, voltage, and EMF, and the unit is radian (rad) or degree (°), which represents the electrical angle of sinusoidal AC at the initial time ($t=0$).

Generally, amplitude (maximum or effective value), frequency (or angular frequency, period), and initial phase are called the three elements of AC. Knowing the three elements of AC, we could write its analytic formula and draw its waveform. On the contrary, knowing the AC analytic formula or waveform could also find out its three elements. Any sinusoidal quantity has these three elements.

To sum up, the basic electrical parameters (physical quantities) representing sinusoidal AC are shown in Table 2-6.

Table 2-6 Physical quantities of sinusoidal AC

Physical quantity	Concept	Symbol
Instantaneous value	Values of current, voltage, EMF and power varying with time at any moment	Expressed by i, u, e, p respectively. For example, the EMF is expressed as $e = E_m \sin \omega t$
Max value	The max value that sinusoidal AC could reach in a cycle, also known as amplitude, peak value, etc.	Expressed by E_m, U_m, I_m respectively
Effective value	The effective value of sinusoidal AC is specified according to the thermal effect of the current. That is, let AC and DC pass through the same resistance respectively. If the heat generated by them is equal within the same time, the value of DC is defined as the effective value of AC.	Expressed by E, U, I respectively
Average value	The ratio of the current passing through the conductor cross-section in the same direction and the half cycle time of the sinusoidal AC in half a cycle	Expressed by I_{PJ}, U_{PJ}, E_{PJ} respectively
Period	The time it takes for the AC to complete a periodic change (or the rotor of the generator rotates for one cycle)	Expressed in T, in second (s)
Frequency	The times that AC completes periodic changes in unit time (1s) (or the number of turns the generator rotates in 1s)	Expressed in f, the unit is Hz. Frequency is also commonly used in kilohertz (kHz) and megahertz (MHz). Their relationship is $1kHz=10^3 Hz$; $1MHz=10^6 Hz$
Angular frequency	Change of electrical angle of AC in 1s (i.e., the geometric angle that the generator rotor rotates in 1s)	Expressed in ω, in radians per second (rad/s)
Phase position	The physical quantity represents the state of sinusoidal AC at a certain time. The phase position not only determines the magnitude and direction of the instantaneous value of sinusoidal AC, but also reflects the changing trend of sinusoidal AC	In the trigonometric function of sinusoidal AC, "$\omega t + \Psi$" is the phase position of sinusoidal AC. Unit: degree (°) or radian (rad)
Initial phase	A physical quantity representing the state at the beginning of the sinusoidal AC. The phase of the sinusoidal AC when t=0 (or the angle between the coil plane and the neutral plane of the generator rotor before rotation) is called the initial phase position, referred to as the initial phase	It is expressed as Ψ_0. The magnitude of the initial phase is related to the selection of the time starting point, and the absolute value of the initial phase is represented by an angle less than π
Phase difference	The phase position difference between two sinusoidal AC with the same frequency at any moment is the phase difference	Expressed in $\Delta\Psi$

① Relation between valid value and maximum value

$$I = \frac{I_m}{\sqrt{2}} = 0.707 I_m, \quad U = \frac{U_m}{\sqrt{2}} = 0.707 U_m, \quad E = \frac{E_m}{\sqrt{2}} = 0.707 E_m$$

② Relation between average value and maximum value

$$I_{pj} = \frac{2}{\pi} I_m = 0.637 I_m, \quad U_{pj} = \frac{2}{\pi} U_m = 0.637 U_m, \quad E_{pj} = \frac{2}{\pi} E_m = 0.637 E_m$$

③ Relationship among period, frequency and angular frequency

$$\omega = 2\pi f = \frac{2\pi}{T}, \quad f = \frac{1}{T} = \frac{\omega}{2\pi}, \quad T = \frac{1}{f} = \frac{2\pi}{\omega}$$

(3) A method of drawing sinusoidal AC waveform

① Horizontal axis: in the waveform of sinusoidal AC, the horizontal axis could be time t, period T and electrical angle ωt, etc. When reading and drawing, pay special attention to what the horizontal axis represents.

② Vertical axis: Look at the letters i, u and e marked on the vertical axis to distinguish whether they represent current, voltage or EMF.

③ Initial phase: It is the most difficult in waveform drawing. When drawing, see whether the initial phase is positive or negative. In the case of negative angle, the starting point of the timing point ($t=0$) should be in the negative half cycle, and its waveform is changed from a negative value to a positive value to the zero point intersected with the horizontal axis. As shown in Fig. 2-8 (a), the electrical angle between point A and origin point O is a negative initial phase, which is on the right side of the vertical axis. In the case of positive angle, the timing starting point $t=0$ should be in the positive half cycle, and its waveform changes from a negative value to a positive value to the zero point intersecting the horizontal axis. As shown in Fig. 2-8 (b), the electrical angle between A' and the origin O is a positive initial phase, which is on the left side of the vertical axis.

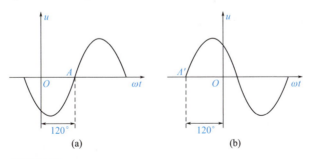

Fig. 2-8 Example of drawing sinusoidal AC waveform

(4) Single phase sinusoidal AC circuit

The basic AC circuits include pure resistance circuit, pure inductance circuit and pure capacitance circuit. See Table 2-7 for the comparison of their properties.

Table 2-7   Properties comparison of three basic AC circuits

Properties			Pure resistance circuit	Pure inductance circuit	Pure capacitance circuit
Voltage and current vector diagram			→i →U	↑$\dot{U}_L$ →i	↑i ↓$\dot{U}_C$
Impedance characteristic	Impedance		Resistance R	Inductive reactance $X_L = \omega L$	Capacitive reactance $X_C = 1/(\omega C)$
	DC characteristic		Showing a certain blocking effect	Conducting DC (equivalent to short circuit)	Cut off DC (equivalent to open circuit)
	AC characteristic		Showing a certain blocking effect	Pass low frequency, resist high frequency	Pass high frequency, resist low frequency
Relation between voltage and current	Magnitude relation		$U_R = RI_R$	$U_L = X_L I_L$	$U_C = X_C I_C$
	Phase position relationship (phase position difference between voltage and current)		u、i are same	u is 90° ahead of i	u is 90° behind of i
Power condition			Energy consuming component, with active power $P_R = U_R I_R = I^2 R$ (W)	Energy storage component ($P_L = 0$), with reactive power $Q_L = U_L I_L = I^2 X_L$(Var)	Energy storage component ($P_C = 0$), with reactive power $Q_C = U_C I_C = I^2 X_C$(Var)
Meet Ohm's Law parameters			Max. value, effective value, instantaneous value	Max. value, effective value	Max. value, effective value

2.3.2 Three phase AC

(1) Formula of three phase AC

There are three groups of coils on the generator stator. Because the geometric dimensions and turns of the three groups of coils are equal, the amplitude of the three EMF e_1, e_2 and e_3, frequencies are the same, and with the same phase difference of $\dfrac{2\pi}{3}$. If the initial phase of e_1 is specified as zero, the instantaneous value formula of three phase EMF is

$$\begin{cases} e_1 = E_m \sin \omega t \\ e_2 = E_m \sin(\omega t - \dfrac{2\pi}{3}) \\ e_3 = E_m \sin(\omega t + \dfrac{2\pi}{3}) \end{cases}$$

(2) Waveform of three phase AC

The waveform of three phase AC power supply is shown in Fig. 2-9. It could be seen from the figure that E_1 is $\dfrac{2\pi}{3}$ ahead of E_2 to the maximum and E_2 is $\dfrac{2\pi}{3}$ ahead of E_3 to the maximum. Normally, the phase sequence of three phase AC power supply is L1–L2–L3.

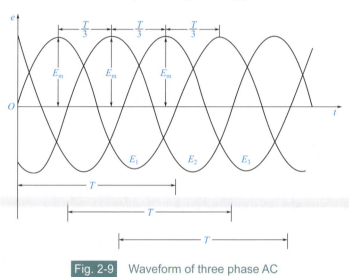

Fig. 2-9　Waveform of three phase AC

(3) Three-phase-four-wire power supply

Each coil that generates three phase EMF is called a phase, and the outgoing line at its head is called a phase wire (commonly known as a live wire). Three coils have three ends, which are connected together and led out with a wire, which is called neutral wire (commonly known as zero wire). The power supply system of this connection mode is called three-phase-four-wire power supply system, as shown in Fig. 2-10, which is represented by the symbol "Y".

Three-phase-four-wire power supply outputs two kinds of voltage, namely phase voltage and wire voltage. The voltage between each phase wire and the neutral wire is called phase voltage, and its effective value is represented by U_1, U_2 and U_3 respectively. The voltage between phase wires is called wire voltage, and U_{12}, U_{23} and U_{31} respectively represent the

effective value. The relationship between them is
The vector diagram is shown in Fig. 2-11.

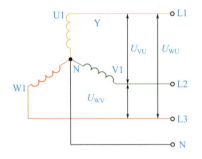

Fig. 2-10 Three-phase-four-wire power supply system

Fig. 2-11 Voltage vector of three-phase-four-wire power supply

It could be seen from the figure that the relationship between wire voltage U_{12} and phase voltages U_1 and U_2:

$$U_{12} = 2U_1 \cos 30° = \sqrt{3} U_1$$

Similarly:

$$U_{23} = \sqrt{3} U_2, \quad U_{31} = \sqrt{3} U_3$$

In general, U_L is used to represent wire voltage and U_Φ is used to represent phase voltage, so the relationship between wire voltage U_L and phase voltage U_Φ is

$$U_L = \sqrt{3} U_\Phi$$

It could also be seen in Fig. 2-11 that U_{12}, U_{23} and U_{31} is $\dfrac{\pi}{6}$ ahead of U_1, U_2 and U_3 respectively, and the phase difference between the three wire voltages is still $\dfrac{2\pi}{3}$, so the phase voltage and wire voltage are centrally symmetric.

 【Special reminder】

> Voltage relationship of three phase power supply in Y-configuration: the wire voltage is 1.732 times ($\sqrt{3}$ times) of the phase voltage; Voltage relationship when three phase power supply is in Δ-configured: the value of wire voltage is equal to that of phase voltage.

(4) Characteristics of three phase AC

① Symmetrical three phase EMF. The three phase EMF has the same amplitude, frequency and phase difference of $\dfrac{2\pi}{3}$. In engineering, it is called symmetrical three phase EMF.

The power supply that could supply symmetrical three phase EMF is called three phase power supply.

② Characteristics of three phase power supply

a. The effective value of symmetrical three phase EMF is equal, the frequency is the same, and the phase difference between each phase is $\frac{2\pi}{3}$.

b. The phase voltage and wire voltage of three-phase-four-wire power supply are centrally symmetric.

c. The wire voltage is $\sqrt{3}$ times of the phase voltage, and the wire voltage phase is $\frac{\pi}{6}$ ahead of the phase voltage phase, i.e.,

$$\begin{cases} U_L = \sqrt{3} U_\Phi \\ \varphi_L - \varphi_\Phi = \frac{\pi}{6} \end{cases}$$

(5) Y-configuration of three phase load

① The Y-configuration of three phase load. Connect the three ends of U-phase, V-phase and W-phase loads together and connect them to the neutral wire of the power supply. The other end of each phase is connected to three phase wires respectively. This configuration is called Y-configuration of three phase loads, and the symbol is "Y". The schematic diagram and actual circuit diagram are shown in Fig. 2-12.

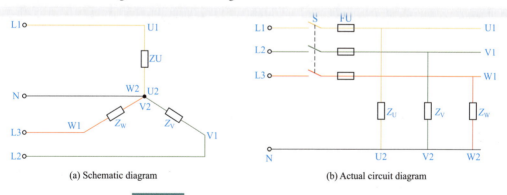

Fig. 2-12 Y-configuration of three phase load

② Y-configuration of three phase symmetrical load, or three-phase-three-wire circuit. When the three phase load is symmetrical, the current in neutral wire is zero, and removing the neutral wire would not affect the normal operation of the three phase circuit. After removing the neutral wire, it becomes a three-phase-three-wire circuit, as shown in Fig. 2-13. Common three phase motors, three phase transformers, three phase electric furnaces, etc., are three phase symmetrical loads, so three-phase-three-wire circuit could be used for power supply.

③ Y-configuration of three phase asymmetrical load —— three-phase-four-wire circuit.

When three phase loads are asymmetric, three-phase-four-wire power supply (as shown in Fig. 2-14) must be adopted to ensure that each phase load could obtain the same phase voltage. Meanwhile, the current in neutral wire is not zero, and the role of the neutral wire is very important. The neutral wire could make the three phase circuit into three independent circuits that do not affect each other. No matter whether the load changes, the load would bear the same phase voltage. In order to prevent accidents, it is stipulated in the three-phase-four-wire system that fuses and switches are not allowed to be installed in the neutral wire. Generally, the neutral wire should be grounded to make it equipotential with the ground to ensure safety.

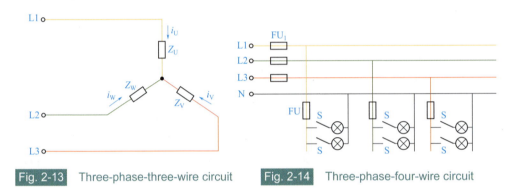

Fig. 2-13 Three-phase-three-wire circuit   Fig. 2-14 Three-phase-four-wire circuit

Once the neutral wire is disconnected, the voltage at both ends of a phase load in the circuit would rise and exceed the rated voltage to damage the electrical equipment; It would also reduce the voltage of a phase load and fail to reach the rated voltage, so that the electrical appliances could not work normally.

(6) △-configuration of three phase load

The three phase loads are connected to each two phase wire of the three phase AC power supply respectively. This connection mode is called delta configuration. It is indicated by the symbol "△", as shown in Fig. 2-15.

When symmetrical loads are configured in △, the wire current is $\sqrt{3}$ times of the phase current. i.e.,

$$I_{\Delta L} = \sqrt{3} I_{\Delta \Phi}$$

When the magnitude and nature of the load are identical, that is $R_U = R_V = R_W$, $X_U = X_V = X_W$, such three phase load is called symmetrical three phase load, otherwise it is asymmetrical three phase load.

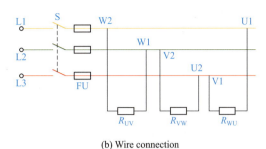

(a) Schematic diagram   (b) Wire connection

Fig. 2-15   △-configuration of three phase load

① Phase voltage and wire voltage of symmetrical three phase load. In a Y-configuration, the voltage at both ends of the load is called the phase voltage, expressed in $U_{Y\Phi}$. When the transmission wire resistance is not considered, the phase voltage $U_{Y\Phi}$ of the load is equal to the phase voltage of the power supply U_Φ; The wire voltage U_{YL} of the load is equal to the wire voltage U_L of the power supply. Therefore, the relationship between load wire voltage U_{YL} and phase voltage $U_{Y\Phi}$ is:

$$\begin{cases} U_{YL} = U_L \\ U_{Y\Phi} = U_\Phi \\ U_L = \sqrt{3}U_\Phi \end{cases}$$

② The phase current and wire current of symmetrical three phase load are symmetrical, so the current flowing through each phase load (phase current) is equal, that is

$$I_{Y\Phi} = I_{U\Phi} = I_{V\Phi} = I_{W\Phi} = \frac{U_{Y\Phi}}{Z_\Phi}$$

The phase difference between the currents of each phase is $\frac{2\pi}{3}$, so it is enough to calculate the current of only one phase.

③ Neutral wire current I_N

According to Kirchhoff's first law, the current flowing into the neutral wire:

$$i_N = i_U + i_V + i_W$$

Rotation vector relation corresponding to the above equation:

$$\dot{I}_N = \dot{I}_U + \dot{I}_V + \dot{I}_W$$

Draw the rotation vector of phase current of symmetrical three phase load, as shown in Fig. 2-16, the sum of the vectors $\dot{I}_U$, $\dot{I}_V$, $\dot{I}_W$ is zero, and the neutral current is zero, that is

$$I_N = 0$$

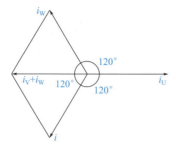

Fig. 2-16　Rotation vector diagram of phase current of three phase load

【Special reminder】

Three phase load could be configured in Y or Δ. The specific configuration should be determined according to the rated voltage of the load and the power supply voltage. The voltage of each phase load must be equal to the rated voltage. For example, for three phase power supply with 380V wire voltage, when the rated load voltage is 220V, the load should be configured into Y; When the rated voltage of the load is 380V, the load should be configured in Δ.

(7) Power of three phase AC circuit

The active power of three phase load is equal to the sum of the power of each phase, namely:

$$P = P_1 + P_2 + P_3$$

In a symmetrical three-phase circuit, no matter whether the load is Y-configured or Δ-configured, the three phase power is:

$$P = 3U_P I_P \cos\varphi = \sqrt{3} U_L I_L \cos\varphi$$

Where, φ is the impedance angle of symmetrical load and the phase difference between load phase voltage and phase current. Apparent power of three phase circuit:

$$S = 3U_P I_P = \sqrt{3} U_L I_L$$

Reactive power of three phase circuit:

$$Q = 3U_P I_P \sin\varphi = \sqrt{3} U_L I_L \sin\varphi$$

Power factor of three phase circuit:

$$\lambda = \frac{P}{S} = \cos\varphi$$

2.4 Fundamentals of electronic technology

2.4.1 Crystal diode and rectifier circuit

(1) Recognizing diode

① Structure and symbol. There is a PN junction inside the crystal diode. A lead wire is led out at each end of the PN junction, and then it is packaged with a housing. The lead from area P is called anode (positive pole), and the lead from area N is called cathode (negative pole). Diode has unidirectional conductivity. Its structure and circuit symbol are shown in Fig. 2-17.

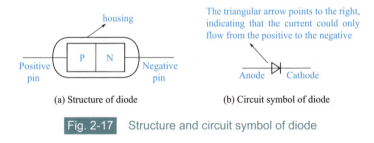

Fig. 2-17 Structure and circuit symbol of diode

② Types

a. According to different structures, diodes could be divided into point contact type and surface contact type. The junction capacitance of the point contact diode is small, and the forward current and the allowed reverse voltage are small. It is commonly used in the detection, frequency conversion and other circuits; The junction capacitance of the surface contact diode is large, the forward current and the allowed reverse voltage are large, and they are mainly used in rectifier and other circuits. The plane-type diode is widely used in the surface contact diode, which could be used as a switch transistor in the pulse digital circuit through higher current.

b. According to different materials, diodes could be divided into germanium diodes and silicon diodes. Compared with the silicon transistor, germanium transistor has the characteristics of lower forward voltage drop (0.2~0.3V for germanium transistor and 0.5~0.7V for silicon transistor), large reverse saturated leakage current, and poor temperature stability.

c. According to different uses, diodes could be divided into common diodes, rectifier diodes, switching diodes, light-emitting diodes, varactor diodes, voltage stabilizing diodes, photodiodes, etc. The diode is turned on by adding forward voltage and turned off by adding reverse voltage. Unidirectional conductivity is the most important characteristic of diode.

③ Identification

a. There are three methods for marking the polarity of the hand-inserted diode pin: direct marking, color ring marking and color dot marking, as shown in Fig. 2-18. The positive and negative polarity of the pin could be identified by observing the marks on the diode package carefully.

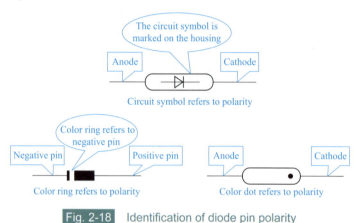

Fig. 2-18   Identification of diode pin polarity

Some diodes manufacturers use the symbols "P" and "N" to mark the diode polarity.

b. There are two types of surface mounting diode (SMD): chip and tube. For the identification of positive and negative polarities of SMD, it is usually sufficient to observe the marking on the housing. Generally, the negative pole is represented by a gray bar or color ring at one end, as shown in Fig. 2-19.

Fig. 2-19   SMD diode polarity identification

c. The positive and negative poles of metal packaged high-power diode could be distinguished according to its shape characteristics, as shown in Fig. 2-20.

d. The positive and negative poles of the light-emitting diodes (LED) could be identified from the length of the pin. The long pin is positive, and the short one is negative. If the pins are the same length, the cathode is the one with a larger area inside the LED, and the anode is the one with a smaller area, as shown in Fig. 2-21. Some LEDs have a small plane, and a lead near the small plane is the negative pole.

Fig. 2-20 Polarity identification of metal encapsulated high power diodes

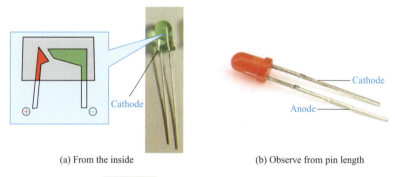

(a) From the inside   (b) Observe from pin length

Fig. 2-21 LED polarity identification

e. Identification of diode polarity on PCB. In PCB, the polarity of diode could be determined by looking at the symbol of silk screen. Several symbols of diode polarity on PCB are shown in Fig. 2-22.

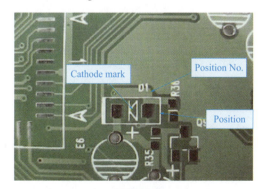 

Fig. 2-22 Several symbols of diode polarity on PCB

The common symbols of diode polarity on PCB are as follows:

- The notched end is the negative pole;
- The end with bar marking is the negative pole;
- The end with dual white bars is the negative pole;
- The end in the direction of the triangle arrow is the negative pole;
- The end with the smaller silk screen circle on the plug-in diode is the negative pole, and the bigger circle is the positive pole.

(2) Main parameters of diode

① The maximum rectification current I_{OM} refers to the maximum average forward DC current value that the diode is allowed to pass when working for a long time.

② The maximum reverse working voltage U_{RM} refers to the maximum reverse working voltage allowed being applied when the diode is used normally.

③ The reverse current I_R refers to the reverse current value when the diode breaks down. The smaller the value, the better the unidirectional conductivity of the diode. The value of reverse current is closely related to temperature, so this parameter should be paid special attention to when using diodes in high temperature environment.

④ The maximum operating frequency f_M is mainly determined by the junction capacitance of the PN junction. Beyond this value, the unidirectional conductivity of the diode would not be well reflected.

(3) Characteristics of common diodes

The characteristics of common diodes are shown in Table 2-8. In the circuit, diodes are often used in rectifier, switch, detection, amplitude limiting, clamping, protection and isolation.

Table 2-8 Characteristics of common diodes

Name	Characteristics	Name	Characteristics
Rectifier diode	The AC is changed into pulsating DC, by using the unidirectional conductivity of PN junction.	Switching diode	The unidirectional conductivity of the diode could be used to control the current in the circuit, which could be turned on or off
Detector diode	Detect the low-frequency signal modulated on the high-frequency electromagnetic wave.	LED	A semiconductor light emitting device, commonly used as an indicating device in electronic appliances
Varactor diode	The junction capacitance varies with the magnitude of the reverse voltage applied to the transistor, which is used to replace the variable capacitor	Zener diode	The voltage at both ends is constant at a certain value, and basically does not change with the current, when the diode is reversely breakdown.

(4) Diode rectifier circuit

The process of changing AC into unidirectional pulsating current is called rectification. The circuit that fulfill this function is called rectifier circuit, also called rectifier. Common diode single-phase rectifier circuits include half wave rectifier circuit, full wave rectifier circuit and bridge rectifier circuit, as shown in Table 2-9.

Table 2-9 Performance comparison of diode single-phase rectifier circuits

Comparison items \ Circuit name	Single-phase half wave rectifier circuit	Single-phase full wave rectifier circuit	Single-phase bridge full wave rectifier circuit
Circuit structure			
Rectified voltage waveform			
Average value of load voltage U_0	$U_0=0.45U_2$	$U_0=0.9U_2$	$U_0=0.9U_2$
Average load current I_0	$I_0=0.45U_2/R_L$	$I_0=0.9U_2/R_L$	$I_0=0.9U_2/R_L$
Average current I_U through each rectifier diode	$I_U=0.45U_2/R_L$	$I_U=0.9U_2/R_L$	$I_U=0.9U_2/R_L$
Maximum reverse voltage U_{RM} of rectifier transistor	$U_{RM}=\sqrt{2}U_2$	$U_{RM}=2\sqrt{2}U_2$	$U_{RM}=\sqrt{2}U_2$
Advantages and disadvantages	Simple, the output rectification voltage fluctuates greatly, and the rectification efficiency is low	Complex, the output voltage fluctuation is small, and the rectification efficiency is high, but the diode withstands high reverse voltage	Complex, the output voltage fluctuation is small, the rectification efficiency is high, and the output voltage is high
Scope of application	The output current is not large and the requirements for DC stability are not high	The output current is large and the requirements for DC stability are high	The output current is large and the requirements for DC stability are high

> **· Tips on diode rectifier circuits ·**
>
> There are two types of rectifier circuits, half wave and full wave ones.
> Half wave rectifier is relatively simple, with output voltage of 0.45 time of the origin.
> Full wave rectifier is more complex, with output voltage of 0.9 time of the origin.

2.4.2 Crystal triode and its amplification circuit

(1) Identification of triode

① The triode has two PN junctions, three regions and three polarities. For example, the semiconductor of PNP type triode is arranged in the order of P, N, P, its middle layer is N-type semiconductor, and its upper and lower layers are P-type semiconductors. The three polarities are collector, emitter and base, as shown in Fig. 2-23.

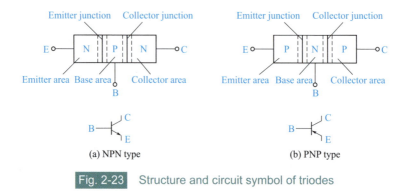

Fig. 2-23 Structure and circuit symbol of triodes

In the symbol of the triode, the arrow marked on the emitter represents its current direction, that is, the current direction when the emitter junction is positively biased.

② The triode is divided into NPN type triode and PNP type triode according to the semiconductor type of the three internal regions; According to semiconductor materials, there are germanium triode and silicon triode.

③ Common triodes could be divided into plastic package triodes, metal package triodes and surface mounting triodes according to different packaging modes. The arrangement of the three pins of triodes in various packaging modes has certain rules to follow, which could be generally identified and judged by the shape, as shown in Fig. 2-24. For the pin judgment of individual special triode, the identification could not completely rely on the shape, and it needs to be tested with the multimeter.

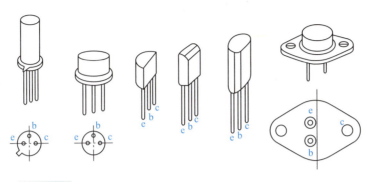

Fig. 2-24　Packaging and pins arrangement of common triodes

④ Main technical parameters of triode

a. AC current amplification factor. Including common emitter current amplification factor β and common base current amplification coefficient, which is an important parameter indicating the amplification ability of transistors.

b. Maximum allowable current I_{CM} of collector. It refers to the collector current when the current amplification coefficient of the triode decreases significantly.

c. Reverse breakdown voltage U_{CEO} between collector and emitter. It refers to the maximum reverse voltage allowed to be applied between the collector and emitter when the base of the triode is open circuit.

d. Maximum allowable dissipation power P_{CM} of collector. It refers to the maximum collector dissipation power when the triode parameter change does not exceed the specified allowable value.

Among the main parameters of the triode, the maximum allowable current I_{CM} of the collector, the reverse breakdown voltage U_{CEO} between collector and emitter, and the maximum allowable dissipated power P_{CM} of the collector are limit parameters. In the triode selection, the actual value in the circuit is not allowed to exceed the limit parameters, otherwise the triode would be damaged. Penetrating current I_{CEO} and current magnification β between collector and emitter are parameters indicating the excellent performance of the triode, especially the requirement on penetrating current I_{CEO} between collector and emitter is the smaller the better.

【Special reminder】

These parameters reflect various performances of triode from different aspects, and are important basis for selecting triode. When using a triode, it is absolutely not allowed to exceed the limit parameters.

⑤ Current relationship of triode

$$I_E = I_B + I_C$$
$$I_C = \beta I_B$$
$$I_E = (1+\beta) I_B$$

(2) Characteristic curve of triode

① Input characteristic curve. It refers to the relationship between the base current I_B and the emitter junction voltage drop U_{BE} of the triode while the U_{CE} remains unchanged.

Since the emitter junction is a PN junction with diode properties, the input characteristics of triode are very similar to the volt ampere characteristics of diode. Generally speaking, the threshold voltage of the silicon transistor is about 0.5V, and when the emitter junction is fully turned on, the U_{BE} is about 0.7V; The threshold voltage of germanium transistor is about 0.2V, and the U_{BE} is about 0.3V when the emitter junction is fully turned on.

② Output characteristic curve. It refers to the relationship between collector current I_C and U_{CE} when the input current I_B of the triode remains unchanged, as shown in Fig. 2-25. It could be seen from the figure that when I_B is unchanged, I_C does not change with the change of U_{CE}; When I_B changes, the relationship between I_C and U_{CE} is a group of parallel curve groups, which have three working areas: cut-off, amplification and saturation.

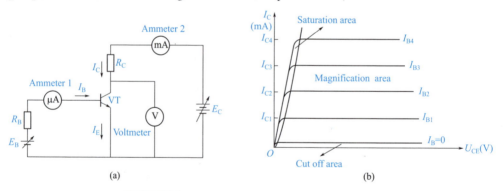

Fig. 2-25　Output characteristic curve of triode

See Table 2-10 for the characteristics of three working states of triode.

Table 2-10　Comparison of three working states of triode

Comparison items \ State	Cut-off	Amplification	Saturation
Position on the output characteristic curve	Areas below $I_B=0$	Parallel and equidistant areas in curves	The area from the steep part on the left side of the curve to the vertical axis

Comparison items \ State	Cut-off	Amplification	Saturation
PN junction offset state	Collector & emitter junctions reverse offset	Collector junctions reverse offset, Positive offset of emitter junction	Positive offset of collector & emitter junctions
Equivalent state between c & e	Equivalent to "switch" off	Constant current source controlled by I_B	Equivalent to "switch" on
Relationship between I_B and I_C	$I_B=0$, $I_C \approx 0$	Controlled $I_C = \beta I_B$	I_B and I_C are large, but I_C is not controlled by I_B

(3) Triode amplifier circuit

A complete amplifying circuit must have a current amplifying element (triode), and must also meet the DC conditions (positive offset of the emitter junction, reverse offset of the collector junction) and AC conditions (the AC path must be unblocked). For the analysis of amplification circuit, the static working point analysis should be carried out first, and then the dynamic analysis. See Table 2-11 for comparison of three configurations of amplifier circuit.

Table 2-11 Comparison of three configurations of amplifier circuit

Comparison items \ Configuration	Common emitter amplifier	Common collector amplifier	Common base amplifying circuit
Circuit form	(circuit diagram)	(circuit diagram)	(circuit diagram)
Voltage magnification A_u	$-\dfrac{\beta R'_L}{r_{be}}$ (high)	Approximately equal to 1 (low)	$\dfrac{\beta R'_L}{r_{be}}$
Input and output signal phase position	Antiphase	In-phase	In-phase
Current magnification A_i	β (high)	$1+\beta$ (high)	Approximately equal to 1 (low)
Input resistance r_i	r_{be} (medium)	$r_{be}+(1+\beta)R'_{L}$ (high)	$\dfrac{r_{be}}{1+\beta}$ (low)
Output resistance r_o	R_c (high)	$\approx \dfrac{r_{be}}{\beta_{(low)}}$	R_c (high)

Comparison items \ Configuration	Common emitter amplifier	Common collector amplifier	Common base amplifying circuit
High frequency characteristic	Poor	Preferably	Good
Stability	Poor	Preferably	Preferably
Scope of application	Intermediate stage and input stage of multistage amplifier	Input stage, output stage, buffer stage of multistage amplifier	High frequency circuit, broadband amplifier

【Special reminder】

When the triode is used in the actual amplification circuit, it needs to add a suitable offset circuit.

2.4.3 Voltage stabilizing circuit

(1) Voltage stabilizing transistor circuit

The simplest voltage-stabilizing circuit could be formed by using the voltage stabilizing characteristics of the voltage stabilizing transistor, as shown in Fig. 2-26. In the figure, VD is a voltage-stabilizing transistor, which plays a voltage-stabilizing role in the circuit; R is the current limiting resistance, which plays the role of voltage reduction in the circuit and could limit the load current. When the current flowing through the load exceeds the maximum current allowed by R, R will burn off.

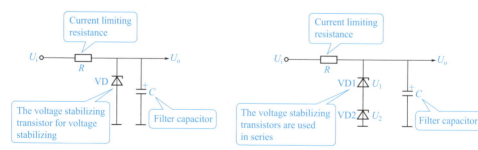

Fig. 2-26 Voltage stabilizing circuit of voltage stabilizing transistor(s)

(2) Series voltage stabilizing circuit

The basic structure of the series voltage stabilizing circuit is shown in Fig. 2-27, and the triode VT is the adjusting transistor. Because the regulating transistor is connected in series with the load, this circuit is called a series voltage stabilizing circuit.

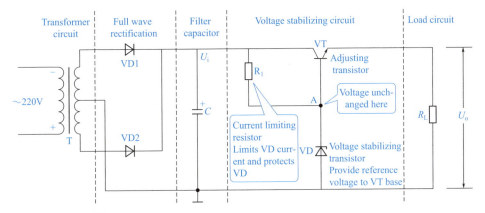

Fig. 2-27 Basic structure of series basic voltage stabilizing circuit

The voltage stabilizing transistor VD provides the base voltage for the regulating transistor, which is called the reference voltage.

The voltage stabilization process of the circuit is:

$U_I \uparrow \to U_o \uparrow \to U_{BE} \uparrow \to I_B \to$ VT weakened continuity $\to U_{CE} \uparrow \to U_o \downarrow$

(3) Three terminals Integrated voltage stabilizing circuit

The three terminal integrated voltage stabilizing circuit is composed of a three terminals svoltage regulator as the core. The three terminals voltage regulator is an integrated voltage stabilizing circuit, which integrates all the components in the voltage stabilizing circuit to form a voltage stabilizing integrated block. It only leads out three pins externally, namely input pin, ground pin and output pin, as shown in Fig. 2-28 (a).

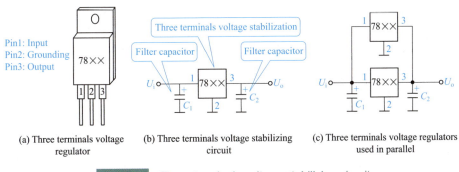

(a) Three terminals voltage regulator
(b) Three terminals voltage stabilizing circuit
(c) Three terminals voltage regulators used in parallel

Fig. 2-28 Three terminals voltage stabilizing circuit

After using a three terminals regulator, the regulator circuit could be very simple, as shown in Fig. 2-28 (b). It only needs to add a filter capacitor on the input and output terminals respectively. In order to obtain greater output current and improve load capacity, three terminals regulators could also be used in parallel, as shown in Fig. 2-28 (c). Table 2-12 shows the application of commonly integrated regulators.

Table 2-12 The application of common integrated voltage regulator

Integrated voltage regulator		Pin function	Output voltage/ V	Applied circuit
Non-adjustable	CW78×× Positive voltage	78×× Input/Grounding/Output	Voltage grades: 5、6、9、12、15、18、24	79×× with C_1 0.33μF, C_2 0.1μF, U_i, U_o
	CW79×× Negative voltage	78×× Grounding/Input/Output	Voltage grades: −5、−6、−9、−12、−15、−18、−24	79×× with C_1 0.33μF, C_2 0.1μF, U_i, U_o
Adjustable	CW317 Positive voltage	W317 Adjusting/Output/Input	Adjustment range: 1.2~37	W317 with 240Ω R_1, C_1 0.1μF, R_P, C_2 10μF, U_i, U_o
	CW337 Negative voltage	W337 Adjusting/Input/Output	Adjustment range: −1.2~−37	W337 with 240Ω R_1, C_1 0.1μF, R_P, C_2 10μF, C_3 1μF, U_i, U_o

【Special reminder】

The common products of the three terminals voltage regulators are 78 series (positive power supply) and 79 series (negative power supply). The output voltage is represented by the last two figures in the specific model, and there are 5V, 6V, 8V, 9V, 12V, 15V, 18V, 24V and other grades. The output current is distinguished by 78 (or 79) followed by a letter. L stands for 0.1A, M stands for 0.5A, and no letter stands for 1.5A. For example, 78L05 stands for 5V 0.1A.

Electrical tools and safety signs

3.1 Use of electrical tools and safety tools

3.1.1 Use of common electrical tools

(1) Use and precautions of common electrical tools

Proper use of tools could not only improve work efficiency and construction quality, but also reduce fatigue, ensure operation safety and extend the service life of tools. Therefore, electricians must pay attention to the reasonable selection and correct use of tools. See Table 3-1 for the purpose and use of common electrical tools.

Table 3-1 Usage and use of common electrical tools

Name	Schematic	Purpose and specification	Use and precautions
Test pencil		Tools used to test whether electrical appliances and electrical equipment such as wires, switches, sockets are live	When using, hold the test pencil body with fingers, touch the metal body (tail) of the electroscope body with index finger, and the small window of the test pencil is facing eyes for observation. The range of voltage measured by the test pencil is 60~500V. It is strictly prohibited to measure high voltage. At present, test pencils (digital) are widely used. The use method of test pencil is the same as that of light-emitting tube. The highest displayed value is the measured value.
Wire cutters		Common tools for clamping and cutting electrical materials (such as wires), with specifications of 150, 175 and 200mm, all equipped with rubber insulated sheath, could be used for live working below 500V	The wire cutters are composed of two parts: tong head and tong handle. The tong head is composed of four parts: jaw, tooth, knife edge and cutting edge. The jaw is used to bend or clamp the wire head; The tooth is used to fasten or loosen the nut; The knife edge is used to cut wires or cut the insulation layer of flexible wires; The cutting edge is used to cut hard metals such as wire cores. Note (1) The wire cutters cannot be used as a hamming tool. (2) Pay attention to protect the insulated sheath of the tong handle to avoid electric shock caused by sleeve damage

Name	Schematic	Purpose and specification	Use and precautions
Needle nose pliers		The tong head of the needle nose pliers is slender and long, which could work in a narrow place, such as the lamp holder, the wire head in the switch, etc. Common specifications are 130mm, 160mm and 180mm	The precautions for use are basically the same as those for wire cutters. Pay special attention to protecting the tong head. The object to be clamped should not be too large. Do not use too much force
Diagonal pliers		Diagonal pliers, also known as wire cutting pliers, are specially used to cut thick wires and cables. Common specifications are 130mm, 160mm, 180mm and 200mm	Precautions for use are basically the same as those for wire cutters
Screwdriver		A tool used to tighten or loosen screws. There are usually two types of screwdrivers, namely, slot type and cross type. The specifications are 75, 100, 125 and 150mm	Attention during use: (1) Select a screwdriver of corresponding size according to the size and specification of the screw, otherwise the screw and screwdriver would be damaged easily; (2) Piercing screwdriver should not be used during live line operation; (3) The screwdriver cannot be used as a chisel; (4) The handle of the screwdriver should be kept dry and clean to avoid electric leakage during live operation
Electrician's knife		It is used to cut the insulation layer of the wire, cable insulation, wooden trough board, etc., in the installation and maintenance of electricians, and the specifications could be divided into large and small ones; The large blade is 112mm long and the small blade is 88mm long	The knife edge should face outward for operation; When cutting the wire sheath, the knife edge should be flat to avoid cutting the wire core; After use, fold the blade into the handle in time to prevent the blade from being damaged or endangering people or cutting skin
Wire stripper		The handle of the special tool for stripping the insulation layer of small diameter wires is insulated, and the withstand voltage is 500V. The specifications are 140mm (applicable to aluminum and copper wires, with diameters of 0.6mm, 1.2mm and 1.7mm) and 160mm (applicable to aluminum and copper wires, with diameters of 0.6mm, 1.2mm, 1.7mm and 2.2mm)	After the length of the insulation to be stripped is fixed with a ruler, the wire could be put into the corresponding cutting edge (slightly larger than the diameter of the wire), and the clamp handle could be held by hand. The insulation layer of the wire would be cut and automatically kick out. Note that the wires with different wire diameters should be placed on the cutting edges with different diameters of the wire stripper

Name	Schematic	Purpose and specification	Use and precautions
Adjustable wrench		Tools used by electricians to tighten or remove hexagon screws (nuts) and bolts. There are four common adjustable wrenches: 150 × 20 (6 inches), 200 × 25 (8 inches), 250 × 30 (10 inches) and 300 × 36 (12 inches)	(1) It could not be used as a hammer; (2) Adjustable wrenches of corresponding specifications should be selected according to the size of nuts and bolts; (3) The opening adjustment of adjustable wrench should be able to not only clamp the nut, but also easily remove the wrench and change the angle
Hammer		Special tools used to hammer cement nails or other objects during installation or maintenance	There are two types of hammer grip: tight grip and loose grip. There are three ways to swing the hammer: wrist swing, elbow swing and arm swing. The right hand is usually used to hold the end of the wooden handle. When hammering, it should be aligned with the workpiece, with even force and accurate drop point

Memory tips

There are many kinds of pliers for electricians, and different usages should be mastered.
The insulated handle should be in good condition for easy operation.
The electrician's knife could not be used for live operated when the handle is not well insulated.
There are two types of screwdrivers, the right one must be chosen.
When using a test pencil to test electricity, it is easy to misjudge in wrong holding method.
When using a wrench to tighten the bolts, the force direction could not be reversed.
When hammering a workpiece, aim at the drop point.

(2) Common sense of common tools maintenance

The basic requirements for users to use common electrical tools are safety, good insulation and moving parts should be flexible. Based on these basic requirements, attention should be paid to the maintenance of electrical tools, which is briefly described below.

① Common electrical tools should be kept clean and dry.

② Before using the electrical pliers, ensure that the insulation performance of the insulated handle is good to ensure the personal safety during live working. If the insulating sheath of the tool is damaged, it should be replaced in a timely manner and should not be used reluctantly.

③ Oil the moving parts of wire cutters, needle nose pliers, wire strippers and other tools frequently to prevent rust.

④ After using the electrician knife, fold the knife body into the handle in time to avoid damage to the knife edge or endangering personal safety.

⑤ The wooden handle of the hammer should not be loose, so as not to affect the drop point or drop the hammer head during hammering.

3.1.2 Use of other electrical tools

Different tools should be used in different electricians working cases. Other electrical tools mentioned here mainly include high-voltage electroscope, hand hacksaw, micrometer, tachometer, electric soldering iron, blowtorch, hand winding machine, puller, foot buckle, pedal, ladder, chisel, tightener, etc. See Table 3-2 for details.

Table 3-2   Other electrical tools and precautions

Name	Schematic	Purpose and specification	Use and precautions
High-voltage electroscope		Used to test electrical equipment with voltage higher than 500V	During use, insulating gloves should be worn, and the hand holding part should not exceed the protective ring; Gradually approach the tested object to see whether the neon tube emits light. If the neon tube does not light all the time, the tested object is not electrified; When using high-voltage electroscope for testing, at least one person should be on site for monitoring
Hand hacksaw		Electricians use it to cut objects	When installing the saw blade, the saw teeth should face forward and the saw bow should be tightened. Saw blades are generally divided into coarse teeth, medium teeth and fine teeth. Coarse teeth are suitable for sawing copper, aluminum and wood materials, medium teeth are for normal materials, while fine teeth could generally saw hard iron plates, threading iron pipes and plastic pipes
Micrometer		Used to measure the outer diameter of enamelled wire	When in use, the enamelled wire to be measured should be straightened and placed between the micrometer anvil and the micrometer rod, and then the micro screw should be adjusted so that it just clamps the enamelled wire, and then the reading could be made. When reading, first look at the integer reading on the micrometer, and then look at the decimal reading on the micrometer. The sum of the two is the diameter of the copper enamelled wire. The integral scale of the micrometer is generally 1mm in one small scale, and the division value on the movable scale is generally 0.01mm in each scale

Name	Schematic	Purpose and specification	Use and precautions
Tachometer	Photoelectric speed measurement / Contact speed measurement	Used to test the rotating speed and linear speed of electrical equipment	When using, first observe the motor speed with eyes, roughly judge its speed, and then turn the speed dial of the tachometer to the speed range to be measured. If it is not sure to judge the motor speed, adjust the speed dial to the high position for observation. After determining the speed, adjust it to the low gear to make the test result accurate. When measuring the speed, the hand-held tachometer should be kept balanced, the test shaft of the tachometer should be concentric with the motor shaft, and the contact force should be gradually increased until the test pointer is stable, and then the data should be recorded
Electric soldering iron		Soldering wire connectors and components	When using an externally heated electric soldering iron, always remove the copper head and remove the oxide layer to avoid burning the copper head over time; After the electric soldering iron is powered on, it cannot be knocked to avoid shortening the service life; After using the electric soldering iron, unplug the plug and place it in a dry place after it cools down to avoid moisture and electric leakage
Blowtorch		Solder the lead coating of lead-sheathed cable, tin lining at the connection of copper core wire with large cross-sectional area, and tin plating of other electrical connections	Before using the blowtorch, check whether there is oil leakage from the oil drum carefully, whether the nozzle is blocked, whether there is air leakage, etc. Fill the corresponding fuel oil for the blowtorch. The oil volume should not exceed 3/4 of the capacity of the oil drum. After filling, tighten the plug at the filling point. When the blowtorch is ignited, no one is allowed to stand in front of the nozzle, and there should be no flammable materials in the workplace. During ignition, add a proper amount of fuel oil into the ignition bowl and ignite it with fire. After the nozzle is heated, open the oil inlet valve slowly; When pumping and pressurizing, close the oil inlet valve first. At the same time, pay attention to keeping a safe distance between the flame and the charged body

Name	Schematic	Purpose and specification	Use and precautions
Manual winding machine		It is mainly used to wind the windings of motors, low-voltage equipment and small transformers	When using the manual winding machine, pay attention to: ① fix the winding machine on the operating platform; ② The number of turns indicated by the pointer at the beginning of winding should be recorded, and the number of turns should be subtracted after winding
Puller		Used to remove pulley, coupling, motor bearing and fan blade of motor	When the puller is used to pull the motor pulley, the puller should be aligned, the screw rod should be aligned with the center of the shaft, and then the lead screw of the puller should be tightened with a wrench with even force. When using the puller, if there is rust between the pulled parts and the motor shaft, soak some gasoline or bolt loosening agent in the joint of the shaft, and then hit the outer circle of the pulley or the top of the screw rod with a hammer, and then pull the pulley outward with force
Foot buckle		Used for climbing power poles and towers	Before use, check whether the arc-shaped snap ring is cracked or corroded, and whether the foot buckle belt is damaged. If it is damaged, repair or replace it immediately. Do not use ropes or wires instead of the straps foot buckle. Before climbing the pole, the foot buckle should be subject to human impact test, and the foot buckle belt should be checked for firmness and reliability
Pedal		Used for climbing power poles and towers	It is used for climbing power poles and towers. Before use, check the appearance for cracks and corrosion, and use it after passing the human impact test; The climbing operation should be stable and correct, and it is prohibited to throw the pedal down from the pole at will; The pedal rope should be subject to static tension test once a year, and could only be used after passing the test

Name	Schematic	Purpose and specification	Use and precautions
Ladder		Electrician climbing tools	There are herringbone ladders and straight ladders. The method of use is relatively simple. The ladders should be stable and anti-skid; At the same time, the ladder should be placed at a safe distance from the electrified body
Chisel		It is used for drilling or chipping the rusted small bolts	When using, hold the chisel with your left hand (note that the tail of the chisel should be exposed about 4cm), hold the hammer with your right hand, and then hit it hard
Tightener		A tool used to tighten electric wires in overhead lines	When in use, the galvanized steel wire rope is wound on the pulley at the right end, hung on the cross arm or other fixed parts, the wire is clamped by the collet at the other end, and the pulley is rotated by the crank to gradually draw the steel wire rope into the pulley, and the wire is tightened and contracted to an appropriate extent

◆ Memory tips ◆

The straight ladder should be antiskid and the herringbone ladder should not open automatically. Foot buckle and pedal must be used to climb the pole, and the hand & foot should be in good coordination.
Tightener should be used for wire tightening, operate slowly without slipping.
Although the blowtorch is small but at high temperature, it could melt the lead coating of the cable. Electric soldering iron could be used to joint components and parts, and the specifications should be selected according to needs.

3.1.3 Use of manual electric tools

Manual electric tools commonly used by electricians mainly include electric hand drills and hammers, as shown in Table 3-3.

Table 3-3 Use of manual electric tools

Name	Schematic	Purpose and specification	Use and precautions
Electric hand drill		For drilling	When installing the drill bit, pay attention to keeping the drill bit and drill clamp on the same axis to prevent the drill bit from swinging back and forth when rotating. During use, the drill bit should be perpendicular to the object to be drilled, and the force should be uniform. When the drill bit is stuck by the object to be drilled, the drilling should be stopped immediately to check whether the drill bit is too loose, and the drill bit can be used after being tightened again. In the process of drilling metal holes, if the temperature of the bit is too high, it is likely to cause the bit to be annealed. Therefore, proper lubricating oil should be added when drilling
Electric hammer		For drilling	Before using the electric hammer, it should be powered on and idling for a while to check whether the rotating part is flexible. The electric hammer can be used only after it is checked that there is no fault; When working, the drill head should be placed on the working surface first, and then the switch should be started to avoid empty drilling as far as possible; In the process of drilling, if the hammer does not rotate, the switch should be released immediately, and the hammer should be started after the cause is found out. When drilling on the wall with an electric hammer, it should know whether there is a power line in the wall first, so as to avoid electric shock caused by drilling through the wire. When drilling in concrete, pay attention to avoid reinforcement

Pay attention to the following points when using manual electric tools such as electric hand drills and electric hammers.

① Before use, check whether the insulation of the power wire is good. If the wire is damaged, wrap it with electrical insulating tape. It is better to use three core rubber wire as the power wire for electric tools, and reliably ground the shell of electric tools.

② Check whether the rated voltage of the electric tool is consistent with the power supply voltage, and whether the switch is flexible and reliable.

③ After the electric tool is connected to the power supply, use a test pencil to test whether the shell is charged. If not, it could be used. If it is necessary to touch the metal shell of electric tools during operation, wear insulating gloves, wear insulating shoes, and stand on the insulating board.

④ When disassembling and assembling the drill bit of the electric hand drill, use the special key, and do not knock the drill clamp with a screwdriver or hammer, as shown in Fig. 3-1.

⑤ When installing the drill bit, it should be noted that the drill bit and drill clamp should be kept on the same axis to prevent the drill bit from swinging back and forth when rotating.

Fig. 3-1　Method of replacing the drill bit of electric hand drill

⑥ In the process of use, if abnormal sound is found, the drilling should be stopped immediately. If the electric tool is hot due to long continuous working, the work should be stopped immediately to allow it to cool naturally, and water should not be poured.

⑦ After drilling, the wire should be wound on the manual electric tool and placed in a dry place for next use.

3.1.4　Use of electrical safety appliances

Electrical safety appliances refer to all kinds of special tools or appliances for electricians that must be used to ensure the safety of operators and prevent electric shock, falling, burning and other industrial accidents during electrical operation. Electrical safety appliances could be divided into basic safety appliances, auxiliary safety appliances and general protective safety appliances according to their purposes.

(1) Basic safety appliances

Basic safety appliances are tools that could contact live parts directly and could reliably withstand the working voltage of equipment for a long time. Commonly used are insulating rods, insulating clamps, etc.

① Insulating rod is a unified name of insulating tools specially used in power system, which could be used for live working, live overhaul and live maintenance tools.

② Insulating clamp is a tool used to install and remove high-voltage fuse or perform other similar work, mainly used in 35kV and below power systems.

(2) Auxiliary safety appliances

Auxiliary safety appliances are used to further enhance the reliability of basic safety appliances and prevent the danger of contact voltage and step voltage. Common ones are insulating gloves, insulating boots, insulating pads, insulating platforms, etc.

① Insulating gloves are five fingered gloves made of rubber. They are mainly used for electrical work, which could protect hands or human body from electricity, water, acid and alkali, chemicals and oil.

② Insulating boots are also called high-voltage insulating boots and mining boots. Insulation means that the electrified body is enclosed with insulating materials to isolate the electrified body or conductors with different potentials, so that the current can flow through a certain path.

③ Insulating rubber pad is also called insulating blanket, insulating mat, insulating rubber plate, insulating rubber mat, insulating ground rubber, insulating rubber, insulating gasket, etc. It has a large volume resistivity and is resistant to electric breakdown, used as insulating materials for worktops or floor coverings in power distribution and other workplaces.

(3) General protective safety appliances

General protective safety appliances include temporary grounding wires, barriers, safety signs, signboards, etc.

3.2 Use of common electrical instruments

3.2.1 Use of multimeter

Multimeter is mainly used to measure resistance, AC and DC voltage and current. Some multimeters could also measure the main parameters of transistors and the capacitance of capacitors.

Multimeters are the most basic and commonly used electrical instruments, mainly including pointer multimeter and digital multimeter.

(1) Use of pointer multimeter

Take MF7 multimeter as an example to introduce the use of pointer multimeter.

① Resistance measurement. The resistance measurement must use the DC power supply inside the multimeter. Open the battery box cover on the back. On the right is the low-voltage battery compartment. Install a 1.5V type C battery; On the left is the high-voltage battery compartment, which is equipped with a 15V laminated battery, as shown in Fig. 3-2. Currently, some manufacturers use 9V laminated batteries for R×10k range in MF47 multimeters.

Fig. 3-2　Battery installation

The method of measuring resistance with pointer multimeter could be summarized as the following tips.

Operation tips

Select the range when measuring the resistance. Short circuits for zero adjustment.
Replace the battery in case of the pointer does not point to zero even the knob to the limit, and then reset it to zero.
Disconnect the power supply and measure again. The contact must be good.
Measure the resistance with two hands hanging in the air to prevent parallel connection from impacting the accuracy.
The pointer should be in the middle to get accurate value.
Don't forget to multiply the reading value with magnification, and the range must be located at the voltage finally.

Range selection for resistance measuring - when measuring resistance, select the appropriate range first, and try to make the measured value between 0.1 ~ 10 of the ohmic scale line, so that the reading would be accurate.

Generally, range "R×1" is selected when measuring resistance below 100 Ω; range "R×10" is for resistance from 100 Ω to 1k Ω; range "R×100" is for resistance from 1k Ω ~ 10k Ω; range "R×1k" is for resistance from 10k Ω ~ 100k Ω; resistance larger than 10k Ω should select range "R×10k".

Two probes short circuits before zero adjustment- ohmic zero adjustment should be carried out for the pointer after proper measuring range is selected. Note that ohmic zero adjustment should be performed after each range change, as shown in Fig. 3-3. The operation time should be as short as possible when the ohm is adjusted to zero. If the two probes connect for a long time, the battery inside the multimeter would consume too quickly.

If the pointer does not point to zero even the knob to the limit, replace the battery and then adjust to zero — if the ohm zero adjustment knob has been turned to the end, the pointer is always on the left side of the 0 Ω line and could not point to the "0" position, which means that the battery voltage in the multimeter is low and could not meet the requirements, and the above adjustment needs to be made after a new battery is replaced.

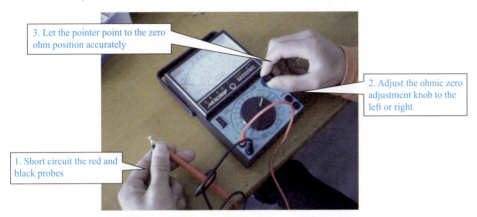

Fig. 3-3   Operation method of ohm zero adjustment

Disconnect the power supply and then measure. The contact must be good. If the resistance value of the resistor is measured on the circuit, the power supply must be disconnected before the measurement. Otherwise, the multimeter might be damaged, as shown in Fig. 3-4. In other words, the resistance could not be measured live. During measurement, make sure that the probe contacts well (when measuring other parameters of the circuit with a multimeter, it is also required that the probe contacts well).

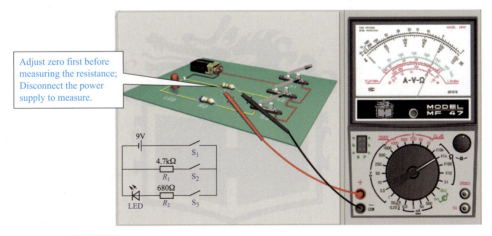

Fig. 3-4   Disconnect the power supply for resistance measurement

Measure the resistance with two hands hanging in the air to prevent parallel connection from impacting the accuracy — during measurement, two hands could not touch two pins of the resistor at the same time, for it is equal to that a resistance (human body resistance) is connected in parallel at both ends of the measured resistor, so the measured value would be less than the actual value of the measured resistance, impacting the measurement accuracy.

The pointer should be in the middle to get accurate value. — the range should be appropriate, if it is too large, it is not easy to read; If it is too small, it would be difficult to measure. Only when the pointer is in the middle of the scale, the reading is accurate.

Don't forget to multiply the reading value with magnification — The reading value should multiply the magnification (the selected range, such as R × 10、R × 100, etc.) is the actual resistance value of the resistance. For example, select range R×100 for measurement, if the pointer indicates 40, the measured resistance value is:

$$40 \times 100 \, \Omega = 4000 \, \Omega = 4k \, \Omega$$

The range must be located at the voltage finally — After the measurement is completed, set the range switch to the highest AC voltage range, that is, the AC 1000V range.

② When measuring the AC voltage below 1000V, set the range switch to the required AC voltage range. When measuring the AC voltage of 1000~2500V, set the range switch in the "AC 1000V" range, and insert the positive probe into the "AC/DC 2500V" special slot.

The following operation tips could be summarized for the method and precautions of measuring AC voltage with pointer multimeter.

Operation tips

The range switch should be at AC, and the range should meet the requirements.
To ensure safety against electric shock, the probes insulation is particularly important.
The probes should be connected to both ends of the circuit, regardless of phase wire or neutral wire.
Measure the effective value of the voltage, and change the slot when measuring high voltage.
Don't touch the front of the probe, and don't forget to power off before shifting.

The range switch should be at AC, and the range should meet the requirements. For AC voltage measurement, proper range must be selected. If resistance measuring range, current measuring range or other measuring ranges are misused, the multimeter might be damaged. In case of that, the internal fuse is generally damaged and could be replaced with a new one of the same specification.

To ensure safety against electric shock, the probes insulation is particularly important — pay attention to safety when measuring AC voltage, which is critical. Since the distance between human body and charged body is relatively close when measuring AC voltage, special attention should be paid, as shown in Fig. 3-5. If the probe or the wire is damaged, it should be completely treated before use.

Fig. 3-5　Measurement of AC voltage

The probes should be connected to both ends of the circuit, regardless of phase wire or neutral wire — the wiring method for measuring the AC voltage is the same as that for measuring the DC voltage, that is, the multimeter is connected in parallel with the circuit to be measured, but no matter which probe is connected to the live wire or neutral wire.

Measure the effective value of the voltage, and change the slot when measuring high voltage. The voltage measured with the multimeter is the effective value of the alternating current. If it is necessary to measure the AC voltage higher than 1000V, insert the red probe into the 2500V slot. However, it is not easy to encounter this situation in practical work.

③ When measuring DC voltage below 1000V, the range switch is set to the required DC voltage range. When measuring DC voltage of 1000~2500V, place the range switch in the "DC 1000V", and insert the positive probe into the "AC/DC 2500V" slot.

The methods and precautions for measuring DC voltage with pointer multimeter could be summarized as the following operation tips.

Operation tips

Determine the positive and negative polarities of the circuit, and select the range first. The red probe should be connected to the high potential, and the black one to the lower side. The probes are connected to both ends of the circuit in parallel. If the pointer rotates reversely, the polarities should be switched. Cut off the power before shifting the range.

Determine the positive and negative polarities of the circuit, and select the range first. Before measuring the DC voltage with a multimeter, it must distinguish the positive and negative polarities of the circuit (or the high potential end and the low potential end), and pay attention to select the appropriate range.

The standard for proper range of voltage range is: the pointer should be at more than 2/3 of the full scale as far as possible (this is different from the standard for proper magnification of resistance range, and it must be noted).

The red probe should be connected to the high potential, and the black one to the lower side. When measuring the DC voltage, the red probe should be connected to the high potential end (or the positive pole of the power supply), and the black one should be connected to the lower end (or the negative pole of the power supply), as shown in Fig. 3-6.

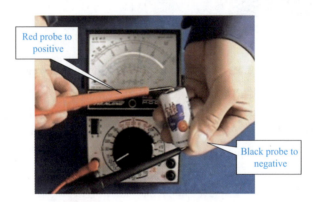

Fig. 3-6　Measurement of DC voltage

The probes are connected to both ends of the circuit in parallel. If the pointer rotates reversely, the polarities should be switched. When measuring DC voltage, two probes are connected to both ends of the circuit (or power supply) in parallel. If the probe rotates reversely, it means that the positive and negative polarities are wrong. In case of that, the red and black probes should be switched for measurement.

Cut off the power before shifting. During the measurement, if it is necessary to shift the range, be sure to remove the probe, and then shift the range after powering off.

④ Generally speaking, the pointer multimeter only has the DC current measurement function and could not be used to measure the AC current directly.

In case of measures DC current below 500mA with MF47 multimeter, set the range switch at "mA". When measuring the DC current of 500mA~5A, place the range switch at "500mA", and insert the positive probe into the "5A" slot.

The methods and precautions for measuring DC current with pointer multimeter could be summarized as the following operation tips.

> **Operation tips**
>
> Set the range switch at DC current, determine the positive and negative poles of the circuit.
> The red probe should be connected to positive, and the black one to negative.
> The probes are connected to the circuit in series, the high and low potentials must be correct.
> The ranges should be changed from large to small, and it should be decided before measurement.
> If the pointer rotates reversely, the positive and negative polarities should be switched.

Set the range switch at DC current, determine the positive and negative poles of the circuit. Generally, the pointer multimeter has the function of measuring DC current, but not AC current. Before measuring the DC current of the circuit, it is necessary to determine the positive and negative polarities of the circuit.

The red probe should be connected to positive, and the black one to negative. When measuring, the red probe should be connected to the positive polairty of the power supply and the black probe to the negative, as shown in Fig. 3-7.

The probes are connected to the circuit in series, the high and low potentials must be correct. Before measurement, the circuit to be measured should be disconnected, and then the multimeter should be connected to the circuit to be measured in series. The red probe should be connected to the high potential end of the circuit (or the positive pole of the power supply), and the black probe should be connected to the low potential end of the circuit (or the negative pole of the power supply), which is exactly the same as the connection method of the probe when measuring DC voltage.

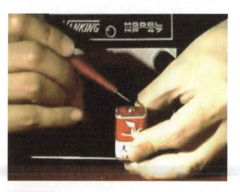

Fig. 3-7　Measurement of battery current

When the multimeter is set in the DC current range, it is equivalent to a DC meter, and the internal resistance would be very small. In case of the multimeter is connected in parallel with the load by mistake, it would cause short circuit and burn the multimeter.

The ranges should be changed from large to small, it should be decided before measurement. Before measuring the current, estimate the circuit current. If the circuit current could not be roughly estimated, the best way is to change the ranges from large to small. If the pointer rotates reversely, the positive and negative polarities are reversed, and the connection positions of the red and black probes should be switched immediately.

⑤ Precautions

a. Check the range before measuring. The measurement category and range switch should be set to the correct position before measurement. For safety, it should be a habit.

b. Do not shift the range during measurement, and shift to the neutral range after measurement. During measurement, the range switch could not be shifted at will, especially when measuring high voltage (such as 220V) or large current (such as 0.5A), to avoid arcing and burning the selector switch's contacts. After measurement, the range switch should be turned to the highest AC range or range "OFF".

c. The dial should be horizontal and the reading should be aligned. When using the multimeter, place it horizontally, and read after the pointer is stable. When reading, the eyesight should be directed to the pointer.

d. The measuring range should be appropriate, and the pointer should be more than half of the full scale. Select the range. If the measured size could not be estimated in advance, try to select a larger range, and then gradually change to a smaller one according to the deflection angle until the pointer deflects to about 2/3 of the full scale.

e. The resistor should not be electrified when it is to be measured, and capacitor must be discharged before measurement. It is forbidden to measure the resistance when the circuit under test is live. When measuring the capacitor with large capacitance on electrical equipment, the capacitor should be short circuited and discharged before measurement.

f. Zero adjustment should be carried out before measurement for resistor, and it is also necessary after range shift. When measuring the resistance, turn the range switch to the resistance range first, short circuit the two probes, and turn the "Ω" zero potentiometer to make the pointer point to zero ohm before measuring. The ohmic zero point should be readjusted every time when the resistance range is shifted.

g. The black probe is connected to positive ("+") internally, which should be memorized clearly. The red probe of the multimeter is positive, and the black probe is negative. But at the resistance ranges, the black probe is connected to the positive pole of the internal battery.

h. Connect the multimeter in series to measure current, while in parallel for voltage. When measuring the current, connect the multimeter in series in the circuit to be measured; When measuring the voltage, connect the multimeter in parallel at both ends of the circuit to be measured.

i. The polarities should not be reversed, single hand operation should become a habit. When measuring DC current and voltage, pay special attention that the polarities of red and black probes could not be reversed. When using a multimeter to measure voltage and current, it is necessary to develop the habit of holding the probes with one hand to ensure safety, as shown in Fig. 3-8 (a). Small components could be measured by holding the probes with both hands when the power is off, as shown in Fig. 3-8 (b).

(a) Single hand operation (b) Double hands operation

Fig. 3-8　Multimeter measurement operation

(2) Use of digital multimeter

① Measurement of resistance

a. Insert the black probe into the COM slot and the red probe into the V/Ω one.

b. Select the range according to the nominal value of the resistance to be measured (which could be observed from the color ring of the resistor), and the selected range should be slightly larger than the nominal value of the resistor.

c. Connect the probes of the digital multimeter in parallel with the measured resistor, and read the measurement results directly from the display screen.

As shown in Fig. 3-9, the measured nominal resistance is 12kΩ, and the actual measured resistance is 11.97kΩ.

Read directly according to the selected range and magnification, which is different from the reading method of a pointer multimeter.

Fig. 3-9　Measurement of resistance

② Measurement of DC voltage

a. Insert the black probe into the COM slot, and the red probe into the V/Ω one.

b. Set the range switch to the appropriate range of DC voltage "DCV" or " V⎓".

c. The two probes are connected in parallel with the circuit to be tested (generally, the red probe is connected to the positive pole of the power supply, and the black probe is connected to the negative pole of the power supply), so that the DC voltage value could be measured. If the polarities of the probes are reversed, the polarity of the red probe is indicated with a "-" sign on the display screen when the voltage value is displayed. As shown in Fig. 3-10, "-3.78" is displayed, indicating that the measured voltage is 3.78V. The negative sign indicates that the red probe is connected to the negative pole of the power supply.

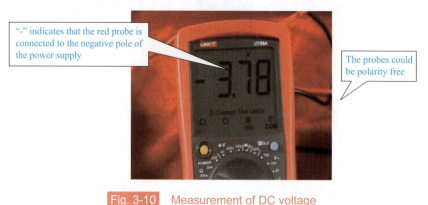

Fig. 3-10 Measurement of DC voltage

③ Measurement of AC voltage

a. Inserted the black probe into the COM slot, and the red probe into the V/Ω one.

b. Set the range switch to the appropriate range of AC voltage "ACV" or "V ~".

c. During measurement, the probe is connected in parallel with the circuit to be measured, and the probes are polarities free. Read the actual value directly from the display screen, as shown in Fig. 3-11.

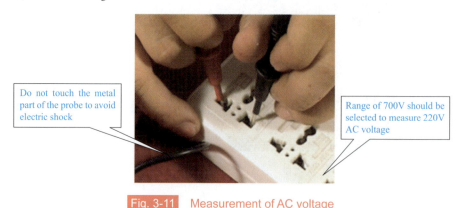

Fig. 3-11 Measurement of AC voltage

④ Measurement of current

a. Insert the black probe into the COM slot. When the measured current is below 200mA, insert the red probe into the "mA" slot. When measuring the current of 0.2~20A, insert the red probe into the "20A" slot.

b. The range switch is set at the appropriate range of "ACA" or "A -".

c. During measurement, the circuit must be disconnected first, and the probes should be connected to the circuit to be measured in series, as shown in Fig. 3-12. When the display screen displays the current value, it would also indicate the polarity of the red probe.

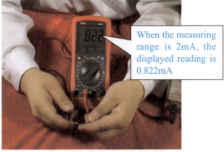

Fig. 3-12   Measurement of current

⑤ Precautions

a. The selection of appropriate range should be in accordance with the magnitude of the U/I to be measured. If the magnitude of the measured voltage or current could not be estimated in advance, it is necessary to switch to the highest range for measurement first, and then reduce the range to an appropriate position gradually as appropriate, as shown in Fig. 3-13. After measurement, turn the range switch to the highest voltage and turn off the power.

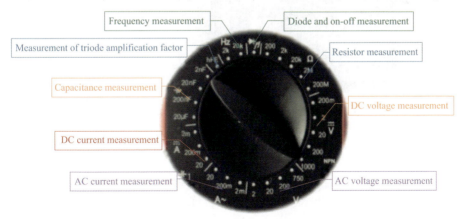

Fig. 3-13   Range selection of digital multimeter

b. The number "1" displayed on the screen has different meanings. For some models of digital multimeter, the number "1" is also displayed on the screen when no measurement is made after powering on. When measuring with a digital multimeter (such as resistance, voltage and current), the screen only displays the number "1" at the highest position, and other digits disappear. "1" means the value "overflows", indicating that the actual value has exceeded the measured maximum value of the range, and the range needs to be shifted to a higher step. Namely, when the range switch is misplaced, "1" appears on the screen.

If the range switch is at the buzzer range (the icon is a diode schematic), and "1" is displayed. If the number "1" is always displayed, the reason is the measured part between the two probes is not connected (or the resistance is much greater than 1000 Ω).

c. The measuring range should be properly selected, and the range should not be shifted during measurement. It is forbidden to change the measuring range when measuring high voltage (above 220V) or high current (above 0.5A) to prevent arcing and burning the range switch's contacts. Generally speaking, the current range of a digital multimeter is in the milliampere range at the smaller range and in the 20A range at the maximum.

d. If the battery power is insufficient, replace it with a new one in time. When "BATT" or "LOWBAT" is displayed, it means that the battery voltage is lower than the working voltage, and a new battery should be replaced. If the digital multimeter is not used for a long time, the battery should be taken out to avoid corrosion of components in the meter due to battery leakage, as shown in Fig. 3-14.

Take out the battery and place it separately to prevent battery leakage from damaging the multimeter, if it is not used for a long time.

Fig. 3-14　Battery compartment of digital multimeter

3.2.2　Use of clamp ammeter

Clamp ammeter is a portable instrument that could measure the AC current on low-voltage lines without interrupting the load operation (i.e., disconnecting the current carrying conductor). Its greatest feature is that it could measure the current without disconnecting the circuit under test, so it is especially suitable for measurement occasions where it is not convenient to disconnect the line or power off is not allowed.

(1) Inspection before use

① Mainly checking whether the insulating materials (rubber or plastic) on the jaw peel off, crack, etc., and whether the whole case, including the glass cover of the meter head, is intact, which are directly related to the measurement safety and the performance of the instrument.

② Check the opening and closing of the jaw. It is required that the jaw should open and close freely, as shown in Fig. 3-15. The two joint surfaces of the jaw should be in good contact. If there are oil stains and debris on the jaw, it should be cleaned with gasoline; If there is rust, it should be lightly wiped off.

Fig. 3-15   Inspection of jaw opening and closing

③ Check whether the zero point is correct. If the pointer is not at the zero point, it could be adjusted through the adjustment knob.

④ The multifunctional clamp ammeter should also be checked whether the wire and probes are damaged. It is required that the conductivity and insulation are in good condition.

⑤ The digital clamp ammeter should also be checked whether the battery in the meter is sufficient, and must be replaced with (a) new one(s) if necessary.

(2) Usage method

① Before measurement, the measured current value should be estimated according to the load current, and the appropriate range should be selected, or the ammeter with a larger range should be selected and then the range should be reduced according to the measured current to make the reading exceed 1/2 of the full scale, so as to obtain a more accurate value.

② During the measurement, pinch the ammeter to open the jaw, and place the measured current carrying conductor in the center of the jaw to reduce the measurement error, as shown in Fig. 3-16. Then, loosen the ammeter to close the jaw (iron core), and the value would be displayed. Note that it is not allowed to clamp all polyphase wires into the jaw for measurement.

③ When measuring the current below 5A, if the range of the clamp ammeter is too large, the wire could be wound on the jaw for several more turns if possible, as shown in Fig. 3-17, and then measured and read. The actual current value in the wire is the displayed value divided by the number of wire turns passing through the inner side of the jaw.

Fig. 3-16 Place the current carrying conductor in the center of the jaw

Fig. 3-17 Method of measuring current below 5A

④ When judging whether the three phase current is balanced, if possible, put the three phase wires of the three phase circuit to be measured into the jaw in the same direction at the same time. In case of the value is zero, it indicates that the three phase load is balanced; If not, it indicates that the three phase load is unbalanced.

(3) Precautions

① Some types of clamp ammeters are equipped with AC voltage measurement. When measuring current and voltage, they should be measured separately, not simultaneously.

② The jaw of the clamp ammeter should be closed tightly during measurement. If there is noise after closing, open the jaw to reclose once it is recommended. If the noise could not be eliminated, check whether the joint surfaces on the magnetic circuit are smooth and clean, and wipe them clean when there is dust.

③ The voltage of the circuit to be measured should not exceed the value indicated on the clamp ammeter, otherwise it is easy to cause grounding accidents or electric shock.

④ On the measuring site, all kinds of equipment should be in order, and the measuring personnel should wear insulating gloves and shoes. The safe distance between each part of the personnel body and the electrified body should not be less than 0.1~0.3m for low-voltage system. When reading the value, it's normal to lower the head or bend the waist. So, special attention must be paid to the safety distance between the limbs, especially the head and the live part.

⑤ When measuring the circuit current, the conductor with insulating layer should be selected, and a safe distance should be kept from other live parts to prevent interphase short circuit accidents. It is forbidden to change the current range during measurement.

⑥ When measuring the current of low-voltage fuse or horizontally arranged low-voltage bus, the fuse or bus should be isolated from each other with insulating materials to avoid short circuit. At the same time, care should be taken not to touch other live parts.

⑦ For digital clamp ammeter, although the battery power has been checked before use, the battery power should also be paid attention to at any time during the measurement process. If the battery voltage is insufficient (such as low voltage indicating symbol appears), the measurement must be continued after the new battery is replaced. The accuracy of measurement is directly related to whether the measurement data could be read correctly. If there is electromagnetic interference on the measuring site, it would certainly interfere with the normal measurement, so try to eliminate it.

⑧ For the pointer clamp ammeter, the selected range should be identified first, and then the scale used should be identified. When observing the scale value indicated by the pointer, the eye should be directly on the watch pointer and scale to avoid squint and reduce parallax. Although the digital ammeter is relatively direct, the effective viewing angle of the LCD screen is very limited. It is easy to misread numbers when the eyes are too slanted. Attention should also be paid to the decimal point and its location, which should not be ignored.

⑨ After the measurement, be sure to place the range switch at the max current range position to avoid damage to the instrument due to carelessness during the next use.

The basic usage and precautions of clamp ammeter could be summarized as the following operation tips.

Operation tips

Keep the power circuit on to detect the current for measurement
It is convenient and fast to measure current with clamp ammeter.
The appearance and insulation of the jaw must be checked before use.
The jaw should be opened and closed freely, and oil dirt and debris should be removed.
The measuring range should be appropriate, and the clamp ammeter could not be used to measure high voltage.
Wind the wire around the jaw for several turns, in case of measuring small current.
Safety distance must be kept on live line measurement.

3.2.3 Use of insulation resistance meter

Insulation resistance meter, commonly known as megger. It is mainly used to check the insulation resistance of electrical equipment, household appliances or electrical wires to the ground and between phases, to ensure that these equipment, appliances and wires work in a normal state, and to avoid accidents such as electric shock casualties and equipment damage.

(1) Usage

① Disconnect the tested equipment from the power supply, discharge it, and then clean the equipment (double circuit lines and double buses, when one is live, the insulation resistance of the other should not be measured).

② Before measurement, the megger should be calibrated, that is, an open circuit test (the measuring line is open circuit, rotate the handle, and the pointer should point to " ∞ ") and a short circuit test (the measuring line is short circuited, rotate the handle, and the pointer should point to "0") should be conducted. The two measuring lines of the megger should not be intertwined with each other, as shown in Fig. 3-18.

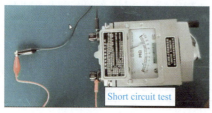
(a) Short circuit test

(b) Open circuit test

Fig. 3-18 Calibration of megger

③ Correct wiring. Generally, there are three terminals on the megger. One is the line terminal with the label of "L", one is the ground terminal with the label of "E", and the other is the protection or shielding terminal with the label of "G". During the measurement, "L" is connected to the conductor part of the equipment insulated from the ground, and "E" is connected to the shell or other conductor parts of the equipment. Generally, only two terminals "L" and "E" are used for measurement. However, when the surface leakage of the equipment under test is serious, which has a great impact on the measurement results and is difficult to eliminate, for example, high air humidity, and the surface of the insulating material is corroded and could not be wiped clean, the "G" terminal must be connected, as shown in Fig. 3-19. Meanwhile, it should also be noted that double stranded wires should not be used during wiring, and single wires with good insulation and different colors should be used, especially the wires connecting the "L" terminal must have good insulation.

Fig. 3-19 Wiring example of megger

④ During measurement, the megger must be laid flat. As shown in Fig. 3-20, hold the megger body with left hand, rotate the handle with right hand with a constant speed of 120r/min to make the pointer rise gradually to a stable value, and then read the insulation resistance value (it is strictly prohibited to measure on the equipment with operator working).

⑤ For equipment with large capacitance, after the measurement, it must be discharged to the ground (Do not touch the megger when it does not stop rotating and the equipment is not fully discharged).

Fig. 3-20 Method of rotating the handle of megger

(2) Precautions

The megger itself generates high voltage during working. In order to avoid personal and equipment accidents, the following precautions must be paid attention to.

① It is not allowed to measure the insulation resistance when the equipment is live. Before measurement, the equipment to be tested must be powered off and discharged; In the case of the equipment that has been measured with megger needs to be measured twice, it must also be grounded and discharged in advance.

② The megger should be far away from the high current conductor and external magnetic field when measuring.

③ The connection wire to the equipment to be tested should be a special measuring wire for megger or two single-core-multi-strand flexible wires with high insulation performance. The two wires must not be twisted together to avoid affecting the measurement accuracy.

④ During the measurement, if the pointer points to "0", it means that the tested equipment is short circuited, and the handle should be stopped immediately.

⑤ If there are semiconductor devices in the tested equipment, the plug-in board should be removed first.

⑥ The part of the equipment being measured should not be touched during the measurement to prevent electric shock.

⑦ When measuring the insulation resistance of capacitive equipment, fully discharge the equipment after measurement.

⑧ During the measurement, the hand or other parts of the body should not touch the part of the equipment being measured or the line terminals of the megger, that is, the operator should keep a safe distance from the measured equipment to prevent electric shock.

⑨ Digital meggers are mostly powered by AA batteries or 9V batteries. The power supply current required during operation is large, so it is necessary to turn it off when not in use. Even meggers with automatic turn-off function are recommended to turn off manually after use.

⑩ Record the temperature of the equipment to be tested and the weather conditions for operation, which is positive for analyzing whether the insulation resistance of the equipment is normal.

3.2.4 Use of electric energy meter

(1) Types

Electric energy meter is an instrument used to measure electric energy, also known as watt hour meter, or kilowatt hour meter.

According to its phase wire(s), AC electric energy meter could be divided into single-phase type, three-phase-three-wire type and three-phase-four-wire type.

According to its working principle, electric energy meters could be divided into electromechanical type and electronic type (also known as static type and solid-state type). In which, electronic type could be divided into full electronic electric energy meters and electromechanical electric energy meters.

Electric energy meters could be divided into active power type, reactive power type, maximum demand type, standard type, multi rate time-sharing type, prepaid type, energy loss type and multi-function type according to their purposes.

(2) Use of single phase electric energy meter

Mechanical single-phase electric energy meter mainly consists of nameplate, voltage coil, current coil, counter, junction box, rotary plate, etc. The structure and wiring schematic diagram of mechanical single-phase electric energy meter are shown in Fig. 3-21.

(3) Use of three phase energy meter

The structure of a three phase active energy meter is the same as that of a single-phase energy meter, except that the three phase energy meter has two groups (three-wire system) or three groups (four-wire system) of voltage coils and current coils. The wiring schematic diagram of the three phase active energy meter is shown in fig. 3-22.

a. The electric energy meter could be directly connected to the circuit for measurement, under the condition of low voltage (no more than 500V) and low current (tens of amperes). In the case of high voltage or large current, the electric energy meter could not be directly connected to the line, so it needs to be used with voltage mutual inductor or current mutual inductor, as shown in Fig. 3-23.

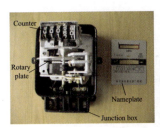
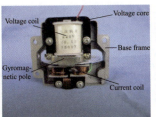

(a) Structure

(b) Wiring diagram

Fig. 3-21　Mechanical single-phase electric energy meter

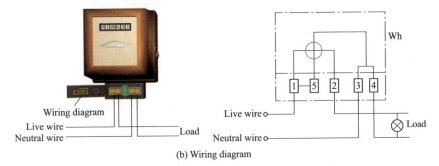

Fig. 3-22　Wiring diagram of three phase active electric energy meter

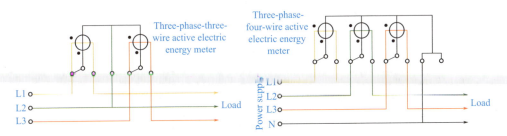

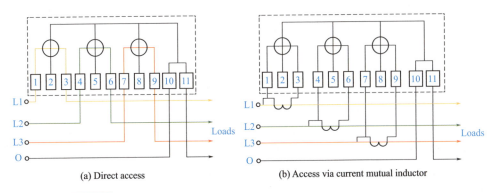

(a) Direct access　　　　(b) Access via current mutual inductor

Fig. 3-23　Wiring diagram of three-phase-four-wire electric energy meter

b. When measuring three phase active energy, three-phase-three-wire active electric energy meter or three-phase-four-wire active electric energy meter should be used according to load conditions. When three phase load is balanced, three-phase-three-wire active electric energy meter could be used; When three-phase load is unbalanced, three-phase-four-wire active electric energy meter should be used.

 【Special reminder】

> The electric energy measured by the direct access three phase electric energy meter could be directly calculated from the difference between the two values on the window of the meter. When the indirect access three phase electric energy meter is used to measure electric energy, the actual measured electric energy should be the value obtained by multiplying the difference between the two reading values by the ratio of current mutual inductor and voltage mutual inductor.

3.2.5 Use of ammeter

Ammeter is a kind of instrument specially used to measure circuit current. The commonly used ammeters include pointer ammeter and digital ammeter.

(1) Measurement of DC current

The instrument used to measure DC current is called DC ammeter. The DC ammeter could be divided into different types, according to the measuring range of A, mA and μA. There are two kinds of DC ammeters: fixed and portable. The fixed ammeters are square and round in shape. The ammeter has two terminals, with "+" and "−" symbols beside the terminals. The "+" terminal of the ammeter is connected to the high potential end of the circuit, and the "−" terminal is connected to the low one. The current flows from the "+" pole of the ammeter to the "−" pole.

The allowable current of the ammeter is small, generally designed as 50μA to 5mA in range. When measuring DC current below several mA, the meter could be used for measurement directly. For DC ammeters measuring large current, additional resistors should be connected in parallel at both ends of the meter. These parallel resistors are called a shunt. Generally, the shunt resistors are installed inside the ammeter. The connection method of DC ammeter is shown in Fig. 3-24.

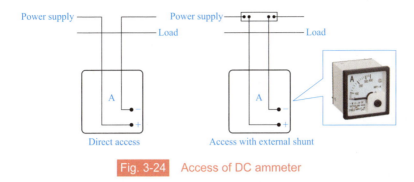

Fig. 3-24 Access of DC ammeter

 【Special reminder】

When measuring DC current, attention must be paid to the polarities of the instrument. The pointer would be damaged due to reverse rotation in case of incorrect polarities connection.

The ammeter must be connected in series with the load, not in parallel. Because the internal resistance of the ammeter is very small, parallel connection is equivalent to short circuiting the positive and negative poles of the power supply. The current is very large, which would damage the power supply and the ammeter.

(2) Measurement of AC current

Because the measuring mechanism of AC ammeter is different from that of DC ammeter, its measuring range is larger than that of DC ammeter. The commonly used AC ammeter in power system is 1T1-A electromagnetic AC ammeter with a maximum range of 200A. Therefore, within this range, the ammeter could be connected in series with the load, as shown in Fig. 3-25 (a).

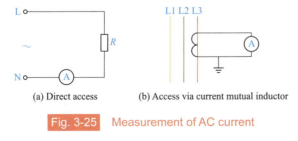

Fig. 3-25 Measurement of AC current

In low-voltage lines, when the load current is greater than the range of the ammeter, the current mutual inductor should be used. The primary winding of the current mutual inductor is connected in series with the load in the circuit, and the secondary winding is connected to an ammeter, as shown in Fig. 3-25 (b). In the high-voltage circuit, the wiring method of the ammeter is the same as that in Fig. 3-25, but the current mutual inductor must be high-voltage.

【Special reminder】

DC ammeter and AC ammeter are quite different. They could not replace each other.

3.2.6 Use of voltmeter

A voltmeter is an instrument that measures the voltage in a circuit. In the field of strong current, AC voltmeter is often used to measure the voltage of the monitoring circuit.

Voltmeters could be divided into DC voltmeter and AC voltmeter according to the waveform of measured current. The voltmeter is divided into microvolt meters (marked with "μV"), millivoltmeter (marked with "mV"), voltmeter (marked with "V"), kilovoltmeter (marked with "kV"), according to different measuring ranges.

(1) Measurement of DC voltage

When measuring DC voltage, connect the voltmeter in parallel with the circuit to be tested. The positive pole of the voltmeter is connected to the "+" terminal of the circuit to be tested, and the negative pole is connected to the "-" one.

In order to prevent the operation of the circuit under test from being affected by the voltmeter, the internal resistance of the voltmeter should be large. When the internal resistance of the voltmeter is not large enough by comparing it to the circuit to be measured, it is necessary to connect a large resistor in series at the voltmeter side for measurement.

(2) Measurement of AC voltage

The wiring of AC voltmeters is polarity free, but in a system, the wiring of all voltmeters should be the same, especially for 220V voltmeters or voltmeters using mutual inductor, this rule must be followed.

When measuring high voltage, voltage mutual inductor must be used. The range of the voltmeter should be consistent with the secondary rating of the mutual inductor. AC voltmeter and its voltage mutual inductor circuits must be equipped with fuses to prevent short circuit.

The wiring method for three phase AC voltage measurement is shown in Fig. 3-26.

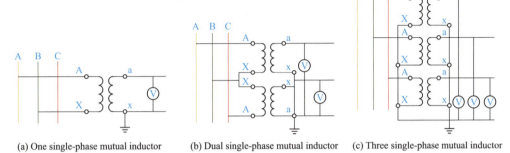

(a) One single-phase mutual inductor   (b) Dual single-phase mutual inductor   (c) Three single-phase mutual inductor

Fig. 3-26   Measurement of three phase AC voltage

3.3   Electrical safety signs

3.3.1   Electrical safety color

(1) Function

Safety color is a specific color to indicate safety information. Color is often used as a sign to enhance safety and prevent accidents. The safety color should be eye-catching and easy to identify.

The use of safety colors could help people form and fix their sensory adaptability in long-term life, which is positive for life and work. The purpose is to enable people to find and distinguish safety signs quickly through bright colors, remind people to pay attention and prevent accidents.

(2) Meaning and purpose

There should be uniform regulations on safety colors. The International Standardization Organization recommends that red, yellow and green be used as safety colors and blue as auxiliary colors. Safety Colors (GB 2893—2008) stipulates that red, blue, yellow and green are safety colors. See Table 3-4 for its meaning and purpose.

Table 3-4   Meaning and use of safety colors

No.	Color	Meaning	Use
1	Red	Indicates prohibition, stop, fire fighting and danger	Prohibited, stopped and dangerous devices, equipment or environment should be painted with red marks

No.	Color	Meaning	Use
2	Blue	Indicates instructions and regulations that must be observed	Generally used for instruction signs, such as wearing personal protective stuffs are painted with blue marks
3	Yellow	Indicates warning and caution	The device, equipment or environment warning people to pay attention should be painted with yellow marks
4	Green	Indicating prompt information, safety and passage	It is used for prompt signs, pedestrian and vehicle passing signs, etc. For example, the machine start button and safety signal flag are painted with green marks

(3) Application of safety color

In practical applications, safety colors often use other colors (i.e., contrast colors) as background colors to make them more eye-catching and improve the discrimination of safety colors.

Contrast color is to make the safety color more eye-catching, including black and white. If the safety color needs to use contrast color, the following methods should be used: red and white, blue and white, green and white, yellow and black. Red and white, blue and white, yellow and black stripes also could be used to indicate the meaning of emphasizing.

According to relevant regulations of power industry, phase L1 of substation buses is painted yellow, L2 phase in green, and L3 in red. In the running state of the equipment, the green signal flash indicates that the equipment is ready for operation, and the red indicates that the equipment is in operation, reminding the staff to concentrate and pay attention to safe operation.

 【Special reminder】

Safety colors do not include lights, fluorescent colors, colors used in aviation, navigation, inland navigation and colors used for other purposes.

3.3.2 Electrical safety sign board

(1) Types

The electrical safety sign board is composed of safety colors, geometric figures and graphic symbols to express specific safety information. There are four types: prohibition signs, warning signs, command signs and prompt signs, as shown in Fig. 3-27.

(a) Prohibition sign
No switching on

(b) Warning sign
Danger! High Voltage

(c) Command sign Safety helmet must be worn when entering the construction site

(d) Prompt sign
Upwards and downwards

Fig. 3-27　Example of electrical safety sign boards

① Prohibition signs are used to prohibit people's unsafe behaviors. Round in shape, white background, red round edge, a red slash in the middle, with black image. Generally, "no fireworks" and "no starting" are commonly used.

② Warning signs are used to remind people to pay attention to the surrounding environment and avoid possible dangers. Equilateral triangle in shape, yellow background, with black edges and images. "Beware of electric shock" and "Pay attention to safety" are commonly used.

③ Command signs are used to force people to take certain actions or take certain preventive measures. Round in shape, blue background, with white image and text. The commonly used ones are "safety helmet must be worn" and "goggles must be worn".

④ Prompt signs are used to provide people with certain information, such as indicating safety facilities or safe places. Rectangle in shape and the background is green, with white image and text.

(2) Regulations on use

Safety signs are generally made of steel plate, plastic and other materials, and there should be no reflection.

The safety sign should be installed at the place at obvious position and with sufficient light; The height should be slightly higher than people's sight to make it easy to find; Generally, it should not be installed on doors and windows, movable parts, or easily accessible parts of other objects; Safety signs should not be used too much in large areas or in the same place. They should usually be used at places with white light sources. Lighting should be added in places with insufficient light.

The Code for Safety Work in Electric Power Industry (electrical part of power plants and substations) clearly stipulates the use of hanging signboards and installing barriers on different occasions:

① On the operating handles of switches that could be powered to the working place once they are switched on, the signboards of "No Switching on, operator is working" with red characters on white background should be hung. If someone is working on the line, the signboard "No switching on, someone is working in line" should be hung on the switch operating handle.

② On the fence of live equipment at the construction site; On the fence of outdoor workplace; On the forbidden aisle; High voltage test site and outdoor structure; The sign board of "Stop, High Voltage Danger!" with red arrows on white background, red edge and black characters should be hung on the beam near the live equipment at the work site.

③ Hang a sign board in circle-shaped with a diameter of 210mm on the green background on the outdoor and indoor work sites or construction equipment, with "Work Here" written on it.

④ Hang the "Upwards and Downwards" signboard with a 210mm diameter white circle and black characters on the green bottom on the steel frame and ladder where the staffs go up and down.

⑤ When the steel frame that the staffs go up and down is close to another one that may go up and down, the ladder of the transformer in operation should be hung with a sign of "No Climbing, High Voltage Danger!" with white background, red edge and black characters.

Low-voltage electrical apparatus

4.1 Basic knowledge of low-voltage apparatus

4.1.1 Classification of low-voltage apparatus

(1) Classified by the use and control object

According to different use and control objects, low-voltage electrical apparatus could be divided into two categories: power distribution apparatus for power grid systems, and control apparatus for electric drive and automatic control systems. See Table 4-1.

Table 4-1　Low-voltage apparatus classified by the use and control object

Name		Main variants	Use	Graphic
Power distribution apparatus	Knife switch	High current knife switch Fuse type knife switch Knife switch for switch board Load switch	It is mainly used for circuit isolation and also for connecting and breaking rated current	
	Change-over switch	Combination switch Reversing switch	Used for shifting and switching on-off circuits of more than two power supplies or loads	
	Circuit breaker	Universal circuit breaker Molded case circuit breaker Current limiting breaker Leakage protection circuit breaker	Used for line overload, short circuit or undervoltage protection, as well as infrequently connecting and breaking circuits	

	Name	Main variants	Use	Graphic
Power distribution apparatus	Fuse	Filled fuse Unfilled fuse Fast acting fuse Self recovery fuse	For short circuit and overload protection of wires or electrical equipment	
Control apparatus	Contactor	AC contactor DC contactor	Mainly used for remote frequent starting or control of motors, as well as connecting and breaking the circuit in normal operation	
	Control relay	Current relay Voltage relay Time relay Intermediate relay Thermal relay	Mainly used for remote frequent starting or control of other electrical appliances or protection of main circuit	
	Starter	Magnetic starter Step-down starter	Mainly used for motor startup and positive and negative direction control	
	Controller	Cam controller Plane controller	Mainly used for connection of conversing main circuit or excitation circuit in electrical control equipment, to achieve the purpose of motor starting, reversing and speed regulation	

Name		Main variants	Use	Graphic
Control apparatus	Master apparatus	Button Limit switch Micro switch Universal change-over switch	Mainly used for connecting and breaking control circuit	
	Resistor	Iron base alloy resistor	Used to change circuit, parameters such as voltage, current and others; or convert electric energy into thermal energy	
	Rheostat	Excitation rheostat Starting rheostat Frequent rheostat	It is mainly used for generator voltage regulation, and step-down start & speed regulation for motor	
	Electromagnet	Lifting electromagnet Drive electromagnet Brake electromagnet	For lifting, operating or driving mechanical devices	

(2) Classification of low-voltage apparatus according to operation mode

According to different operation modes, low-voltage apparatus could be divided into automatic apparatus and manual apparatus, as shown in Table 4-2.

Table 4-2 Classification of low-voltage apparatus according to operation mode

Name		Use
Automatic apparatus	Contactor Relay	The apparatus that completes the actions such as connecting, breaking, starting, reversing and stopping through electromagnetic (or compressed air) work is called automatic apparatus
Manual apparatus	Knife switch Change-over switch Master switch	The apparatus that complete the actions of connecting, braking, starting, reversing and stopping through manual work is called manual apparatus

(3) Classification of low-voltage apparatus according to actuator

According to different actuators, low-voltage apparatus could be divided into contacted apparatus and contactless apparatus, as shown in Table 4-3.

Table 4-3　Classification of low-voltage apparatus according to actuator

Name		Use
Contacted apparatus	Contactor Relay	It has detachable movable contact(s) and static contact(s), and the contact and separation of the contact(s) is used to realize the connection and disconnection control of the circuit
Contactless apparatus	Proximity switch Solid state relay, etc.	There is no detachable contact, and the on-off control of the circuit is mainly realized by the switching effect of semiconductor components

【Special reminder】

Electrical equipment could be divided into high-voltage apparatus and low-voltage apparatus according to the working voltage. Internationally recognized boundary of high and low voltage electrical apparatus: AC 1kV (DC 1.5kV). High-voltage apparatus is above 1kV AC and low-voltage apparatus is below 1kV AC.

Generally, 380/220V and 6kV voltage are used for industrial electrical apparatus. The former is a low-voltage apparatus, and the latter is a high-voltage apparatus.

4.1.2　Common terms of low-voltage apparatus

To select low-voltage electrical apparatus correctly, the meaning of their common terms must be understood, as shown in Table 4-4.

Table 4-4　Meaning of common terms of low-voltage apparatus

No.	Term	Meaning
1	Short circuit connecting capacity	The connecting capacity, including the short circuit at the outlet line end of the switching apparatus under specified conditions
2	Short circuit breaking capacity	The breaking capacity, including short circuit at the outlet line end of the switching apparatus under specified conditions
3	Operating frequency	The maximum number of cycle operations that could be achieved by the switching apparatus per hour
4	Power on duration rate	The ratio of on load time and working cycle of electrical apparatus, usually expressed in percentage
5	Electrical life	The number of load operation cycles of mechanical switching apparatus without repair or replacement of parts under specified normal working conditions

No.	Term	Meaning
6	On-off time	The time interval from the moment when the current starts to flow through one pole of the switching apparatus to the moment when the arc at all poles is finally extinguished
7	Arcing time	The time interval from the moment that electric arc occurs when the contact is disconnected (or the fusant melts) to the time when the electric arc is completely extinguished
8	Breaking capacity	The expected breaking current value that the switching apparatus could break at a given voltage under specified conditions
9	Connecting capacity	The expected connecting current value that the switching apparatus could connect under the given voltage and the specified conditions
10	Connecting and breaking capacity	The expected current value that the switching apparatus could connect and break at a given voltage and the specified conditions

【Special reminder】

Low-voltage electrical apparatus is widely used in power transmission and distribution systems, electric drive and automatic control systems. Electricians must be familiar with the structure and principle of commonly used low-voltage electrical apparatus, and be able to correctly select and maintain them. As shown in Fig. 4-1, it is an application example of low-voltage apparatus.

Fig. 4-1 Operation of low-voltage apparatus controlled motor

4.1.3 Selection and installation of low-voltage electrical apparatus

(1) Selection principles

The main low-voltage apparatus used in the electric drive and transmission system have different uses and use conditions, so there are different selection methods, but the general requirements should follow the following two basic principles.

① Safety principle. Safe and reliable use is the basic requirement for any switching device. Ensuring the reliable operation of circuits and electrical apparatus is an important guarantee for normal production and life.

② Economic principle, under the condition that the safety standards are met and the required technical specifications could be reached, the electrical apparatus with higher cost performance should be selected as far as possible. In addition, it should be selected according to the service time, maintenance & replacement cycle and convenience of maintenance of low-voltage apparatus.

(2) Precautions for selection

① Select appropriate low-voltage electrical apparatus according to the type of control objects (motor control, machine tool control, electrical control of other apparatus), control requirements and environment for use.

② Select appropriate low-voltage electrical apparatus according to the normal working conditions, such as working altitude, relative humidity, corrosivity of harmful gas and conductive dust, allowable installation azimuth, impact resistance, indoor and outdoor working conditions, etc.

③ Determine the electrical technical specification (such as rated voltage, rated current, operating frequency, working system, etc.)according to the technical requirements of the controlled object.

④ The capacity of the selected low-voltage apparatus should be greater than that of the controlled equipment. For some equipment with special control requirements, special low-voltage electrical apparatus should be selected (such as speed requirements, pressure requirements, etc.).

⑤ When selecting low-voltage electrical apparatus that is consistent with the controlled equipment, factors such as "connecting" and "braking" capacity, service life and process requirements of electrical appliances should also be considered.

(3) Installation requirements

① Low-voltage apparatus should be installed vertically. The installation position should be convenient for operation and not easy to be damaged.

② Low-voltage electrical apparatus should be installed in a place without severe vibration and at a proper height from the ground. If the low-voltage apparatus is installed in places with severe vibration, vibration reduction measures should be taken.

③ The metal shell or metal support of low-voltage apparatus must be grounded (or connected to neutral wire). The exposed part of the electrical apparatus should be provided with a protective cover, and the opening position of the double knife switch should be equipped with a device to prevent self-closing.

④ In the workshop with flammable, explosive gas or dust, the electrical apparatus should be installed outdoors in a sealed manner, and rain proof measures should be taken. Explosion proof electrical appliances must be used in places with explosion hazards.

⑤ The contact surface of the electrical apparatus should be kept clean, smooth and in good contact during use. The contact(s) should have sufficient pressure, the action of each phase contact should be consistent, and the arc extinguishing device should be kept intact.

⑥ The single pole switch must be connected to the phase wire. The bottom of the floor mounted low-voltage apparatus should be 100mm higher than the ground. The name of the installed equipment and the circuit number or circuit type should be marked on the plate where the low-voltage electrical apparatus is installed.

4.2 Low-voltage circuit breaker and leakage protector

4.2.1 Low-voltage circuit breaker

(1) Functions of low-voltage circuit breaker

Low-voltage circuit breaker is mainly used for infrequent circuit connecting and breaking, and could automatically break the circuit in case of overload, short circuit and voltage loss.

(2) Type and structure of low-voltage circuit breaker

Low-voltage circuit breakers mainly include frame type (universal type) and molded case type (device type), and their structures are shown in Fig. 4-2.

(a) Universal low-voltage circuit breaker

1—Arc shield; 2—Switch body; 3—Drawer base; 4—Closing button; 5—Opening button; 6—Intelligent release; 7—Insertion position of shaking handle; 8—Connection/test/separation indication

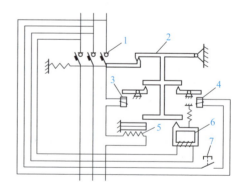

(b) Molded case low-voltage circuit breaker

1—Main contactor; 2—Free stripper; 3—Overcurrent stripper; 4—Shunt release; 5—Thermal release; 6—Loss of voltage release; 7—Button

Fig. 4-2 Type and structure of low-voltage circuit breaker

(3) Selection of low-voltage circuit breaker

Low-voltage circuit breakers are mainly used to control three types of loads: distribution lines, motors and lighting. The following principles should be generally followed when selecting low-voltage circuit breakers.

① The setting current of low-voltage circuit breaker should not be less than the normal working current of the circuit. The setting current of the circuit breaker is also called the current setting value of the overload release, which refers to the current value of the release adjusted to action. See Table 4-5 for selection of setting current of low-voltage circuit breaker.

Table 4-5 Selection of setting current of low-voltage circuit breaker

Type of load	Relation between setting current and load working current
Lighting	6 times of the load current
Motor (1)	Current for device type breaker should be 1.7 times of the staring current of motor Current for universal breaker should be 1.35 times of the staring current of motor
Motor (>1)	It is the sum of 1.3 times of the starting current of one motor with the largest capacity plus the rated current of other motors
Distribution lines	It should be equal to or greater than the sum of the rated current of the load in the circuit

② The setting current of the thermal release should be consistent with the rated current of the controlled load, otherwise, it should be adjusted manually, as shown in Fig. 4-3.

Fig. 4-3 Adjustment of the setting current

③ When selecting low-voltage circuit breakers, the protection characteristics of upper and lower switches should be matched with the type, grade, specification, etc. It is not allowed to skip the aimed switch due to the failure of the protection of its level, so as to expand the scope of power failure. Fig. 4-4 shows the types of low-voltage circuit breakers commonly used in families and similar places.

(a) 1P circuit breaker

The live wire enters the circuit breaker, but the neutral wire (Zero wire) does not enter

1P with DPN, the live wire and neutral wire (Zero wire) enter the circuit breaker at the same time, with higher safety

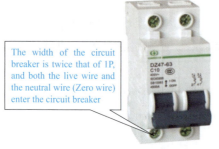

(b) 2P circuit breaker

The width of the circuit breaker is twice that of 1P, and both the live wire and the neutral wire (Zero wire) enter the circuit breaker

2P with DPN, the live wire and neutral wire (Zero wire) enter the circuit breaker at the same time, and the width of the circuit breaker with leakage protection is twice that of 1P with DPN

Fig. 4-4   Types of common low-voltage circuit breakers

4.2.2 Leakage protector

(1) Functions of leakage protector

The leakage protector has the protection functions against leakage, electric shock, overload, short circuit, etc. It is mainly used to protect the direct and indirect electric shock of low-voltage power grid effectively, and also could be used as the phase loss protection of three phase motors.

Like other circuit breakers, the leakage protector could connect or disconnect the main circuit, and has the function of detecting and judging the leakage current. In case of leakage or insulation damage occurs in the main circuit, the leakage protection switch could disconnect the main circuit according to the judgment result.

(2) Type of leakage protector

There are single-phase and three phase leakage protectors.

(3) Structure of leakage protector

The leakage protector is mainly composed of test button, operating handle, leakage indicator and wiring terminals, as shown in Fig. 4-5.

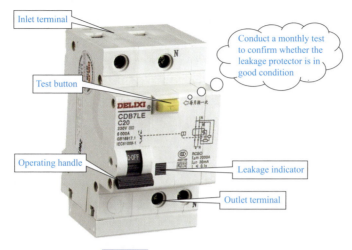

Fig. 4-5   Leakage protector

(4) Selection of leakage protector

Leakage protectors for residential and power use (generally referred to as 400V system) are mainly selected based on the leakage current value, as shown in Table 4-6.

Table 4-6   Selection of leakage protector

Applicable places	Reason to select
Family and similar places	Generally, small leakage protectors with action current not exceeding 30mA and action time not exceeding 0.1s are selected
Bathrooms, swimming pools and other places	The rated action current of leakage protector should not exceed 10mA
Places where electric shock may cause secondary accidents	The rated action current of leakage protector should not exceed 10mA

【Special reminder】

In the industrial power distribution system, the leakage protector, together with the fuse and thermal relay, could form a low-voltage switching element with complete functions.

4.3　Contactor and Relay

4.3.1   Contactor

(1) Function of contactor

Contactors are mainly used to frequently connect or disconnect AC and DC circuits.

They have large control capacity and could be operated remotely. They could be used with relays to achieve timing operation, interlocking control, various quantitative control, voltage loss and undervoltage protection. They are widely used in automatic control circuits. Their main control objects are motors, and they could also be used to control other electrical loads, such as electric heaters, lighting, electric welding machines, capacitor banks, etc.

【Special reminder】

The AC contactor uses the main contact to connect and disconnect the circuit, and uses the auxiliary contact to execute the control command. For the industrial electrical apparatus, there are many types of contactors, and the current varies from 5 to 1000A, which are widely used.

(2) Structure of AC contactor

The contactor is mainly composed of electromagnetic system, contact system, arc extinguishing device, etc. See Table 4-7. Its shape and structure are shown in Fig. 4-6.

Table 4-7 Structure of AC contactor

Device or system	Composition and description
Electromagnetic system	Movable iron core (armature), static iron core, electromagnetic coil, reaction spring
Contact system	Main contact (used to connect and cut off the large current of the main circuit), auxiliary contact (used the small current to control the circuit); Generally, there are three pairs of main contacts and several pairs of auxiliary contacts
Arc extinguishing device	It is used to quickly cut off the electric arc generated when the main contact is disconnected, so as to avoid the main contact being singed and fused. The large capacity contactor (above 20A) adopts the gap arc extinguishing chamber and arc extinguishing grid to extinguish the arc, while the small capacity contactor adopts the double break contact arc extinguishing, electric power arc extinguishing, interphase arc plate for arc isolating and ceramic arc extinguishing chamber to extinguish the arc

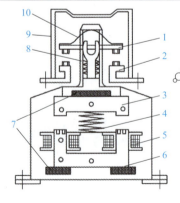

The structure of AC contactor is basically the same as that of DC contactor

Fig. 4-6 Appearance and structure of AC contactor

1—Moving contact bridge; 2—Static contact; 3—Armature; 4—Buffer spring; 5—Solenoid coil; 6—Iron core; 7—Mat; 8—Contact spring; 9—Arc extinguishing chamber; 10—Contact pressure spring

Tips on the structure and principle of AC contactor

AC contactor is composed of three parts.
Electrical energy becomes mechanical energy in the electromagnetic force drive device.
Contact(s) is/are the actuator(s), with two types of dynamic disconnecting and dynamic connecting.
Arc is generated during contact disconnection, and the arc extinguishing device is used to extinguish the arc.
There are many other parts, and each has its own role.
The armature moves for the suction of coil conducting, and circuit is connected because of contact(s) closing.
If the voltage is too high or too low, the coil might be burnt.

(3) Selection of contactor

See Table 4-8 for selection method of contactor.

Table 4-8 Selection method of contactor

Key points for selection	Method and description
Type of contactor	Select according to the type of load current in the circuit. AC contactor should be selected for AC load and DC contactor should be selected for DC load. If the control system is mainly composed of AC load, and DC motor or DC load with small capacity, AC contactor could also be selected for control, but the rated current of contact(s) should be larger
Rated voltage of main contact(s)	The rated voltage of the main contact(s) of the contactor should be equal to or greater than the rated voltage of the load. The rated voltage of AC contactor mainly includes: 127, 220, 380, 500V; The rated voltage of DC contactor mainly includes: 110, 220, 440V
Rated current of main contact(s)	The rated current of the main contact(s) of the selected contactor should be greater than the rated current of the load circuit. It could also be selected according to the maximum power of the controlled motor. If the contactor is used to control the frequent starting, positive and negative rotation, or reverse braking of the motor, the rated current of the main contact(s) of the contactor should be used at reduced grade. The rated current of AC contactor mainly includes: 5, 10, 20, 40, 60, 100, 150, 250, 400, 600A; The rated current of DC contactor mainly includes: 40, 80, 100, 150, 250, 400, 600A
Rated voltage of suction coil and capacity of auxiliary contact(s)	If the control circuit is relatively simple and the number of contactors used is small, the rated voltage of AC contactor coil is generally 380V or 220V. If the control circuit is complex and there are many electrical appliances used, for safety reasons, the rated voltage of the coil could be selected to be lower. Meanwhile, a control transformer needs to be added. The rated voltage of AC contactor suction coil mainly includes: 36, 110 (127), 220, 380V; Rated voltage of DC contactor suction coil mainly includes: 24, 48, 220, 440V

Tips on selection of AC contactor

The load requirements should be met when selecting AC contactor(s).
The selection of operating frequency depends on the operation times and current.
The motor power is the basis for the selection of rated current.
The rated voltage should be equal to or greater than the load voltage.
The selection of coil voltage depends on the complexity of the circuit.

4.3.2 Relay

(1) Types of relay

Relay is an automatic switching device with isolation function. Its contact(s) is/are usually connected in the control circuit, instead of directly controlling the main circuit with large current, it controls the main circuit through contactors or other electrical apparatus.

There are many kinds of relays. See Table 4-9 for common relays.

Table 4-9 Types of relays

Classification method	Types
According to the nature of input signal	Current relay, voltage relay, speed relay, pressure relay
According to the working principle	Electromagnetic relay, electric relay, inductive relay, transistor relay and thermal relay
According to the output mode	Contacted relay and contactless relay
According to the overall dimensions	Micro relay, ultra-micro relay, mini relay
According to protection characteristics	Sealed relay, plastic sealed relay, dust cover relay, open relay

【Special reminder】

The rated current of relay is generally not greater than 5A.

(2) Voltage relay

① Features. The coils are connected in parallel in the circuit, with many turns and thin wires.

② Function. The voltage relay is mainly used to monitor the voltage changes in the electrical circuit. The voltage relay acts, when the voltage value of the circuit changes beyond the set value, and the contact state switches to send a signal, as shown in Fig. 4-7.

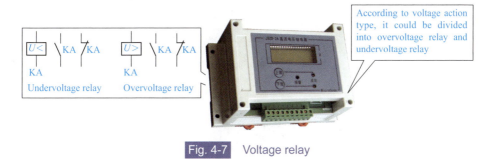

Fig. 4-7 Voltage relay

③ Selection.

a. The selection of overvoltage relay mainly depends on the rated voltage, action voltage and other parameters. The action value of overvoltage relay is generally set according to 1.1~1.2 times of the rated voltage of the system;

b. The rated voltage of the voltage relay coil could generally be selected according to the rated voltage of the circuit.

 【Special reminder】

The coil turns of voltage relay are large and the wires are thin. When in use, the electromagnetic coils of voltage relay are connected in parallel to the monitored circuit, in parallel with the load, and the action contact(s) is/are connected in series to the control circuit.

Memory tips of voltage relay

There are two types of voltage relay, overvoltage and undervoltage.
The coil turns are numerous and thin, and the setting range could be diverse.
The coils are connected in parallel in the circuits, to monitor the electrical changes closely.
The rated voltage should be suitable, and several tests must be conducted after installation.

(3) Current relay

① Features. The coil is connected in series in the circuit, with thick wire, few turns and small impedance.

② Function. Current relay is a control apparatus that reflects current changes and is mainly used to monitor current changes in electrical circuits. The current relay acts when the change of circuit current exceeds the set value, and the contact state would switch to send a signal, as shown in Fig. 4-8.

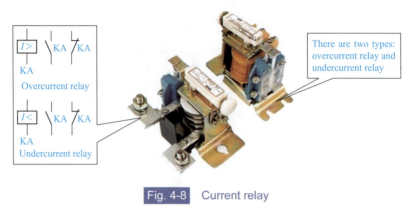

Fig. 4-8　Current relay

③ Selection.

a. The rated current of overcurrent relay coil could generally be selected according to the rated current of the motor for long-term operation. For motors that start frequently, take the thermal effect of the starting current in the relay into account, the rated current could be one grade higher.

b. The action current of the overcurrent relay could be set according to the working conditions of the motor, generally 1.1~1.3 times of the starting current of the motor, and 2.25~2.5 times for frequent starting motors. Generally, the starting current of common wound rotor induction motor is considered as 2.25~2.5 times of the rated current, and the starting current of cage induction motor is selected with 5-8 times of the rated current.

c. Undercurrent relay is often used for weak magnetic protection of DC motor magnetic field, which must be set according to actual needs.

【Special reminder】

The coil turns of the current relay are little and the wire is thick. When in use, the solenoid coil is connected in series to the monitored main circuit, in series with the load, and the action contact is connected in series to the auxiliary circuit.

(4) Intermediate relay

① Characteristics. The intermediate relay is essentially a voltage relay with the same structure and working principle as the contactor. However, it has pretty many contacts, mainly expanding the number of contacts in the circuit. In addition, the rated current of its contacts is large.

② Function. Intermediate relay is a low-voltage electrical device that transmits or converts signals. It could transmit, amplify, flip, shunt, isolate and memorize control signals to achieve the goal of one point controlling multiple points, and small power controlling large power, as shown in Fig. 4-9.

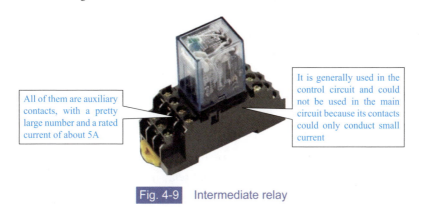

All of them are auxiliary contacts, with a pretty large number and a rated current of about 5A

It is generally used in the control circuit and could not be used in the main circuit because its contacts could only conduct small current

Fig. 4-9　Intermediate relay

③ Selection. There are many variants and specifications of intermediate relays, including J27 series, J28 series, JZ11 series, JZ13 series, JZ14 series, JZ15 series, JZ17 series and 3TH series. The intermediate relay should be selected mainly according to the voltage level of the controlled circuit, the contact number, type, capacity and other requirements of the required contacts.

(5) Thermal relay

① Function. Thermal relay is used to protect motor or other electrical equipment and electrical circuit from overload. It is mainly used for overload protection of motor and the control of heating state of other electrical devices. Some types of thermal relays also have protection against phase failure and unbalanced current operation.

【Special reminder】

Thermal relay is mainly used together with fuse.

② Structure. Thermal relays are mainly of bimetallic sheets type, thermistor type and fusible alloy type, as shown in Table 4-10.

Table 4-10　Structure of thermal relay

Structural style	Protection principle
Bimetallic sheets type	Use bimetallic sheets to bend by heating to push the lever to make the contact act, as shown in Fig. 4-10
Thermistor type	Made of the characteristic that the resistance value changes with temperature
Fusible alloy type	Use the overload current to heat to melt the fusible alloy and make the relay act

③ Selection and use. The thermal element of the thermal relay is used in series with the main circuit of the protected motor, and the contact(s) of the thermal relay is/are connected in series with the control circuit where the contactor coil is located.

a. Generally, the two phase thermal relay could be selected in case of the motor starts with light load or works in a short time; When the balance of power supply voltage and working environment are poor or the power difference of multiple motors is significant, the three phase thermal relay could be selected; For the motor in Δ-configuration, the thermal relay with phase failure protection device should be selected.

b. The rated current of thermal relay should be greater than that of the motor.

c. Generally, the setting current of thermal relay is adjusted to be equal to the rated current of the motor; For motors with poor overload capacity, the setting value of thermal element could be adjusted to 0.6~0.8 times of the rated current of the motor; For motors that have a long starting time, drive impact loads or are not allowed to stop, the setting current of thermal element should be adjusted to 1.1~1.15 times of the rated current of the motor. It is absolutely not allowed to bend the bimetallic sheets.

Tips on thermal relay

It is mainly consists of three parts, contact, bimetallic sheets and thermal element.
Connected to the main circuit of the motor in series to detect overload or disconnection.
The setting current needs to be adjusted, and the action time is critical.
The protection object is the motor, and it is difficult to realize frequent starting.

(6) Time relay

① Function. The time relay is essentially a timer. After the timing signal is sent out, the time relay turns on and off the controlled circuit according to the preset time and timing delay. In short, the circuit is turned on and off according to the setting time.

② Categories. According to the composition principle, it could be divided into electromagnetic type, electric type, air damping type, transistor type and digital type. According to the delay mode, it could be divided into power on delay type and power off delay type.

③ The graphical symbols are shown in Fig. 4-11.

④ Selection and use

a. The working voltage of time relay should be within the rated working voltage range.

b. When the load power is greater than the rated value of the relay, intermediate relay should be added.

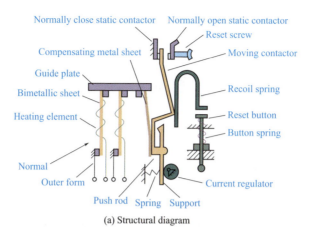

(a) Structural diagram

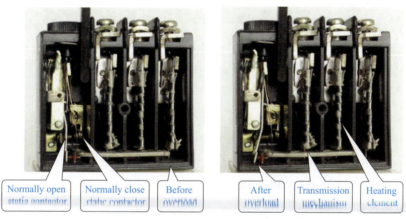

(b) Working principle

Fig. 4-10　Structure and working principle of bimetallic sheets thermal relay

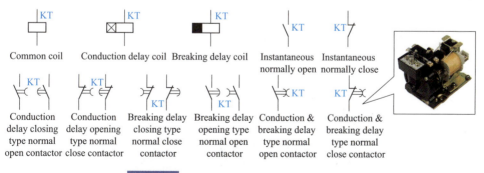

Fig. 4-11　Graphical symbols of time relay

　　c. It is strictly forbidden to install or remove the time relay when the power is on.

　　d. For equipment that may cause significant economic losses or personal safety, it is necessary to make enough allowance for the technical specifications and performance values during design, and safety measures such as dual circuit protection should be adopted.

4.4 Fuse and master switch

4.4.1 Fuse

(1) Function of fuse

Low voltage fuse, commonly known as fuse, which breaks the circuit by melting the fusant when the current exceeds the limit value. It is an electrical apparatus used for overload and short circuit protection of lines or equipment.

【Special reminder】

Most fuses are non-recoverable products (except self recovery fuse). Once damaged, they should be replaced with fuses of the same specification.

(2) Structure of common fuses

There are four types of commonly used fuses: ceramic plug-in type, screw type, closed tube type and powder-filled cartridge type. Their structures are shown in Fig. 4-12.

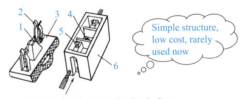

(a) RC1A series ceramic plug-in fuse
1—Fusant; 2—Moving contactor; 3—Ceramic cover; 4—Cavity; 5—Static contactor; 6—Ceramic base

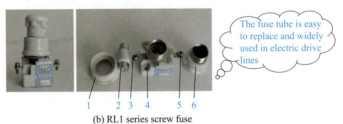

(b) RL1 series screw fuse
1—Ceramic case; 2—Fuse tube; 3—Lower wiring base; 4—Ceramic base; 5—Upper wiring base; 6—Ceramic cap

(c) RM10 series powder-filled cartridge fuse
1—Clamp base; 2—Fuse tube; 3—Vulcanized paper tube; 4—Brass sleeve; 5—Brass cap; 6—Fusant; 7—Knife clamp

(d) RT0 series powder-filled cartridge fuse
1—Fuse indicator; 2—Quartz sand filler; 3—Indicator fusant; 4—Clamp; 5—Clamp base; 6—Base; 7—Fusant; 8—Fuse tube; 9—Tin bridge

Fig. 4-12   Structure of common fuses

(3) Characteristics and application of common fuses (see Table 4-11)

Table 4-11   Characteristics and application of common fuses

Type	Characteristics	Application	Graphic
Ceramic plug-in type	It has the advantages of simple structure, low price and convenient replacement. When using, insert the ceramic cover into the base, and then remove the ceramic cover to replace the fuse	In the terminal or branch circuit of low-voltage line with rated voltage of 380V and below and rated current of 5~200A, it could be used for short circuit protection of lines and electrical equipment, as well as overload protection in lighting lines	
Screw type	The fuse tube is equipped with quartz sand, fusant and melting indicator with small red dots. The quartz sand is used to enhance the arc extinguishing performance. There is obvious indication after the fuse is blown	As a short circuit protection device in the circuit with AC rated voltage of 500V and rated current of 200A and below	

Type	Characteristics	Application	Graphic
Closed tube type	The fuse tube is made of steel, and its two ends are detachable caps made of brass. The fusant in the tube is a melting piece with variable cross section, and it is convenient to replace the fusant	Used in power lines with rated AC voltage of 380V and below, DC voltage of 440V and below, and current of 600A and below	
Powder-filled cartridge type	The fusant is two reticulated copper sheets, which are connected by a tin bridge. Fill quartz sand around the fusant for arc extinguishing	Used for short circuit protection and overload protection of lines and electrical equipment in power transmission and distribution systems with AC 380V and below and large short circuit current	

Tips on type and application of fuse

Edison invented the simple fuse.
Faults should be strictly avoided, and it would melt when the current exceeds the limit.
There are four kinds of commonly used fuses, and ceramic plug-in type is used for residential power distribution.
Screw type is commonly used for machine tool power distribution.
The powder-filled free type fuse is commonly used for equipment cables.
The powder-filled type is commonly used for rectifying elements.

4.4.2 Master switch

(1) Function and types

The master switch is used to switch on and off the control circuit to issue commands, or to program the production process.

The master switches include control button (button for short), travel switch, universal change-over switch, master controller, etc. In addition, there are foot switch, proximity switch, reversing switch, emergency switch, toggle switch, etc.

(2) Button switch

There are many types of button switches, including ordinary button type, mushroom head type, self-locking type, self resetting type, rotary handle type, indicator type, symbol type and key type. There are single button, double buttons, three buttons and different combinations. Common button switches are shown in Fig. 4-13.

In order to avoid misoperation, button caps are usually made into different colors to show differences

The building block structure is generally adopted, which is usually made into a composite type, with a pair of dynamic breaking contact and dynamic closing contact

Fig. 4-13　Button switch

(3) Limit switch

Limit switches, also known as position switches, are commonly divided into two categories. One is the travel switch and microswitch driven directly by mechanical stroke as the input signal; The other is proximity switch with electromagnetic signal (non-contact) as input action signal. As shown in Fig. 4-14.

(a) Travel switch　　　　　　　　　(b) Proximity switch

Fig. 4-14　Common limit switch

(4) Universal change-over switch

The universal change-over switch is a master switch with multi grades, multi segments, or control multi loops. When rotating the operating handle, the cam inside the switch would be driven to rotate, so that the contacts are connected or disconnected in the specified order, as shown in Fig. 4-15. The universal transfer switch is mainly used for the conversion of various control lines, the commutation measurement and control of voltmeter and ammeter, the conversion and remote control of power distribution device lines, etc. The universal change-over switch could also be used to control the starting, speed regulation and commutation of small capacity motors directly.

Fig. 4-15　Universal change-over switch

Chapter 5

Asynchronous motor

A motor is a device that converts electrical energy into mechanical energy. It is indicated by the letter M in the circuit. Its main function is to generate driving torque, as the power source of electrical appliances or industrial and agricultural production machinery. According to the structure and working principle, motors could be divided into synchronous motors and asynchronous motors. During operation, the motor whose speed is slower than the rotating magnetic field formed by the input voltage is called asynchronous motor, which could be divided into single-phase asynchronous motor and three phase asynchronous motor.

5.1　Basic knowledge of asynchronous motor

5.1.1　Brief introduction to motor

(1) Type of motors

① According to the difference of power supply, it could be divided into DC motor and AC motor, and there are many types of them, as shown in Table 5-1. In addition, there is a single-phase series motor, which could use either DC or AC power.

Table 5-1　Classification of motors by power supply

DC motor	Brushless DC motor		
	Brushed DC motor	Permanent magnet DC motor	
		Electromagnetic DC motor	Separately excited DC motor
			Shunt DC Motor
			Series excited DC motor
			Compound DC motor
AC motor	Asynchronous motor	Three-phase asynchronous motor	Cage rotor
			Wound rotor
		Single-phase asynchronous motor	Split phase motor
			Capacitor starting motor
			Capacitor run motor
			Capacitor starting running motor
			Shaded pole motor
	Synchronous motor (three-phase, single-phase)		

② According to the structure and working principle, it could be divided into synchronous motor and asynchronous motor.

a. During running, the motor whose speed is slower than the rotating magnetic field formed by the input voltage is called asynchronous motor, which could be divided into three-phase asynchronous motor and single-phase asynchronous motor.

b. During running, a motor whose speed is consistent with the rotating magnetic field formed by the input voltage is called synchronous motor. Synchronous motor could also be divided into permanent magnet synchronous motor, reluctance synchronous motor and hysteresis synchronous motor.

③ It could also be classified according to the form of housing protection, cooling mode, installation form, insulation grade, working mode, center height of motor dimensions, stator core outer diameter and other characteristics, as shown in Table 5-2.

Table 5-2　Classification of motor

Classification criteria	Type
According to the form of housing protection	Open type, protective type, closed type, dust-proof type, explosion-proof type, etc.
According to cooling mode	Self-cooling type, self-fan cooling type, other fan cooling type, duct ventilation type, etc.
According to the installation form	Horizontal, vertical, flanged (with or without feet)
According to insulation grade	Class A, E, B, F, H
According to working mode	Continuous, short-time, periodic and aperiodic
According to the center height of motor dimensions and outer diameter of stator core	Large, medium, small

(2) Motor model

The motor model consists of the motor type code, feature code and design serial number.

Motor type code: Y-asynchronous motor; T-synchronous motor. For example, the model identification of a motor is Y2-160M2-8, and its meaning is shown in Table 5-3.

Table 5-3　Meaning of motor model Y2-160M2-8

Identification	Meaning
Y	Model, indicating asynchronous motor
2	Design serial number, "2" refers to the product with improved design based on the first time
160	Center height is the height from the shaft center to the plane of the motor base
M2	Specification of base length: M refers to medium, in which "2" is the second specification of M-type iron core, and iron core of type "2" is longer than type "1"
8	Number of poles, "8" refers to 8-pole motor

(3) Nameplate of motor

The nameplate shows some parameters of the motor. To use the motor correctly, the nameplate must be fully understood. The nameplate of the motor is shown in Fig. 5-1.

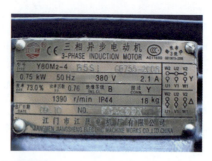

Fig. 5-1　Example of motor nameplate

The meanings of the main technical parameters marked on the nameplate of AC asynchronous motor are shown in Table 5-4.

Table 5-4　Meaning of each item on motor nameplate

Item	Meaning
Model	It refers to the serial variants, performance, protective structure form, rotor type and other product codes of the motor
Rated power	It refers to the mechanical power at the output end of the motor when it runs under the rated conditions specified by the manufacturer. The unit is generally kW or HP. 1HP=0.736kW
Voltage	It refers to the line voltage applied to the stator windings during the rated running of the motor, and the unit is volt (V). Generally, the working voltage of the motor should not be higher than or lower than 5% of the rated value
Current	The three-phase line current of the stator windings when the motor outputs rated power at rated voltage and frequency
Configuration	It refers to the configuration method of the stator three-phase windings, which should be consistent with the configuration method specified on the motor nameplate. Generally, the three-phase asynchronous motor should be configured into a star (Y) for the power below 3kW; For 4kW or above, configure into a delta (Δ)
Rated frequency	It refers to the frequency of the AC power supply connected to the motor, which is specified as 50Hz ± 1Hz in China
Rotating speed	It refers to the speed of motor per minute (r/min) under rated voltage, rated frequency and rated load. The formula of motor speed and frequency is $n=60f/p$ Wherein, n-motor speed (r/min); 60-every minute (second); f-power frequency (Hz); P-number of polar pair of rotating magnetic field of motor
Rated efficiency	It refers to the efficiency of the motor when running under rated conditions, and is the ratio of rated output power to rated input power. The rated efficiency of asynchronous motor is about 75%~92%
Isolation grade	It refers to the heat resistance grade of insulating material used for motor windings. The commonly used insulating materials of motors are divided into five grades: A, E, B, F and H according to their heat resistance
Working mode	It refers to the running mode of the motor. It is generally divided into "continuous" (code: S1), "short-time" (code: S2), "intermittent" (code: S3)
LP value	It refers to the total noise level of the motor. The lower the LP value, the lower the noise of motor running. Noise unit: dB

(4) Protection grade of motor

There are two types of motor housing protection, the first is the protection against solid foreign matters entering internal and the human body touching the internal live parts or moving parts; The second is the protection of water entering the internal.

The marking method of motor housing protection grade is shown in Fig. 5-2. Wherein, the first digit represents the grade of the first protection form; The second digit indicates the level of the second protection form, as shown in Table 5-5. When only one protection is considered, the other digit is replaced by "X". The in-front additional letter is the additional letter of the motor product, W represents the weather protection motor, R represents the duct ventilation motor; The following additional letter is also the additional letter of the motor product. S represents the motor for which the second protection form test is conducted under static state, and M represents the motor for which the second protection form test is conducted under running state. Additional letters may be omitted without special instructions.

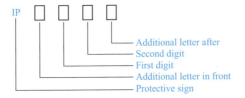

Fig. 5-2 Marking method of motor housing protection grade

Table 5-5 Housing protection grade of motor

The 1st digit	Classification of protection against human body and solid foreign matters	The 2nd digit	Classification of protection on water proof
0	Unprotected	0	Unprotected
1	Semi-protective type (to prevent solid foreign matters with a diameter greater than 50mm from entering)	1	Anti-dripping type (prevent vertical dripping)
2	Protective type (to prevent solid foreign matters with a diameter greater than 12mm from entering)	2	Anti-dripping type (to prevent vertical dripping with angle $\theta \leqslant 15°$ to perpendicular)
3	Sealed type (to prevent solid foreign matters with a diameter greater than 2.5mm from entering)	3	Anti-drenching type (to prevent vertical dripping with angle $\theta \leqslant 60$ to perpendicular)
4	Fully sealed type (to prevent the entry of solid foreign matters with a diameter greater than 1mm)	4	Splash proof type (to protect water from splashing in any direction)
5	Dustproof type	5	Water spray proof type (to protect against water spray in any direction)
		6	Anti sea wave type or enforced spraying
		7	Anti-flooding type
		8	Merged type

Take the housing protection level of IP44 for instance, where the first digit "4" indicates the protection level against human contact and solid foreign matters (that is, the motor housing could prevent solid foreign matters with a diameter greater than 1mm from touching or approaching the live or rotating parts in the housing); The second digit "4" indicates the degree of protection level against water entering the motor (that is, the motor housing could withstand water splashing in any direction without harmful effects).

【Special reminder】

The most commonly used protection levels of motors are IP11, IP21, IP22, IP23, IP44, IP54, IP55, etc.

5.1.2 Structure and principle of asynchronous motor

(1) Structure of asynchronous motor

① The basic structure of single-phase asynchronous motor consists of fixed part (stator), rotating part (rotor) and supporting part (end cover and bearing), as shown in Fig. 5-3.

Fig. 5-3 Basic structure of single-phase asynchronous motor

② The structure of a three phase asynchronous motor is basically the same, which is usually composed of a magnetic circuit part, an electric circuit part and other parts, as shown in Fig. 5-4.

a. Magnetic circuit part

Stator iron core: It is made of silicon steel sheets coated with insulating paint with 0.35~0.5mm in thickness, which reduces the eddy current loss of the iron core caused by the alternating magnetic flux. The inner circle of the iron core has evenly distributed notches, which are used to embed the stator windings.

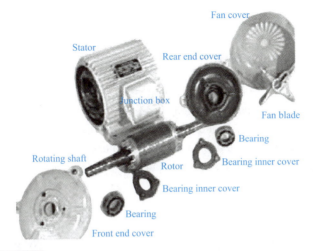

Fig. 5-4 Basic structure of three-phase asynchronous motor

Rotor iron core: It is made of silicon steel sheets with thickness of 0.5mm and is embedded on the shaft, with the same function as the stator iron core. On the one hand, it is used as a part of the magnetic circuit of the motor, and on the other hand, it is used to place the rotor winding.

b. Electric circuit part

Stator winding: three-phase winding is composed of three independent windings, and each winding is composed of several coils connected together. The coil is wound by insulated copper wire or insulated aluminum wire. The winding of three-phase motor has plenty variants, such as single-layer winding, double-layer stacked winding, single-layer and double-layer mixed winding, etc.

The junction box is an important part for the connection between the motor windings and the external power supply.

c. Other parts

Base: used to fix the motor.

End cover: it could be divided into front and rear covers.

Rotating shaft: generates electromagnetic torque under the induction of stator rotating magnetic field, rotates in the direction of rotating magnetic field, and outputs power to drive machinery running.

Bearing: It is a component that ensures the high-speed running of the motor at the central position.

Fan, fan cover and fan blade: used for cooling, dust prevention and safety protection.

Junction box: used for wiring between windings and three-phase power supply.

 【Special reminder】

The air gap between stator and rotor is generally 0.2~2mm. The size of the air gap has a great impact on the running performance of the motor. The larger the air gap is, the larger the excitation current supplied by the grid is, and the power factor cosφ the lower. To improve the power factor, the air gap should be reduced as much as possible; However, due to assembly requirements and other reasons, the air gap could not be too small.

(2) Working principle of asynchronous motor

The rotor of asynchronous motor is a rotatable conductor, usually in the form of a cage. The stator is the non-rotating part of the motor, whose main task is to generate a rotating magnetic field. Rotating magnetic field is not realized mechanically. Instead, AC current passes through several pairs of electromagnets, which makes the magnetic poles' properties change circularly, so it is equivalent to a rotating magnetic field.

Through the relative movement between the rotating magnetic field generated by the stator (its speed is synchronous speed n_1) and the rotor winding, the rotor winding cuts the magnetic induction line to generate an induced electromotive force, thus generating an induced current in the rotor winding. The induced current in the rotor winding acts with the magnetic field to generate electromagnetic torque and make the rotor rotate. As the rotor speed gradually approaches the synchronous speed, the induced current gradually decreases, and the electromagnetic torque generated correspondingly decreases. When the asynchronous motor runs in working state, the rotor speed is less than the synchronous speed.

The slip ratio s of the asynchronous motor is equal to the relative value of the difference between the actual speed n and the synchronous speed n_1 expressed in percentage.

$$s = \frac{n_1 - n}{n_1} \times 100\%$$

$$n = \frac{60f_1}{p}(1-s)$$

Where, p is the number of magnetic poles of the motor, and f_1 is the frequency of AC.

【Special reminder】

Slip ratio s is an important parameter of asynchronous motor, which could reflect various running conditions and speed of asynchronous motor. The greater the load of asynchronous motor, the lower the speed, and the greater the slip ratio; On the contrary, the smaller the load, the higher the speed, and the smaller the slip ratio. When the asynchronous motor is under rated load, its rated speed is very close to the synchronous speed, so the slip ratio is very small, generally 2%~6%.

(3) Configuration of three phase asynchronous motor windings

The stator windings of three phase asynchronous motor are the electric circuit part of the asynchronous motor, which is composed of three phase symmetrical windings and embedded in the stator slots according to certain spatial angles.

Generally, there are 6 outlet lines in the junction box of three phase cage motor, marked with A, B, C, X, Y and Z. Where: A & X are the two ends of the first phase winding; B & Y are the two ends of the second one; C & Z are the two ends of the third one. If A, B and C are the starting ends (heads) of three phase windings respectively, X, Y and Z are the corresponding ends (tails). The six outlet terminals must be correctly connected to each other before being connected to the power supply.

Three phase stator windings could be configured into star (Y) or delta (Δ) according to different power supply voltages and requirements on motor nameplate, as shown in Table 5-6.

Table 5-6 Configuration of three phase windings of asynchronous motor

Configuration style	Actual wiring	Wiring diagram	Wiring schematic diagram
Y-configuration			
Δ-configuration			

① Y-configuration. Connect the tail ends X, Y and Z of the three phase windings together in a short circuit, and connect the head ends A, B and C to the three phase power supply respectively.

② Δ-configuration. Connect the head and tail ends of each phase winding of the three phase coil in turn. That is, the tail end X of the first phase is connected to the head end B of the second phase, Y to C, Z to A, and then the three contacts are connected to the three-phase power supply respectively.

Memory tips

There are two types of motor configuration, Y and Δ.
Y-configuration is generally adopted for rated voltage of 220V.
One end of each three phase winding is short circuit, and the other ends are connected to the power supply respectively.
The three phase windings are connected head to tail to be Δ-configuration for rated voltage of 380V.
The 3 contacts are connected to the power supply respectively, which is also called Δ-configuration.
The motor configuration is determined by the manufacturer and could not be changed at will.

【Special reminder】

Whether the three phase asynchronous motor is Y- configured or Δ- configured, the reverse direction (forward or reverse) could be obtained by switching any two phases of the three phase power supply.

Either Y-configuration or Δ-configuration, the wire voltage and wire current are the same. The difference is that the current and voltage of the coil windings are different. In Y-configuration, the voltage passed by the coil is the phase voltage (220V), characterized by low voltage and large current; When configured in Δ, the voltage at the coil is 380V, which is characterized by high voltage and low current.

5.1.3 Characteristics of asynchronous motor

(1) Mechanical characteristics of asynchronous motor

The mechanical characteristics of the motor refer to the characteristics of the relationship between the running speed n of the motor and the generated torque M under certain conditions, which is expressed by the function $n=f(M)$, as shown in Fig. 5-5. It is an important characteristic of motor operation. If the load changes and the speed changes little, it is called hard characteristic; in case of a large speed change, called soft characteristic.

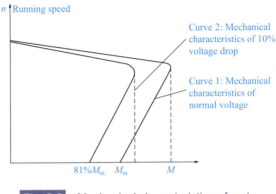

Fig. 5-5   Mechanical characteristics of motor

The artificial mechanical characteristics of AC asynchronous motor could be obtained by changing the stator voltage, the number of magnetic pole pairs, the series impedance in the stator circuit, the series resistance in the rotor circuit and the power frequency.

Torque and speed are two basic requirements of production machinery for motor. Different production machinery has different torque-speed relationship, which requires the mechanical characteristics of motor to adapt to it. For example, if the speed is required to be constant when the load changes, the synchronous motor should be selected; Series excited or compound excitation DC motors should be selected for those requiring large starting torque and soft characteristics, such as trams, electric locomotives, etc.

(2) Running characteristics of asynchronous motor

The running characteristics of asynchronous motor refer to the torque generated by the action between the rotor magnetic potential and the stator magnetic potential formed by the rotor permanent magnet of the motor in the air gap when the motor operates at rated voltage and frequency. When the motor is close to the synchronous speed, the alternating frequency gradually decreases, and a stable synchronous electromagnetic torque is formed with the stator magnetic potential during synchronous running.

The running state of asynchronous motor is closely related to the speed range of rotor, as shown in Fig. 5-6. In which, N and S represent the air gap rotating magnetic potential, the arrow of n_1 represents the direction of rotation, the middle two small circles represent a short circuit coil of the rotor, and f represents the EMF.

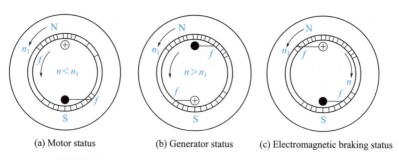

(a) Motor status   (b) Generator status   (c) Electromagnetic braking status

Fig. 5-6   Three running states of asynchronous motor

① When the rotor of the motor rotates forward and the speed is lower than the synchronous speed, that is, $0 < n < n_1$, or $1 > s > 0$, the motor runs in the running state (s is the slip rate, the same below), as shown in Fig. 5-6 (a).

② When the rotor of the motor rotates forward and the speed is higher than the synchronous speed, that is, $n > n_1$, or $s < 0$, it runs under the generator state, as shown in Fig. 5-6 (b).

③ When the rotor of the motor rotate reversely, that is, $n < 0$, or $s > 1$, it operates in the electromagnetic braking state, as shown in Fig. 5-6 (c).

Motor working mode is a description of the load that the motor bears, including starting, electric braking, load, no-load, energy cut-off and shutdown, as well as the duration and sequence of these stages. The working mode is divided into 10 categories from S1 to S10, as shown in Table 5-7.

Table 5-7   Working modes of motor

Working mode		Definition	Description
S1	Continuous mode	Working mode with constant load for a long time without specified period	Continuous running under constant load to reach thermal stability
S2	Short-term mode	Running under constant load for a specified time, stop and cut-off power before reaching thermal stability, and the time is enough to make the motor or cooler cool to the temperature difference with the cooling medium finally within 2K	—
S3	Intermittent cycle mode	Running according to a series of the same working cycles, and each cycle consists of a period of constant load running time and a period of shutdown & power cut-off time. However, the running time in each cycle is short, which is not enough to make the motor reach thermal stability, and the starting current in each cycle has no obvious impact on the temperature rise	The load duration should be indicated at S3 working mode, such as S3 25%

Working mode		Definition	Description
S4	Intermittent cycle with startup mode	Running according to a series of the same working cycles, and each cycle consists of a period of startup time, a period of constant load running time and a period of shutdown & power cut-off time	The starting and running time is short in each cycle, which is not enough to make the motor reach thermal stability
S5	Intermittent periodic with electric braking mode	Running according to a series of the same working cycles, and each cycle consists of a period of starting time, a period of constant load running time, a period of fast electric braking time and a period of shutdown & power cut-off time	The starting, running and braking time is short in each cycle, which is insufficient for the motor to reach thermal stability
S6	Continuous periodic mode	Running according to a series of the same working cycles, and each cycle consists of a period of constant load running time and a period of no-load running time	The load running time is short in each cycle, which is insufficient for the motor to reach thermal stability
S7	Continuous periodic with electric braking mode	Running according to a series of the same working cycles, and each cycle consists of a starting time, a constant load running time and an electric braking time	No shutdown & power cut-off time
S8	Continuous periodic with corresponding change of load-speed mode	Running according to a series of the same working cycles. Each cycle consists of a period of constant load running time at a planned speed, followed by other constant load running time at one or several different speeds	The feature is that there are 3 constant loads in each cycle, such as the application of multi-speed asynchronous motor
S9	Non-periodic changes in load and speed mode	Running mode of non-periodic change of load and speed within an allowable range	Including frequent overloads, which could be far more than full load
S10	Discrete constant load mode	The working mode includes no more than 4 discrete load values (or equivalent loads), and the running time of each load should be sufficient to make the motor reach thermal stability	The minimum load value in a running cycle could be zero (no-load or shutdown & power cut-off)

5.2 Starting and running of asynchronous motor

5.2.1 Starting of asynchronous motor

(1) Starting mode of single-phase motor

There are three starting modes for 220V AC single-phase motor.

① Split phase starting mode. The auxiliary starting winding is used for starting, and the starting torque is not large. The running speed is approximately maintained at a constant value. It is mainly used as motors in electric fans, air conditioning fans, washing machines, etc.

② Centrifugal switch-off mode. When the motor is stationary, the centrifugal switch is on. After power is supplied, the starting capacitor engages in starting. When the rotor speed reaches 70%~80% of the rated value, the centrifugal switch would automatically break away. The starting capacitor completes the task and is disconnected. The starting winding does not engage in the running, while the motor continues to run via the running winding coil.

③ Dual value capacitor mode. When the motor is stationary, the centrifugal switch is on. After power is supplied, the starting capacitor engages in starting. When the rotor speed reaches 70%~80% of the rated value, the centrifugal switch would automatically break away. The starting capacitor completes the task and is disconnected. The running capacitor is connected to the starting winding in series to engage in the running. This connection method is generally used in places with large and unstable loads such as air compressors, cutting machines and woodworking machines.

(2) Starting of three phase asynchronous motor

① Full voltage starting refers to the starting by implementing all the power supply voltage to the motor windings, which is also called direct starting.

In the full voltage starting of the asynchronous motor, the starting current is 4-7 times of the rated current. In production, whether the cage asynchronous motor could start directly depends on the following conditions.

a. The motor is allowable to direct starting. For motors with large inertia, long starting time or frequent starting, excessive starting current would cause motor aging or even damage.

b. The driven mechanical equipment could withstand the impact torque of direct starting.

c. The normal running of other equipment on the power grid would not be impacted by the grid voltage drop caused by the direct starting of the motor. The specific requirements are as follows: the power grid voltage drop caused by frequent starting motors should not be greater than 10%; The power grid voltage drop caused by infrequent starting motors should not be greater than 15%; When the starting torque required by the production machinery could be ensured, and the voltage fluctuation caused in the power grid would not impact the running of other electrical equipment, the grid voltage drop caused by the motor startup is allowed to be 20% or more; In case of a transformer supplies power to multiple loads with different characteristics and some loads require small voltage fluctuation, the power of asynchronous motors that could be directly started should not be too big.

d. The motor should not be started too frequently. Because the more frequent the starting, the more impact it would have on other loads on the same grid.

② Step-down starting. The step-down starting of the motor is to reduce the voltage implemented to the motor stator windings during starting, limit the starting current, and restore the working voltage to the rated value when the motor speed is basically stable with unchanged power supply voltage.

When the load has no strict requirements on the starting torque of the motor, but also limits the starting current of the motor, and the motor meets the 380V and the rated running state in Δ-configuration, the step-down starting could be used.

Common step-down starting methods include rotor with the resistor in series step-down starting, Y/Δ starting, reactance step-down starting, extended Δ starting, soft starting and autotransformer step-down starting. Only the most commonly used Y/Δ starts are described below.

The feature of Y/Δ step-down starting is that for motors with large capacity and rated running with Δ-configuration, the stator windings are in Y-configuration during starting. When the speed rises to a certain value, it is changed to Δ-configuration to achieve the purpose of step-down starting. This starting mode is called Y/Δ step-down starting of a three-phase asynchronous motor.

Y/Δ step-down starting refers to the method of windings in Y-connection when the motor is started, and then changed to Δ-connection after the motor is successfully started. The Y/Δ step-down starting method is simple, economical and reliable. The starting voltage of Y-configuration is only $\frac{1}{\sqrt{3}}$ of that of Δ-configuration, the starting current and the starting torque are both only 1/3 of that under normal operation of Δ-configuration. Therefore, Y/Δ starting is only applicable to no-load or light load conditions.

As shown in Fig. 5-7, there are several commonly used Y/Δ step-down starting control circuits for motors.

In the circuit shown in Fig. 5-7 (a), the six wiring terminals of the three-phase windings of motor M are connected with the contactors KM1, KM2 and KM3 respectively. When starting, close the power switch QS, the upper part of the main contact of contactor KM1 is energized, and the control circuit is also energized. Press the start button SB2, the coils of contactor KM1 and KM2 are powered on at the same time (KM2 is powered on through the dynamic breaking contact of time relay KT and the dynamic breaking contact of KM3). Meanwhile, the asynchronous motor is in the starting state in Y-configuration, and the motor starts to start; Since the auxiliary contact (interlock contact) of KM2 and KM3 connected in series is disconnected, the contactor KM3 is not energized at this time.

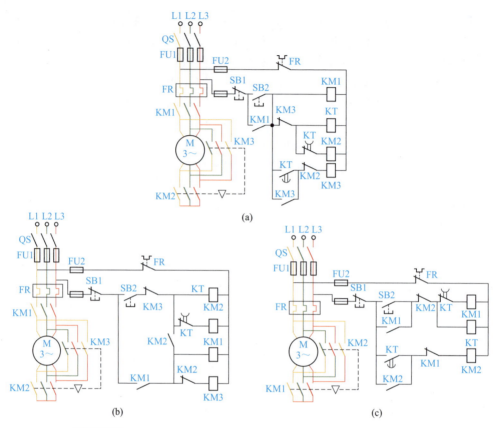

Fig. 5-7 Common Y/Δ step-down starting control circuits of motor

After KM1 acts, the time relay KT coil starts to delay after being energized. During the set delay time of KT, the asynchronous motor starts and accelerates. After the delay time of relay KT expires, all contacts of KT change state, KM2 coil is powered off, and the main contact is disconnected, so that the center point of Y-configured asynchronous motor is disconnected; After the KM2 coil is powered off, the dynamic breaking auxiliary contact KM2 connected in series to the KM3 coil circuit is closed to release the interlock. After KM2 is closed, the coil circuit of contactor KM3 is connected, KM3 acts and all its contacts change state. After the KM3 coil is energized, the main contact is closed. Meanwhile, the motor switches to Δ-configuration automatically for running, and the secondary starting is carried out; The dynamic contact in parallel with KT dynamic contact is closed and self-locking; The dynamic breaking auxiliary contact (interlock contact) in series with the KT and KM2 coils is disconnected, the time relay KT and the contactor KM2 coil are de-energized, and the starting process is ended.

In the Y/Δ step-down starting circuit as shown in Fig. 5-7 (a), since the main contact of KM2 is closed with rated voltage, the contact capacity is required to be large, while KM2 does not work when the asynchronous motor runs normally, which would cause certain waste. At the same time, If the main contact of contactor KM3 is melted and stuck for some reason, the asynchronous motor would be directly connected in Δ-configuration instead of Y-configuration for step-down starting, and the step-down starting function would be lost. Therefore, the circuit is not reliable enough. If a restart relay is added between KT and KM3 (the restart relay actually has the same meaning as the intermediate relay. Generally, the fast intermediate relay is selected, which is mainly used for electrical isolation between two circuits and provides more contact capacity), so the circuit would be more reliable.

In comparison, the control circuit shown in Fig. 5-7 (b) has high reliability. The coil of time relay KT and contactor KM2 could be energized only when KM3 dynamic breaking contact is closed (no fusing fault exists) and start button SB2 is pressed. The delay starts after the KT coil is energized. All contacts change state after KM2 coil is powered on. The main contact connects the asynchronous motor into Y-configuration without voltage; The closing of auxiliary contact KM2 energizes the coil of contactor KM1; The dynamic breaking auxiliary contact (interlock contact) in series with KM3 coil is disconnected. After the KM1 coil is energized, the main contact KM1 is closed, and the main circuit is connected. Meanwhile, the motor has been connected to Y-configuration, so the motor is energized and started; The closing auxiliary contact (self-locking contact) of KM1 is closed and connected with the stop button SB1 to form a self-locking.

When the delay time set by KT is up, the auxiliary contact KT of the dynamic breaker would be disconnected, the KM2 coil would lose power, and the main contact KM2 would disconnect the center point of the Y-configured asynchronous motor to prepare for the Δ-configuration; The dynamic breaking auxiliary contact (interlock contact) in series with the KM3 coil is reset and closed to energize the KM3 coil of the contactor. After KM3 is powered on, the asynchronous motor is in Δ-configuration for secondary starting, and the interlocking contact in series with the start button SB2 is disconnected, so the starting process is ended. Since the main contact of KM2 is closed without power, a contactor with relatively small contact capacity could be selected for KM2. However, in actual use, if the selected contact capacity is too small, when the time delay setting of the time relay is also short, it is easy to cause the main contact of KM2 to burr or damage, which should be noted in actual use.

The control circuit shown in Fig. 5-7 (c) only uses two contactors. In fact, it is obtained by renumbering the contactors after removing KM1 from the circuit shown in Fig. 5-7 (a). This circuit is suitable for occasions where the control requirements are relatively low and the capacity of asynchronous motor is relatively small.

【Special reminder】

Y/Δ step-down starting is a common starting method for three phase asynchronous motors. During starting, the motor stator windings are in Y-configuration, and during running, it's Δ-configuration, as shown in Fig. 5-8.

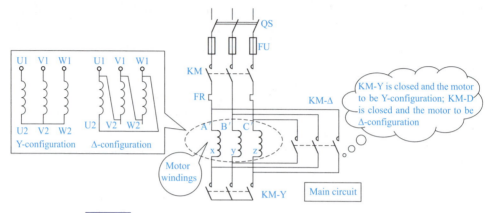

Fig. 5-8 Winding configuration during Y/Δ step-down startup

Y/Δ step-down starting is not allowed for motors with rated running status of Y-configuration.

5.2.2 Running of asynchronous motor

(1) Conditions for safe running of asynchronous motor

① Running parameters. Voltage, current, frequency, temperature rise and other running parameters of the motor should meet the requirements.

② All insulation indexes of insulated motor should meet the requirements. In any case, the insulation resistance of the motor should not be less than 1000 Ω per volt of running voltage.

③ Protection facilities. Motor protection should be complete. For example, when fuse is used for short circuit protection, the rated current of fuse should be 1.5 times of the rated current of asynchronous motor (step-down starting) or 2.5 times of the rated current of asynchronous motor (full voltage starting); When the thermal relay is used for overload protection, the current of the thermal element should not be greater than 1.1~1.25 times of the rated current of the motor; The motor should be equipped with voltage loss protection device; Important motors should be equipped with phase failure protection unit, and the motor housing should be reliably connected to neutral or grounded according to the running mode of the power grid.

④ Maintenance. The motor should keep its main body complete, spare parts complete, undamaged and clean. The motor should be maintained regularly. The daily maintenance work includes removing the external dust and oil stain, checking the bearing and replacing the lubricating oil, checking the slip ring and commutator and replacing the brush, checking the grounding (zero) wire, tightening each bolt, checking the connection of the outlet wires and the insulation resistance. The starting equipment should be overhauled together with the motor. After the motor is overhauled, the insulation resistance of each part should be measured (absorption ratio and AC/DC withstand voltage test should also be measured and carried out for large motors), the polarities of stator windings should be measured, and no-load test should be conducted.

⑤ Technical data. In addition to the original technical data provided by the manufacturer, the running records and maintenance records of the motor should also be made. If possible, running procedures could also be composed to ensure the safe running of the motor better.

(2) Abnormal running of asynchronous motor

In the actual running of asynchronous motor, the power supply voltage or frequency is not equal to the rated values, and the possibility of three phase voltage or three phase load asymmetry exists. For example, the frequency fluctuation would be caused by the active power imbalance, and the voltage fluctuation would be caused by the reactive power imbalance. The motor would be in an asymmetric running state when there is a single-phase load or a phase failure, or when there is a two-phase short circuit or a single-phase short circuit to ground, these are called the abnormal running state of the motor.

In case of abnormal running, the motor would be burnt seriously, causing losses.

① Running at non-rated voltage. In order to make full use, the iron core is always close to saturation when the motor is running at rated voltage. When the voltage changes, the saturation degree of the motor core changes accordingly, which causes changes in the excitation current, power factor and efficiency. If the difference between the actual voltage and the rated voltage does not exceed ± 5%, there is no significant impact on the running of the motor; If the voltage change exceeds ± 5%, the running of the motor would be greatly affected.

If the motor runs under $U_1 > U_N$, the main flux Φ_m would increase. Because the saturation degree of the magnetic circuit also increases simultaneously, the excitation current I_m would greatly increase when the magnetic flux increases little, which would reduce the power factor of the motor, and the iron core loss would increase with the increase of Φ_m, which leads to the decrease of motor efficiency and the increase of temperature rise. Meanwhile, in order to make the motor run safely, the load must be reduced.

If the motor runs under $U_1 < U_N$, the main flux Φ_m would decrease, the excitation current I_m would decrease, and the iron core loss would also decrease as the decreases of Φ_m. If the load is constant, the motor speed would decrease, the slip, the rotor current and the rotor copper loss would increase.

【Special reminder】

The increase in voltage would increase the iron core loss, decrease the efficiency and the power factor; The decrease of voltage would increase the copper loss, decrease the efficiency and the power factor when the voltage is about the rated load. If the voltage is decreased too much, it might also cause stalling or even burning out. It is beneficial to decrease the voltage under light load.

In actual production, the motor is started by Y/Δ step-down starting, and is often connected to Y-configuration under light load to improve the power factor and efficiency of the motor.

② Asymmetrical running. The phase failure running of three phase asynchronous motor is one of the main reasons for burning out the motor winding.

a. Three phase motor runs with one phase failed. Running of three phase motor without one phase would result in a large starting current, reduced speed, vibration, buzzing sound, and windings burning during long-term operation.

b. Three-phase motor runs in two-phase with neutral mode. The two-phase with neutral running of three phase motor is caused by the wrong connection between a phase wire and the protective neutral wire connected to the metal housing. Meanwhile, the motor housing is charged, and there is a great risk of electric shock. If the load torque is small, the motor could still be started in the positive direction when the power is connected; During running, the speed change is very small, and the abnormal sound is not obvious.

【Special reminder】

In order to ensure the safe and economical running of the motor, the running regulations stipulate that the voltage fluctuation should not exceed ± 5% of the rated value, and the frequency fluctuation should not exceed ± 1% of the rated value.

When the motor runs asymmetrically, its performance becomes worse. Asymmetric running has disadvantages but no advantages for motors.

5.3 Fault inspection and maintenance of asynchronous motor

In the long-term running process of three phase asynchronous motor, various faults would occur. The faults could be summarized into electrical faults and mechanical faults. In electrical aspect, there are mainly faults in stator windings, rotor windings, stator and rotor cores, switch and starting equipment, etc.; Mechanical failures mainly include bearing, rotating shaft, fan, base, end cover, load mechanical equipment, etc.

5.3.1 Fault inspection and maintenance of three phase cage asynchronous motor

It is important work to timely determine the cause of the failure and handle it accordingly to prevent the expansion of the failure and ensure the normal running of the equipment. The common fault phenomena, possible causes and corresponding treatment solutions of three phase cage asynchronous motor are shown in Table 5-8 below for reference when analyzing and handling faults.

Table 5-8 Common fault phenomena and treatment solutions of three phase cage asynchronous motor

Fault phenomenon	Possible causes	Treatment solutions
The motor could not be started after the power is on, but there is no abnormal noise, odor and smoke	(1) Power supply is not on (at least two phases are not on) (2) Fusant melted (at least two phases melted) (3) The set value of overcurrent relay is too small (4) Wiring error of control equipment	(1) Check whether the power switch and junction box are broken and repair them (2) Check the fusant specification and the cause of melting, replace it with a new one (3) Adjust the relay setting value to match with motor (4) Correct wiring
After powering on the motor could not rotate, and then the fusant melted	(1) Lack of one phase power supply (2) Short circuit between stator windings (3) Stator winding grounding (4) Stator winding wiring error (5) Fusant's sectional area is too small	(1) Find out the disconnection point and connect it the power circuit (2) Find out the short circuit point and repair it (3) Find out the grounding point and eliminate it (4) Check the wrong connection and correct it (5) Replace the fusant
The motor does not start, but with a buzzing sound after powering on	(1) One phase of stator/rotor windings or power supply is open circuited (2) The outlet wire of the windings or the internal connection of the windings is wrong (3) The contact of power circuit is loose and the contact resistance is large (4) The motor is overloaded or the rotor is jammed (5) The power supply voltage is too low (6) The bearing is stuck	(1) Find out the break point and repair it (2) Check whether the head and tail ends of the windings are correct, and correct the wrong one(s) (3) Fasten the loose wiring screws, use a multimeter to check whether the contacts are falsely connected, and repair them (4) Reduce load or find out and eliminate the mechanical faults (5) Check whether the Δ-configuration of three phase windings is wrongly configured to Y, and correct it if it is wrong (6) Replace the qualified grease or repair the bearing

Fault phenomenon	Possible causes	Treatment solutions
The motor is difficult to start, and the speed with rated load is much lower than the rated value	(1) The power supply voltage is too low (2) Motor in Δ-configuration is wrongly configured Y (3) The cage rotor is open welded or broken (4) Misconnection of partial coils of stator winding (5) Motor overload	(1) Measure the power supply voltage and try to improve it (2) Correct the connection (3) Check and repair open welds and broken points (4) Check the wrong connection and correct it (5) Reduce load
The motor no-load current is unbalanced, the current in three phases differ too much from each other	(1) Stator windings short circuit internally (2) When rewinding, the turns of three phase windings are not equal (3) Supply voltage imbalance (4) Wiring error of some coils of stator windings	(1) Repair stator windings and eliminate short circuit fault (2) Rewind the coils of stator windings in serious cases (3) Measure the power supply voltage and try to eliminate the imbalance (4) Check the wrong connection and correct it
The pointer of the ammeter is unstable and swings when the motor is unloaded or loaded	(1) The guide bar of cage rotor is open welded or broken (2) One phase of the wound rotor is open-circuited, or the short circuit device of the brush and collector ring is in poor contact	(1) Find out the broken bars or open welds and repair them (2) Check and repair the wound rotor circuit
The motor is overheated and even smokes	(1) Motor overload or frequent starting (2) Power supply voltage is too high or too low (3) Motor running with phase failure (4) Short circuit between turns or phases of stator windings (5) Stator and rotor core friction (6) Bars broken in cage rotor, or open welded in the welding points of wound rotor winding (7) Poor motor ventilation (8) Poor insulation or burrs between silicon steel sheets of stator core	(1) Reduce the load and control the starting according to the specified times (2) Adjust the power supply voltage (3) Find out the open circuit and repair it (4) Repair or replace stator windings (5) Find out the cause and eliminate the friction (6) Find out the cause and re-weld the rotor windings (7) Check the fan and dredge the air duct (8) Maintain the stator iron core and handle the insulation
The motor makes abnormal noise during running	(1) Loose stator and rotor iron cores (2) Stator and rotor iron core friction (3) Bearing is short of oil (4) The bearing is worn or there is foreign matter in the oil (5) The fan rubs against the cover	(1) Overhaul the stator and rotor iron cores and re-tighten them (2) Eliminate the friction, and small the rotor if necessary (3) Add lubricating oil (4) Replace or clean the bearing (5) Reinstall the fan or cover

Fault phenomenon	Possible causes	Treatment solutions
The motor vibrates greatly during running	(1) The motor foundation bolts are loose (2) The motor foundation is uneven or not firm (3) Bent or unbalanced rotor (4) Coupling center is not calibrated (5) Fan imbalance (6) Excessive bearing wear clearance (7) The rotating part of the machine with load on the rotating shaft is unbalanced (8) Partial short circuit or grounding of stator windings (9) Partial short circuit of wound rotor	(1) Tighten the foundation bolts (2) Re-strengthen the foundation and level it (3) Straighten the shaft and perform dynamic balancing of the rotor (4) Recalibrate to make it meet the requirements (5) Overhaul the fan and correct the balance (6) Repair the bearing and replace it if necessary (7) Conduct static balance or dynamic balance test, adjust the balance (8) Find short circuit or grounding point, and repair or replace the windings (9) Repair rotor windings
Bearing overheating	(1) Excessive grease in rolling bearing (2) The lubricating grease is deteriorated or contains impurities (3) Improper fit of bearing and journal or end cover (too tight or loose) (4) The inner hole of the bearing cover is eccentric and rubs against the shaft (5) The belt tension is too tight or the coupling is improperly assembled (6) The bearing clearance is too large or too small (7) Bent shaft (8) The motor has been put aside for too long	(1) Add lubricating grease as required (2) Clean the bearing and replace it with clean grease (3) If it is too tight, turn and grind the journal or end cover inner hole, and if it is too loose, repair it with adhesive (4) Repair the bearing cover to eliminate friction (5) Properly adjust the belt tension and correct the coupling (6) Adjust the clearance or replace it with a new bearing (7) Correct the shaft or replace the rotor (8) No load running, stop when overheating, and try again after cooling several times. If it still fails, disassemble it for maintenance
The no-load current is too large (the normal no-load current is 20%~50% of the rated current)	(1) Power supply voltage is too high (2) Y-configuration is wrongly to be Δ-configuration (3) The internal wiring of the windings is incorrect during repair, such as connecting the series windings in parallel (4) The mechanical loss of the motor is increased due to assembly quality defects, lack of oil or damage of the bearing (5) The stator and rotor iron cores are not complete after maintenance (6) The wire diameter of stator windings is too small during repair (7) Insufficient turns or wrong internal polarities during repair (8) Short circuit, broken wire or grounding fault in windings (9) The iron core does not match the motor during repair	(1) If the power supply voltage exceeds 5% of the rated value of the grid, it could be reported to the power supply department to adjust the tap changer on the regulating transformer (2) Correct wiring (3) Correct internal winding wirings (4) Disassemble for inspection, reassemble, add lubricating oil or replace the bearing (5) Open the end cover for inspection and adjustment (6) Rewinding with specified wire diameter (7) Rewind the windings according to the specified number of turns, or check the windings' polarities (8) Find out the fault point and deal with the insulation at the fault point. If it cannot be recovered, replace the windings (9) Replace with the original iron core

Fault phenomenon	Possible causes	Treatment solutions
The no-load current is too small (less than 20% of the rated current)	(1) The Δ-configuration is wrongly to be Y-configuration (2) The wire diameter of stator windings is too small during repair (3) The internal wiring of the windings is incorrect during repair, such as connecting the parallel windings in series	(1) Correct wiring (2) Rewinding with specified wire diameter (3) Correct internal windings wiring
Y/Δ switch starts, it is normal in Y-configuration, and motor stops or three-phase current is unbalanced in Δ-configuration.	(1) The switch is wrongly connected, and the three phases in Δ-configuration are not connected (2) When in Δ-configuration, the switch has poor contact and is connected in V shape	(1) Correct wiring (2) Repair the connectors with poor contact
The motor housing is charged	(1) Unqualified grounding resistance or open circuit of protective grounding wire (2) Damaged winding insulation (3) The junction box insulation is damaged or too much dust inside (4) Winding damped	(1) Measure the grounding resistance. The grounding wire must be good and reliable (2) Repair insulation, and then dipping and drying (3) Replace or clean the junction box (4) Drying treatment
The insulation resistance is only tens to hundreds of ohms, but the windings are good	(1) The motor is damped (2) Brush powder (winding motor), dust and oil contamination enter the windings (3) Poor windings insulation	(1) Drying treatment (2) Strengthen maintenance, timely remove the accumulated dust and oil stains, and wash the dirty motor with gasoline. After the gasoline volatilizes, conduct paint dipping and drying treatment to restore it to a good insulation state (3) Disassemble for maintenance, strengthen the insulation, paint dipping and drying, and rewind the windings when it could not be repaired
The brush spark is too heavy	(1) The brand or size of the brush does not meet the specified requirements (2) Dirt on slip ring or commutator (3) Improper brush pressure (4) The brush is jammed in the brush holder (5) The slip ring or commutator is oval or grooved	(1) Replace a proper brush (2) Clean the slip ring or commutator (3) Adjust the brush pressure of each group (4) Polish the brush so that it could move up and down freely in the brush holder (5) Handle with lathe

Fault phenomenon	Possible causes	Treatment solutions		
Motor axial displacement	The motor with rolling bearing is poorly assembled	Disassemble for maintenance, the allowable axial displacement of the motor is as follows:		
		Capacity/kW	Allowable axial displacement/mm	
			To one side	To both sides
		10 and below	0.50	1.00
		10~22	0.75	1.50
		30~70	1.00	2.00
		75~125	1.50	3.00
		More than 125	2.00	4.00

5.3.2 Fault inspection and maintenance of three phase wound asynchronous motor

The slip ring and brush are the most vulnerable parts of three phase wound asynchronous motor. See Table 5-9 for common faults and solutions of slip ring and brush.

Table 5-9 Common faults of slip ring and brush and the treatment solutions

Fault phenomena	Possible causes	Treatment solutions
Slight damage to slip ring surface, such as brush marks, spots and small dents	Slightly uneven contact between brush and slip ring	Adjust the contact surface between the brush and the slip ring to make them contact evenly; Rotate the slip ring, gently grind it with an oilstone or a fine file until it is flat, and then polish it with No.0 sand rubber when the slip ring rotates at high speed until the surface of the slip ring shows metallic luster
The slip ring surface is seriously damaged, such as the surface concavity and convexity, groove depth exceeding 1mm, and the damaged area exceeding 20%~30% of the slip ring surface area	(1) The model of the brush is incorrect, the hardness is too high, the size is inappropriate, and the slip ring is damaged due to long-term use (2) There are hard particles such as carborundum in the brush, which makes the slip ring surface appear linear traces of different thickness and length (3) The spark is too heavy, which burns the slip ring surface	(1) Replace with brush of specified model and size (2) Use qualified brushes (3) Find out the cause of heavy spark and eliminate it
The slip ring is oval (it would burn down in severe cases)	(1) The motor is not firmly installed (2) The fit clearance between the inner sleeve of the slip ring and the motor shaft is too large, resulting in irregular swing during operation	(1) Tighten the foundation bolts (2) Check and secure the position of the slip ring on the shaft

Fault phenomena	Possible causes	Treatment solutions
The brush catches fire	(1) Improper maintenance, rough slip ring surface, causing vicious circle and aggravating sparks	(1) Strengthen inspection and maintenance, and deal with problems in a timely manner
	(2) The model and size of the brush are inappropriate, or the brush is worn and too short due to long-term use	(2) Replace the brush with the specified model and size, and replace the short brush
	(3) The brush is stuck in the brush holder	(3) Find out the cause, so that the brush can move up and down freely in the brush holder, but it cannot be too loose
	(4) The brush is poorly ground, the contact surface is uneven, and the contact with the slip ring is poor	(4) Grind the contact surface with fine sand rubber, and ensure that the contact surface is not less than 80%, or replace it with a new brush (the contact surface of the new brush also needs to be polished)
	(5) Uneven or insufficient pressure of brush spring	(5) Adjust the pressure of the bush spring. If the elasticity fails to meet the requirements, replace the brush spring (the pressure should be 15~20kPa)
	(6) The slip ring is not flat or round	(6) Grind the slip ring with sand rubber, and treat it round with lathe if serious
	(7) Oil dirt or sundries fall between the slip ring and the brush, causing poor contact between them	(7) Wipe the brush and slip ring with clean cotton cloth dipped in gasoline to remove the oil around and on the bearing, and take anti contamination measures
	(8) Corrosive medium exists in the air	(8) Improve the use environment and strengthen maintenance
Arc short circuit between brushes or slip rings	(1) The conductive powder falling off the brush covers the insulating part, or flies in the space between the brush holder and the slip ring, forming a conductive circuit	(1) Strengthen maintenance, and timely remove the accumulated brush powder with compressed air or vacuum cleaner; An isolating plate (2mm thick insulating laminate) could be added beside the brush holder, and it could be fixed on the brush holder with a flat-headed screw to separate the brush from the brush holder
	(2) Bakelite gasket or epoxy resin insulation gasket is broken	(2) Replace each insulating washer on the slip ring
	(3) Bad environment, corrosive medium or conductive dust	(3) Improve environmental conditions

Electrical line

6.1 Types and characteristics of electrical line

Electrical line is an important part of power system, which could be divided into power line and control line. The power line is mainly used to complete the task of power transmission, and the control line is used for connection of protection and measurement. According to the laying method, it could be divided into overhead lines, cable lines, indoor lines, etc. According to its insulation properties, it could be divided into insulated lines and bare lines; According to its purpose, it could be divided into bus, trunk line and branch lines.

6.1.1 Overhead line

(1) Composition of overhead lines

Overhead lines refer to high and low-voltage power lines laid by using poles and towers with a spacing of more than 25m. Overhead lines are mainly composed of conducting wires, poles/towers, cross arms, insulators, fittings, stay wires and foundations.

① The conducting wires of overhead lines are used to transmit current, and most of them are steel-cored aluminum strands, hard copper strands, hard aluminum strands and aluminum alloy strands. Insulated conducting wires should be used for low-voltage overhead lines in the plant area (especially in places with fire hazards).

② The poles and towers of overhead lines are used to support conducting wires and their accessories, including reinforced concrete poles, wooden poles and iron towers. According to its functions, the tower is divided into straight line pole/tower, tension pole/tower, corner pole/tower, terminal pole/tower and branch pole/tower, as shown in Table 6-1.

Table 6-1　Pole/tower of overhead line

No.	Tower type	Description
1	Straight line pole/tower	It is located on the straight line section of the line and is only used for supporting conducting wires, insulators and fittings. Under normal conditions, it could bear the wind force at the side of the line, but not the tension in the direction of the line. It is accounting for more than 80% of the total electric quantity
2	Tension pole/tower	It is located between several straight poles/towers on the straight line section of the line or at places with special requirements, such as the intersection with railways, highways, rivers, pipes, etc. This kind of pole/tower could withstand the tension of one side of the conducting wires in case of wire breaking accident and wire tightening during stringing
3	Corner pole/tower	It is located where the line changes direction. This kind of pole/tower might be of tension type or straight type, depending on the size of the corner. It could bear the resultant force of conducting wires on both sides

No.	Tower type	Description
4	Terminal pole/tower	It is located at the head and tail of the line. Under normal conditions, it could withstand all conducting wires tension in the line direction
5	Branch pole/tower	It is located at the branch of the line. There are two types of poles: straight pole/tower and tension pole/tower in the main line direction, and tension pole/tower in the branch direction

③ Insulators are used to insulate conducting wires between each other and conducting wire(s) to the earth, ensure reliable electrical insulation performance of lines, and fix conducting wires to bear vertical and horizontal loads of conducting wires. Pin, suspension and butterfly insulators are mostly used for overhead lines, as shown in Fig. 6-1. In order to ensure the safe running of the line, insulators with cracks or scars on the enamel surface should not be used.

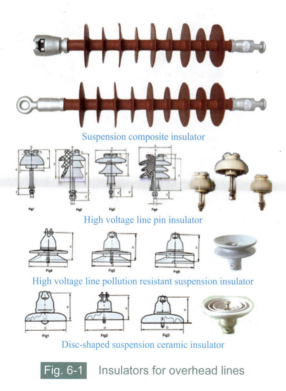

Suspension composite insulator

High voltage line pin insulator

High voltage line pollution resistant suspension insulator

Disc-shaped suspension ceramic insulator

Fig. 6-1 Insulators for overhead lines

④ Cross arms are used to support insulators. The iron cross arm is firm and durable, but its lightning protection performance is poor, and rust prevention treatment must be done. Ceramic cross arm is a combination of insulator and common cross arm. It has a simple structure, convenient installation and good electrical insulation, but its ceramic is fragile and its mechanical strength is poor.

⑤ Fitting is mainly used to fix the conducting wires and cross arm, including wire clamp, cross arm support, hoop, sizing block, connection fitting, etc., as shown in Fig. 6-2.

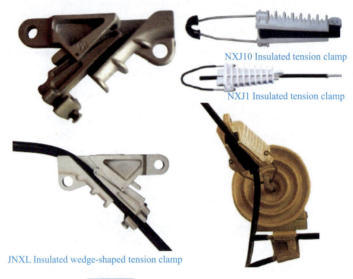

Fig. 6-2 Fittings of overhead line

⑥ Stay wire is used to balance the transverse load and conduct wires tension acting on the pole/tower. On the one hand, improve the strength of the pole/tower and bear the force of external loads on the pole/tower to reduce the material consumption of the tower and reduce the cost of the line; On the other hand, the pole/tower should be fixed on the ground together with the stay rod and the cable tray to ensure that the pole and tower would not tilt and collapse.

⑦ The underground installations of overhead power line pole/tower are generally referred to as foundations. The foundation is used to stabilize the pole/tower, so that the pole/tower would not be pulled up, sunk or collapsed due to bearing vertical load, horizontal load, accidental line-breaking tension and external force.

(2) Technical parameters of overhead lines (Table 6-2)

Table 6-2 Technical parameters of overhead lines

Parameters	Definition
Spacing	Distance between the center lines of two adjacent poles on the same line (40~50m at 10kV and below)
Distance between lines	Distance between conducting wires on the same pole. It is related to line voltage and spacing (the minimum distance between lines of 10kV and below is 0.6m; the minimum distance between low-voltage lines is 0.5m)

Parameters	Definition
Sag	For flat ground, the vertical distance between the lowest point of overhead line and the suspension point of conducting wires on poles at both ends (it cannot be too large or too small, depending on the spacing, conducting wires' material and sectional area. Too large it would be unsafe, and if too small, the tension would be too large)
Distance between conductor and ground	10kV and below, 6.5m for high voltage, 6m for low voltage

(3) Characteristics of overhead lines

The overhead line has a simple structure, convenient to install and low cost; Large transmission capacity and high voltage; Good heat dissipation condition; Easy for maintenance. But when the network is complex and centralized, it is not easy to set up; It is neither safe nor beautiful to install in densely populated urban areas; Poor working conditions, vulnerable to environmental conditions, such as ice, wind, rain, snow, temperature, chemical corrosion, lightning, etc.

6.1.2 Cable line

Cable line refer to the line that uses cables to transmit electric energy. Cable line is generally laid underground, overhead or underwater.

(1) Composition of cable line

The cable line is mainly composed of cable body, cable intermediate joint, cable terminal connector, etc., and also includes corresponding civil facilities, such as cable trench, pipe, shaft, tunnel, etc.

(2) Structure of cable

The cable is mainly composed of wire core, insulating layer and protective layer. The wire core is divided into copper core and aluminum core, and the insulation layer is divided into impregnated paper insulation, plastic insulation, rubber insulation, oil-filled insulation, etc. The protective layer is divided into inner protective layer and outer protective layer. The inner protective layer includes lead coating, aluminum coating, PVC sheath, cross-linked polyethylene sheath, rubber sheath, etc.; The outer protective layer includes jute liner, steel armor, anti-corrosion coating, etc. The structure of the cable is shown in Fig. 6-3.

Cables could be laid in cable trenches and cable tunnels, or directly buried underground as required. In case of being buried underground directly, it is easy to construct, with good heat dissipation. It could not reliably prevent external force damage, and it is easy to be corroded by acid and alkali substances in the soil.

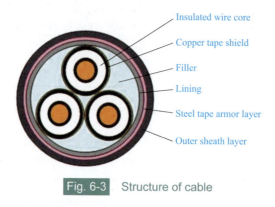

Fig. 6-3　Structure of cable

(3) Cable terminal connector

Cable terminal connectors are divided into outdoor and indoor ones. The terminal connectors for outdoor use are cast iron shell, ceramic shell and epoxy resin; Nylon and epoxy resin terminal connectors are mainly used indoors. The epoxy resin terminal connectors have simple forming process, strong adhesion with the metal sheath of the cable, good insulation and sealing performance, and are most widely used.

Cable intermediate joints mainly include lead-aluminum intermediate joints, cast iron intermediate joints and epoxy resin intermediate joints. Intermediate joints of 10kV and below are usually epoxy resin type.

【Special reminder】

> Cable terminal connectors and cable intermediate joints are the weak links of the whole cable line.

(4) Characteristics of cable line

Compared with overhead line, cable line has the following characteristics: it is not interfered by natural meteorological conditions (such as lightning, wind and rain, smoke, pollution, etc.); It isn't affected by the growth of trees along the lines; It is positive to beautifying the urban environment; It does not occupy the ground space, and the same underground passage could accommodate multiple cable lines; It is beneficial to prevent electric shock and use electricity safely; Maintenance costs are little. Therefore, it is widely used in modern enterprises, especially in corrosive gas and steam, flammable and explosive places.

The cable line also has the following disadvantages: The transmission current is smaller than the overhead lines, even though with the same cross-sectional area of the conducting wires; The investment and construction cost is several times more and increases with the increase of voltage; The fault repair time is also long.

6.1.3 Indoor line

(1) Types of indoor line

There are many types of indoor lines. Buses could be divided into hard bus and soft bus; Trunk line could be clear, hidden and underground pipe styles; The branch lines include sheathed line, direct laying line, ceramic (plastic) plywood line, drum insulator line, pin insulator line, steel pipe line, plastic pipe line, etc.

(2) Principles of indoor line

Due to different indoor lines, the technical requirements are also different. No matter which line it is, it must meet the basic requirements of indoor line, that is, the basic principles that indoor line should follow.

① Security. Indoor line and electrical equipment must ensure safe running.

② Reliability. Ensure the reliability of line power supply and the running reliability of indoor electrical equipment.

③ Convenience. Ensure the convenience of construction, running and maintenance.

④ Beautification. The installation of indoor lines and electrical equipment should contribute to the beautification of buildings.

⑤ Be economic. On the premise of ensuring safety, reliability, convenience and beautification, to be economic should be considered to achieve reasonable construction and save cost.

6.2 Safety of electrical lines

6.2.1 Conductivity

The conductivity of the conducting wires includes requirements for heating, voltage loss and short circuit current.

(1) Heating

In order to prevent the lines from overheating and ensure the normal running of the lines, the maximum operating temperature of the conducting wires should not exceed the limits specified in Table 6-3.

Table 6-3 Maximum temperature limits for conductor operation

Wire type	Max. temperature limit /°C	Wire type	Max. temperature limit /°C
Rubber insulated wire	65	Bare wire	70
Plastic insulated wire	70	Lead or aluminum clad cable	80
Plastic cable	65		

(2) Voltage loss

Voltage loss is the algebraic difference between the voltage at the receiving end and the voltage at the supply end. If the voltage loss is too large, not only the electrical equipment could not work normally, but also heat would be generated at the electrical equipment and electrical lines.

According to relevant national standards, for power supply voltage, the voltage loss of 10kV and below power lines should not exceed ±7% of the rated voltage, and that of low-voltage lighting lines and agricultural user lines should not exceed 7%~−10%.

(3) Short circuit current

In order to ensure reliable quick action of disconnection protection device in case of a short circuit, there must be sufficient short circuit current. This also requires that the wire sectional area should not be too small. On the other hand, due to the large short circuit current, the conducting wires should be able to withstand the impact of the short circuit current without being damaged.

Especially in TN system, the impedance of the circuit between phase wire and protective neutral wire should meet the requirements of protective grounding connection. The single-phase short circuit current should be more than 4 times the rated current of the fusion (more than 5 times for explosive hazardous environment) or 1.5 times the setting value of instantaneous overcurrent release of low-voltage circuit breaker.

6.2.2 Mechanical strength

The conducting wires in running would be subject to their own weight, wind force, thermal stress, EMF and icing gravity. Therefore, sufficient mechanical strength must be ensured.

According to the requirements of mechanical strength, the minimum cross sectional area of the overhead line conducting wires is shown in Table 6-4; See Table 6-5 for the minimum sectional area of low-voltage wires.

Table 6-4 Minimum cross sectional area of overhead line conductor   mm²

Category	Copper	Al and Al alloy	Iron
Single strand wire	6	10	6
Stranded wire	6	16	10

Table 6-5 Minimum cross sectional area of low-voltage wire   mm²

Category		Min. cross sectional area		
		Copper core flexible wire	Copper wire	Al wire
Power wire of mobile equipment	For daily use	0.2	—	—
	For production	1.0	—	—
Pendant lamp wire	Civil building, indoor	0.4	0.5	1.5
	Industrial building, Indoor	0.5	0.8	2.5
	Outdoor	1.0	1.0	2.5
Insulated conducting wire on supports with distance d inbetween	$d \leqslant $ 1m, Indoor	—	1.0	1.5
	$d \leqslant $ 1m, Outdoors	—	1.5	2.5
	$d \leqslant $ 2m, Outdoors	—	1.0	2.5
	$d \leqslant $ 2m, Outdoors	—	1.5	2.5
	$d \leqslant $ 6m, Outdoors	—	2.5	4
	$d \leqslant $ 6m, Outdoors	—	2.5	6
Household wire	$d \leqslant $ 10m	—	2.5	6
	$d \leqslant $ 25m	—	4	10
Wire through pipe		1.0	1.0	2.5
Plastic sheathed wire		—	1.0	1.5

【Special reminder】

The power wire of mobile equipment and pendant lamp wire must use copper core flexible wire. Except for the wires through pipe, other kinds of wires should not be flexible wires in other cases.

6.2.3 Spacing

(1) Spacing between electrical lines and other facilities

The buried depth of overhead line poles should not be less than 2m, and should not be less than 1/6 of the pole height.

(2) Spacing of low-voltage household wires

As the faults of the household wire and the incoming wire are common, the following spacing requirements should be paid attention to when installing the low-voltage household wire.

① If there is an important traffic road below, the minimum height of the household wire from the ground should not be less than 6m; In case of inconvenient access, the minimum height of the household wire from the ground should not be less than 3.5m.

② The household wire should not cross the building. If it must be, the minimum height from the building should not be less than 2.5m.

③ The distance from the household wire to the protruding part of the building should not be less than 0.15m, and the vertical distance to the balcony below should not be less than 2.5m, the vertical distance to the window below should not be less than 0.3m, the vertical distance to the upper window or balcony should not be less than 0.8m, and the horizontal distance to the window or balcony should not be less than 0.8m.

④ When the household wire crosses the communication wire. The vertical distance between the household wire and the communication wire should not be less than 0.6m when the household wire is above; And the vertical distance between them should not be less than 0.3m, when the household wire is below.

⑤ The minimum distance between the household wire and the trees should not be less than 0.3m.

If the above distance requirements could not be met, other protective measures must be taken. In addition to the above requirements for a safe distance, it should also be noted that the length of the household wire should not exceed 25m generally; Insulated conducting wire should be used for the household wire, and the sectional area of copper conducting wire should not be less than $2.5mm^2$, and that of aluminum conducting wire should not be less than $10mm^2$; When the angle between the household wire and the distribution line reaches $45°$, a cross arm should be installed on the pole of the distribution line; The household wire should not have joints.

6.2.4 Identification and connection of conducting wires

(1) Identification of conducting wires

① Conducting wires with sectional area under $25mm^2$. The sectional area and diameter comparison of conducting wires with sectional area under $25mm^2$ is shown in Table 6-6 (the minimum sectional area of aluminum wire is $2.5mm^2$).

Table 6-6 Cross sectional area and diameter comparison of conducting wires

Sectional area /mm²	1.5	2.5	4	6	10	16	25
Diameter /mm	1.37	1.76	2.24	2.73	7×1.33	7×1.68	7×2.11

It could be seen from Table 6-6 that the diameter (1.33mm) of a single strand of conducting wire with 10mm² in sectional area is similar to that of conducting wire with 1.5mm² in sectional area (1.37mm); The diameter (1.68mm) of a single strand of conducting wire with 16mm² in sectional area is similar to that of 2.5mm² in sectional area (1.76mm), And the single strand of conducting wire with 25mm² in sectional area (2.11mm) is similar to that of 4mm² in sectional area (2.24mm); Therefore, the identification of single strand of conducting wire with 1.5mm², 2.5mm² and 4mm² in sectional area is equivalent to the identification of conducting wires with 10mm², 16mm² and 25mm² in sectional area.

② Conversion between sectional area (S) and diameter (D) of single core conducting wire

$$S=\pi R^2=\pi(D/2)^2$$

Namely: section=$\pi \times$ (Diameter/2)²

③ Estimation and selection of the sectional area of the conducting wire could be achieved according to the power (or load power) of the given equipment.

The sectional area of conducting wire could be calculated according to the following formula:

$$S=\frac{I_e}{J \times 0.8}$$

Namely: Sectional area=$\dfrac{\text{Load current(A)}}{\text{Safety current density(A / mm}^2) \times 0.8}$

Tips on estimating safe ampacity of conductors

If the sectional area is less than 10mm², the ampacity is 5 times of the sectional area value. The ampacity with a sectional area of more than 100mm² is twice the value of the sectional area. The section area of 25mm² and 35mm² is the boundary of 4 times and 3 times. The sectional areas of 70mm² and 95mm² are 2.5 times.
In case of wire through pipe, 20% discount should be given after calculation; If the ambient temperature exceeds 25 ℃, 10% discount should be given after calculation; if the temperature exceeds 25 ℃ and through pipe, 20% discount should be given after calculation, or simply 30% discount should be given in general.

The ampacity of bare aluminum wire with the same section could be increased by half compared with aluminum core insulated wire.
The sectional area order of copper conducting wire should be improved by one level, and then calculated according to the corresponding aluminum wire conditions.
The first three sentences of the tips refer to the safe ampacity of aluminum conducting wire, open laying and ambient temperature of 25 ℃. The last three sentences of the tips refer to the safe ampacity when conditions change.

For example, the rated power of a three-phase asynchronous motor is 10kW, the rated current is 20A, and the sectional area of its conducting wire could be known from the tips of safe ampacity. The safe current density per square millimeter of aluminum conducting wire below 10mm² is 5A. It is also known that the ampacity is calculated at a 20% discount when the wire is through pipe:

$$S=20/(5\times0.8)=5(mm^2)$$

As there is no conducting wire with the specification of 5mm², 6mm² aluminum wire or 4mm² copper wire should be used for 10kW three-phase motor.

【Special reminder】

① For through pipe insulated wire, the minimum sectional area of copper wire is 1mm²; The minimum sectional area of aluminum wire is 2.5mm².

② For the secondary circuit of various electrical equipment (except the secondary circuit of current mutual inductor), although the current is very small, in order to ensure the mechanical strength of the secondary wire, insulated copper wire with a sectional area of no less than 1.5mm² is often used.

(2) Twisted connection of conducting wires

The connection of conducting wires has a variety of methods, such as twisted connection, welding, crimping, etc.

The conducting wires connection must be tight. In principle, the mechanical strength of the connection should not be lower than 80% of that of the conducting wire itself; The insulation strength should not be lower than that of the conducting wire itself; The resistance of the joint should not be greater than 1.2 times of the resistance of the conducting wire itself.

① Connection of single copper core wire

It could be divided into linear connection and T-connection. See Table 6-7 for the process and technical requirements.

Table 6-7 Connection process and technical requirements of single copper core wire

Type		Operation schematic	Operation process and technical requirements
Linear connection	Single copper core wire with small sectional area		(1) Cross the two strands of two core wires with insulating layer and oxide layer removed, and twist each other for 2~3 turns; (2) Straighten the free ends of the two wire ends, and wind each free end on the other core wire. The winding length is 6~8 times of the core wire diameter; This is the common stranded connection; (3) Cut off the excess part and trim the burrs
	Single copper core wire with big sectional area		(1) Add a core wire with the same diameter at the overlap of the two strands of two wires to increase the contact surface at the joint; (2) A bare copper wire (binding wire) with a sectional area of about 1.5mm^2 is tightly wound on the joint, and the winding length is about 10 times of the core wire diameter; (3) Use wire cutters to fold back the two core wire ends to be joint respectively, continue to wind with the binding wire for 5~6 turns, then cut off the excess part and trim the burrs; (4) In the case of copper wires with different cross sections to be connected, wind the core wire of the smaller one on the bigger one for 5~6 turns tightly at first, and then fold the bigger one back with wire cutters to make it contact close to the core wire of the smaller cross-section; Finally, wind the smaller one for 4~5 turns, then cut off the excess part and trim the burrs.
	Memory tips	Intersect the two conducting wires in a red cross Twisted their free ends for three turns. Straighten the wire tails and wind six turns tightly. Cut off the excess part	
T- connection	Single copper core wire with small sectional area		(1) Intersect the core wires of the branch and the trunk lines vertically, leave 3~5mm bare wire for the core wire of the branch line, wind it clockwise for 6~8 turns on the core wire of the trunk line, cut off the excess part and remove the burrs. (2) For a T-type connection of core wire with a small sectional area, a winding knot could be made by the branch core wire on the trunk wire, and then wound the branch core wire for 5-8 turns tightly on the trunk wire.

Type		Operation schematic	Operation process and technical requirements
T- connection	Single copper core wire with big sectional area	10 times of the wire diameter	Intersect the core wires of the branch to the trunk lines at the right angle, and fill in a bare copper wire with the same diameter, then wind it with the binding wire by referring to the linear connection of copper core wire with a large sectional area
	Memory tips	Intersect the branch line and the trunk line vertically; Wind the branch core wire clockwisely. After 6 ~ 8 turns of winding, cut off the excess part and trim the burrs.	

② Connection of multi strands copper core wire. The following describes the connection methods with 7 strands copper core wire as an example. See Table 6-8 for the connection process and technical requirements.

Table 6-8 Connection process and technical requirements of 7 strands copper core wire

Type		Operation schematic	Operation process and technical requirements
Linear connection			(1) The stranded wires removed from the insulating layer are dispersed and straightened, and twisted tightly at the position where is about 1/3 of the insulating removed length to the insulating layer; (2) Separate the remaining free ends into two parts with umbrella shape, make the two umbrella-shaped wire heads opposite each other, and cross the strands until the roots are connected; (3) Pinch the loose wire ends on both sides, divide the wires into three groups to 2, 2 and 3, pull the first group vertically, wind it for two turns in a clockwise direction, then bend it down and pull it into a right angle to cling to the opposite core wire; (4) The winding method of the second and third groups is the same as that of the first group (note: during winding, let the latter group of thread heads press the root of the former group, which has been folded into right angles, and the last group of thread heads wind the core for 3 turns), cut off the excess part and remove the burrs
	Memory tips		Tips on straight connection of 7-strand copper core wire The stranded wires removed from the insulating layer should be straightened, they should be dispersed except in the root area. The free ends of the dispersed are in umbrella shapes, When making "umbrella bone", divide them into three groups to 2, 2 and 3. Wind the groups of 2 strands for 2 turns one by one, and wind the group of 3 strands for 3 turns, then cut off the excess part and remove the burrs. The roots should be twisted tightly, and the same method should be applied to the other end.

Type	Operation schematic	Operation process and technical requirements
T-connection		**Method 1**: Bend the branch core to 90° and stick to the trunk line, then fold the branch head back in 2 groups (3 strands and 4 strands) and tightly wind it on the trunk line in different directions, with the winding length 10 times the diameter of the core line; **Method 2**: Further twist the branch core wire tightly at 1/8 of the root along the original twisting direction, divide the remaining wire ends into two groups (3 strands and 4 strands), make an open at the joint area on the trunk core wire (3 strands and 4 strands), insert one group through it. Place the other group in front of the trunk core wire, wind 4~5 turns in the right direction, and wind 4~5 turns in the left direction for the group inserted into the trunk core wire. Cut off the excess part and remove the burrs
	Memory tips	Tips on T-connection of 7 strands copper core wire The core wires are divided into two groups (3 strands and 4 strands) of trunk and branch lines. Wind the 3 strands of branch line on the trunk line with 3 strands for 4 turns, and the other group for 5 turns in a similar way.

③ Connection of cable wires. When two-core sheathed wire, three-core sheathed wire or multi-core cable are connected, the connection method is the same as the twisted connection described previously. The connection points of each core wire should be staggered as far as possible to prevent leakage or short circuit between core wires. The connection of dual-core sheathed wire is as shown in Fig. 6-4 (a); The connection of three core sheathed wire is as shown in Fig. 6-4 (b); The connection of four core power cable is as shown in Fig. 6-4 (c).

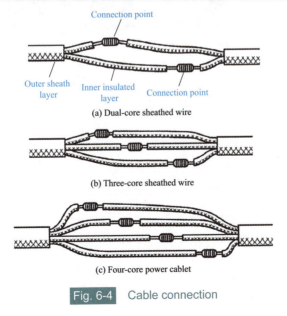

Fig. 6-4 Cable connection

【Special reminder】

When wiring, cut off the power supply and pay attention to safety to prevent electric shock.

(3) Crimping connection of conducting wires

Crimping connection refers to the use of copper or aluminum sleeves to cover the core wire to be connected, and then press the sleeve with crimping pliers or crimping die to keep the core wire connected.

When connecting multiple strands of copper and aluminum conducting wires with a large cross section area, copper aluminum transition connection clamp [Fig. 6-5 (a)] or copper aluminum transition connection pipe [Fig. 6-5 (b)] should be used. When using oval sectional sleeve, insert the core wires of the two wires to be connected from the left and right ends respectively and pass a little through the sleeve, as shown in Fig. 6-5 (c), and then crimp the sleeve, as shown in Fig. 6-5 (d).

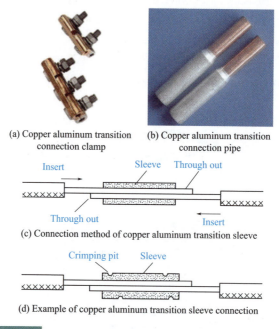

(a) Copper aluminum transition connection clamp

(b) Copper aluminum transition connection pipe

(c) Connection method of copper aluminum transition sleeve

(d) Example of copper aluminum transition sleeve connection

Fig. 6-5　Copper aluminum wires crimping connection

When aluminum conducting wire is connected with copper terminal of electrical equipment, copper aluminum transition nose should be used, as shown in Fig. 6-6.

Fig. 6-6　Copper aluminum transition nose

 【Special reminder】

Generally, copper and aluminum conducting wires could not be connected directly, but transition connection must be adopted.

6.2.5　Line protection

(1) Line protection

Various lines should have sufficient protection capacity against hazards of chemical, thermal, mechanical, environmental, biological and other harmful factors. The line protection design should meet the relevant requirements.

According to the requirements of relevant regulations, the power cables should be protected through pipes at the following positions:

① Cable leading in or out of buildings (including partition walls and floor slabs), trenches and tunnels;

② Where cables pass through railways and roads;

③ Cable leading in or out of the ground in a section 2m above the ground and 0.1~0.25m below the ground should be protected by pipe;

④ Parts of the cable where it may be mechanically damaged;

⑤ The distance between the cable and various pipes or trenches is less than the specified distance.

(2) Line overcurrent protection

When a short circuit occurs in a power line, the basic characteristics are: the current increases significantly, the voltage decreases significantly, and the line impedance decreases significantly.

Overcurrent protection is to judge whether the line is in fault state according to the principle as the first characteristic that current increases significantly in case of short circuit. The overcurrent protection of electrical circuits includes short circuit protection and overload protection.

① Short circuit protection. The short circuit current is very large, which could cause serious consequences if it lasts a little longer. Therefore, the short circuit protection device must act instantaneously in case of short circuit. The electromagnetic overcurrent release (or relay) has the characteristics of instantaneous action, and is suitable to be used as a short-circuit protection device. When the current is 6 times of the rated current of the fusant, the fusing time of the fast fuse generally should not be longer than 0.02s, and it also has good short circuit protection performance.

② Overload protection. The magnitude of current allowed to pass through the electrical circuit continuously without overheating the wire is called safe ampacity or safe current. If the current passing through the conducting wire exceeds the safe ampacity, it is called conducting wire overload. Generally, the maximum allowable working temperature of conducting wire is 65℃. In case of overload, if the temperature exceeds this temperature, the insulation would age rapidly and even the line would burn. Therefore, in the actual power supply line, there is usually an overload protection function to prevent overload from causing potential safety hazards.

The thermal release (or thermal relay) should be used as an overload protection device, but the action of the thermal release is too slow (the action time is still greater than 5s when the setting current is 6 times); Therefore, it could not be used as a short-circuit protection device. In the circuit with no or very small impulse current, the fuse is not only used as a short circuit protection device, but also as an overload protection device.

6.2.6　Line management

Data and documents are necessary for electrical lines, such as construction drawings, test records, etc. The system of patrol, cleaning and maintenance should also be established.

For overhead lines, in addition to the protection against harmful factors in the design, patrol inspection and maintenance must be strengthened, and measures to prevent the expansion of accidents must be considered. Accidents may occur when cables are damaged by external forces, chemical corrosion, flooding, insect bites, and cable terminal joints or intermediate joints are polluted or water enters. Therefore, the management of cable lines must also be strengthened and tests must be carried out regularly.

The corresponding management system should be established for temporary lines. For example, application and approval procedures should be carried out for the installation of temporary lines; The temporary line should be in charge of a specially assigned person; There should be a clear place of use and service life. Safety issues must be considered before installing temporary lines. The movable temporary wire must be rubber sheathed flexible wire with protective core, and the length generally does not exceed 10m. In principle, the height and other spacing of temporary overhead lines should not be less than the limit value specified by the regular line. If necessary, shielding measures should be taken, and the length should not exceed 500m.

Patrol inspection is one of the basic jobs of line running and maintenance. Through patrol inspection, defects could be found in time to take preventive measures to ensure the safe running of the line. The inspector should record the defects found in the record book and report to the superior in time.

(1) Overhead line patrol inspection

Overhead line patrol includes regular patrol, special patrol and fault patrol. Regular inspection is one of the daily tasks. 10kV and below lines should be inspected at least once a quarter. Special patrol refers to the patrol after sudden changes in running conditions, such as thunderstorm, heavy snow, heavy fog, earthquake, etc. Fault patrol refers to the patrol after a fault occurs. Generally, the fault should not be eliminated individually during patrol inspection. Overhead line patrol inspection mainly includes the following contents.

① Whether there are inflammable, explosive or strongly corrosive substances stacked on the ground along the line; Whether there are dangerous buildings near the line; Whether there are buildings and other facilities that may cause harm to the line in thunderstorm or windy weather; Whether there are branches, kites, bird nests and other sundries on the line, if any, try to remove them.

② Whether the electric pole is inclined, deformed, decayed, damaged, and the foundation sinks; Whether the cross arm and fitting are displaced, fixed firmly, the weld is cracked, and nuts are missing.

③ Whether the conducting wire has broken strands, corrosion, scar and be twisted caused by external force; Whether the wire joints are in good condition, and whether there are traces of overheating, serious oxidation and corrosion; Whether the distance between the conducting wire and the earth, adjacent buildings, or adjacent trees meets the requirements.

④ Whether the insulator has cracks, dirt, burns and flashover traces; Deviation degree of insulator string and damage condition of insulator iron parts.

⑤ Whether the stay wire is intact and loose, whether the binding is fastened, and whether the bolt is rusty.

⑥ Whether the size of protective gap (discharge gap) is qualified; Whether the ceramic bushing of lightning arrester has cracks, dirt, burns and flashover traces; Whether the sealing is good and whether the fixing is loose; Whether the lead wire on the arrester is broken and well connected; Whether the lightning arrester downlead is intact, whether the fixing is changed, whether the grounding body is exposed, and whether the connection is good.

(2) Cable line patrol inspection

The regular inspection of cable lines is generally once a quarter, and the outdoor cable terminals are inspected once a month. The cable line patrol inspection mainly includes the following contents.

① Whether the directly buried cable line stakes are intact; Whether garbage and other heavy objects are stacked on the ground along the line, and whether there are temporary buildings; Whether the ground near the line is excavated; Whether there are corrosive emissions such as acid and alkali near the line, and whether there are lime and other corrosive substances stacked on the ground; Whether the cables exposed to the ground are protected through pipes; Whether the protective pipe is damaged or rusted, and whether it is firmly fixed; Whether the plugging of the cable leading into the room is tight; During flood or after rainstorm, inspect whether there is serious scouring or collapse in the vicinity.

② Whether the cable lines in the trench and the cover plate of the trench are complete; Whether there is water seepage in the trench, whether there is water in the trench, and whether there are flammable and explosive substances stacked in the trench; Whether the cable armor or lead coating is corroded; Whether the cable is bitten by rats; During the flood period or after the rainstorm, inspect whether there is water inflow in the indoor trenches, and whether the drainage of outdoor trenches is smooth.

③ Whether the ceramic bushing of cable terminal and intermediate connector terminal has cracks, dirt and flashover traces; Whether the terminal filled with cable glue (oil) has glue overflow (oil leakage); Whether the wiring terminals are well connected; Whether there are signs of overheating; Whether the grounding wire is intact and loose; Whether the intermediate joint is deformed and the temperature is too high.

④ Whether the hooks or supports along the exposed cables are firm; Whether the cable sheath is corroded or damaged; Whether there are inflammable, explosive or strongly corrosive substances stacked near the line.

Electrical lighting

7.1　Electrical lighting mode and type

7.1.1　Modes of electric lighting

Lighting is a way of providing people with sufficient illumination (general lighting) by using artificial light or natural light, or providing good identification (road lighting, advertising signs, etc.), emphasizing features (architectural lighting, key lighting, etc.), or creating a comfortable light environment (residential lighting, etc.), creating a special atmosphere (commercial stage lighting), and other special purposes (biochemical, medical, plant cultivation, etc.).

(1) Classification according to the nature of light source

According to the nature of light sources, electrical lighting could be divided into thermal radiation light source lighting, gas discharge light source lighting and semiconductor light source lighting.

① Thermal radiation light source lighting. The thermal radiation light source is made of the principle of radiation and luminescence when the object is electrically heated to high temperature. This type of lamp has simple structure and is easy to use. It could be used when the rated voltage of the bulb is the same as the power supply voltage, such as incandescent lamp, iodine tungsten lamp and other lighting lamps. Its lighting feature is low luminous efficiency.

② Gas discharge light source lighting. The gas discharge light source is made of the principle of emitting light when the current passes through the gas. This kind of lamp has high luminous efficiency, long service life, and various light colors. Such as fluorescent lamp, high-pressure mercury lamp, high-pressure sodium lamp and others. Its lighting feature is high luminous efficiency, up to 3 times of incandescent lamps.

③ Semiconductor light source lighting. Semiconductor light source lighting, namely light emitting diode (LED for short), is a semiconductor solid light emitting device. It uses solid semiconductor chips as light emitting materials and emits excess energy through carrier recombination in semiconductors to cause photon emission, which directly emits red, yellow, blue, green, blue, orange, purple and white light.

Semiconductor lighting products are lighting appliances made of LED as light source. Semiconductor lighting has the remarkable characteristics of high efficiency, energy conservation, environmental protection, easy maintenance, etc. It is an effective way to achieve energy conservation and emission reduction, and has gradually become another revolution of lighting sources following incandescent lamps and fluorescent lamps in the lighting history.

(2) Classification according to the light scattering mode of lamps

According to the light scattering mode of the lamp, the lighting mode could be divided into 7 types: indirect lighting, semi-indirect lighting, direct indirect lighting, diffuse lighting, semi-direct lighting, wide beam direct lighting and highly focused beam downward direct lighting, as shown in Fig. 7-1. See Table 7-1 for the introduction of common lighting modes.

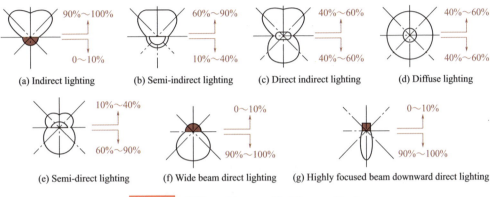

Fig. 7-1 Different types of lighting methods

Table 7-1 Common lighting modes

Lighting mode	Introduction	Remarks
Indirect lighting	Indirect lighting is generated by shading the light source, and 90%~100% of the light is directed to the ceiling, dome or other surfaces, from which it is reflected indoors When indirect lighting is close to the ceiling, almost cause no shadow would be caused, which is the most ideal overall lighting Upward lighting is another form of indirect lighting. The cylindrical upward lighting could be used on many occasions	
Semi-indirect lighting	60%~90% of the light would shine on the ceiling or the upper part of the wall, the ceiling would be used as the main reflection light source, and 10%~40% of the light would shine on the working surface directly The light reflected from the ceiling tends to soften the shadow and improve the brightness ratio. Since the light is downward directly, the brightness of the lighting device is nearly equal to the brightness of the ceiling	In order to avoid too bright ceiling, the upper edge of the suspended lighting device should be at least 305~460mm lower than the ceiling
Direct indirect lighting	Direct indirect lighting devices provide nearly the same illuminance to the ground and ceiling, that is, 40%~60%, while there is only a little light around, so the brightness in the direct glare area must be low. This is a device with both internal and external reflective bulbs. For example, some table lamps and floor lamps could produce direct indirect light and diffuse light	
Diffuse lighting	This kind of lighting device is almost the same for all directions. In order to control glare, the circle of the diffuser should be large and the wattage of the lamp should be low	

Lighting mode	Introduction	Remarks
Semi-direct lighting	In the semi direct lighting device, 60%~90% of the light shines down on the working surface directly, while the remaining 10%~40% of the light shines up, and the percentage of softening shadows by downward lighting is very small	
Wide beam direct lighting	It has a strong contrast between light and dark, and could create interesting and vivid shadows. Because its light shines directly on the target, if the reflective bulb is not used, it would produce strong glare. Gooseneck lamps and guide rail lighting belong to this category	—
Highly focused beam downward direct lighting	The light focus formed by the highly concentrated beam could be used to highlight the effect of light and emphasize the role of focus. It could provide sufficient illumination on the wall or other vertical planes, but high brightness ratio should be prevented	

7.1.2　Types of electrical lighting

(1) Classification according to lighting mode

According to different lighting modes, it could be divided into general lighting, partial lighting and mixed lighting.

(2) Classification according to lighting function

According to different lighting functions, it could be divided into normal lighting, emergency lighting, guard duty lighting, obstacle lighting, colored lights and decorative lighting. See Table 7-2 for lighting types and functions.

Table 7-2　Lighting types and functions

Type		Functions
Normal lighting	General lighting	The lighting generally required for the whole room is called general lighting
	Partial lighting	Lighting should be set near the work site to meet the illumination requirements of a partial work site
	Mixed lighting	It is composed of general lighting and partial lighting. It is applicable to places with high illuminance requirements, low working position density, and unreasonable for general lighting
Emergency lighting		The lighting set for continuous work and personnel evacuation in case of normal lighting is interrupted for some reason is called emergency lighting
Guard duty lighting		The lighting set in duty room, guard room and other places is called guard duty lighting
Obstacle lighting		Lighting installed on buildings for obstacle signs is called obstacle lighting
Colored lights and decorative lighting		Lighting designed to beautify the night scene of the city, as well as festival decoration and interior decoration is called colored lights and decorative lighting

7.2 Installation and maintenance of lighting equipment

7.2.1 Installation of lighting equipment

(1) Selection of lighting switch

Normal lighting switch refers to a low-voltage electrical device designed for families, offices, public entertainment places, etc., which is used to isolate the power supply or could connect or disconnect the current in the circuit or change the circuit connection according to regulations.

There are many kinds of lighting switches. For example, according to the panel type, there are 86, 120, 118, 146 and 75 types; The connection modes of switches are divided into single pole switch, two poles switch, three poles switch, three poles with neutral wire switch, two-way switch with public access line, two-way switch with one disconnection position, two poles two-way switch, two-way reversing switch (or neutral switch), etc. According to the installation mode, there are two types: surface mounted and concealed. Therefore, when selecting the switches, the practicability, aesthetics, cost performance and other aspects should be considered.

【Special reminder】

The size of the switch panel should be consistent with the size of the embedded switch junction box.

(2) Installation requirements of lighting switch

① Control requirements. The lighting switch should be connected in series on the phase wire (live wire), and should not be installed on the neutral wire. If the lighting switch is installed on the neutral wire, although the light does not light up when it is disconnected, the phase wire of the lamp holder is still connected, that people would mistakenly think it is in the power-off state because the light does not light up. In fact, the voltage to ground of each point on the lamp is still a dangerous voltage of 220V. If people touch these actually charged parts when the lights are off, an electric shock accident would be caused. Therefore, the safety of various lighting switches or switches of single-phase small capacity electrical equipment can only be ensured if they are connected in series to the live wire.

Switches in the same room should be controlled orderly without dislocation.

② Location requirements. The installation position of the switches should be convenient for operation and should not be blocked by other objects. The distance between the switch edge and the door frame edge is 0.15~0.2m.

③ Height requirements. The pull switch is generally 2.2~2.8m from the ground and 0.15~0.2m from the door frame; The toggle switch is generally 1.2~1.4m from the ground and 0.15~0.2m from the door frame. The switch height error in the same room should not exceed 5mm. The height error of switches installed side by side should not exceed 2mm. The allowable vertical deviation of the switch panel should not exceed 0.5mm.

④ Aesthetic requirements. The switches installed in the same room should use the same series of products. The on-off positions of the switches should be consistent, the operation should be flexible and the contact should be reliable. The concealed switch panel should be close to the wall, with no gap around it, firmly installed, and the surface should be smooth and tidy, free of cracks and scratches. The spacing between adjacent switches should be consistent.

⑤ Special requirements. In flammable, explosive and special places, lighting switches should be explosion-proof and airtight. Three-way switch (two at bedside and one at entrance) could be considered for the bedroom ceiling lamp, based on the principle that two people do not interfere with each other for rest. The living room ceiling lamp could be considered to be equipped with a dual control switch according to the needs (at the main entrance and at bedroom door area).

(3) Selection of sockets

The socket, also known as power socket, refers to the base where one or more circuit wiring(s) could be inserted, through which various wirings could be inserted. This makes it easier to connect to other circuits. Through the connection and disconnection between the circuit and the copper parts, the final connection and disconnection of this part of the circuit could be achieved.

When selecting, the rated current value of the socket should match the current value of the electrical appliances. If it is overloaded, it is easy to cause accidents. In general, the rated current of the power socket should be 1.25 times greater than the rated current of the known equipment. Generally, the rated current of single-phase power socket is 10A, and that of special power socket is 16A. The distribution circuit and connection mode of special high-power household appliances should be selected according to the actual capacity.

For electrical equipment (such as washing machine) with electric shock risk when plugged into the power supply, the socket with switch should be used to disconnect the power supply. Indoor power sockets should be of safety type, and splashproof sockets should be used in wet places such as toilets.

(4) Socket installation requirements

① Height requirements. In families and similar places, the height from the ground to the surface-mounted socket is generally 1.5~1.8m, and the height from the ground to the concealed socket cannot be less than 0.3m. Split type and wall-mounted type air conditioner sockets should be 1.8m above the ground according to the reserved hole of outlet pipe. The safety socket should be used in children's playgrounds, and the height should not be less than 1.8m.

② Special requirements. The sockets with different voltage levels should be obviously different and should not be mixed.

For sockets used for portable or mobile electrical appliances, single-phase power supply should use three-hole sockets, and three-phase power supply should use four-hole sockets.

In wet places such as kitchen and toilet, waterproof boxes should be installed together with the sockets installation.

③ Wiring requirements. There are two types of single-phase two-hole sockets: horizontal and vertical. During horizontal installation, the right pole facing the socket is connected to the phase wire (L), and the left pole is connected to the neutral wire (N), that is, "left neutral and right phase"; During vertical installation, the upper pole facing the socket is connected to the phase wire, and the lower pole is connected to the neutral wire, that is, "upper phase and lower neutral".

When wiring single-phase three-hole socket, the protective grounding wire (PE) should be connected at the top, the right pole for the phase wire, and the left pole for the neutral wire, that is, "left neutral, right phase and PE at the middle". The wiring method of single-phase socket specified in the national standard is shown in Fig. 7-2.

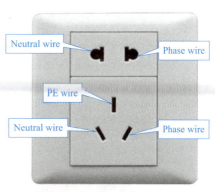

Fig. 7-2 Wiring method of single-phase socket specified in national standard

(5) Installation requirements of lamp

① The most basic requirement for lamp installation is that it must be firm, especially for large lamps.

a. When the lamp mass is more than 3kg, the ceiling lamp should be installed in the masonry structure with embedded bolts, or fixed with expansion bolts, nylon plugs or plastic plugs. Wooden wedges cannot be used, because they are too unstable and easy to decay after a long time.

And the bearing capacity of the above fixing parts should match the weight of the ceiling lamp. To ensure that the ceiling lamp is fixed firmly and reliably, and its service life could be extended.

b. When expansion bolts are used for fixing, the bolt specifications should be selected according to the technical requirements of the lamp products, and the drilling diameter and burial depth should be consistent with the bolt specifications. The number of bolts for fixing the lamp base should not be less than the number of fixing holes on the lamp base, and the bolt diameter should match the hole diameter.

For lamps without fixed mounting holes on the base (self-drilling during installation), each lamp should have at least 2 bolts or screws for fixing, and the center of gravity of the lamp should be consistent with the center of gravity of the bolts or screws.

Only when the diameter of the insulating stand is 75mm and below, it could be fixed with one bolt or screw.

c. The ceiling lamp could not be directly installed on combustible objects. Some families use painted plywood to line the back of the ceiling lamp for beauty. In fact, this is very dangerous, and thermal insulation measures must be taken. If the high temperature part of the lamp surface is close to combustibles, thermal insulation or heat dissipation measures should also be taken.

d. The pendant lamp should be equipped with a hanging box, and each hanging box could only be equipped with one set of pendant lamp, as shown in Fig. 7-3. The surface of the pendant lamp must be well insulated, without joints, and the sectional area of the conducting wire should not be less than $0.4mm^2$. The wiring in the hanging box should take measures to prevent the lamp from falling due to the force on the wire head. The lamps with a mass of more than 1kg should be equipped with lifting chains. When the mass of the pendant lamp exceeds 3kg, it should be fixed with embedded hooks or bolts. The lamp line of the pendant lamp should not be pulled, and the lamp line should be braided with the pendant chain.

Fig. 7-3 One set of pendant lamp is installed in each hanging box

② Attention to safety must be paid on the installation of lamps.

The safety here includes use safety and construction safety.

a. The metal housing of the lamp should be reliably grounded to ensure safe use. As shown in Fig. 7-4, the metal housing of a brand of LED lamp should be grounded.

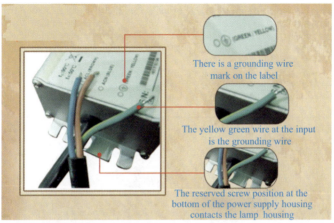

Fig. 7-4　Grounding of metal housing of LED lamps

In the actual application process of Class I lamps, if there is no grounding wire, false grounding or poor grounding, the potential hazards may include the potential danger of electricity leakage, the inability to effectively release similar induced electricity and static electricity generated by lamps, etc. Therefore, when wiring Class I lamps, a grounding conducting wire should be added to the lamp box.

b. When the screw lamp holder is wired, the phase wire (i.e., the live wire connected with the switch) should be connected to the central contact terminal, and the zero wire should be connected to the screw terminal, as shown in Fig. 7-5.

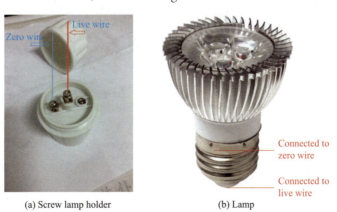

(a) Screw lamp holder　　　　　　(b) Lamp

Fig. 7-5　Screw lamp holder and lamp

c. The two wire ends connected with the incoming wires of the lamp power supply should have good electrical contact, and should be wrapped with electrician waterproof insulating tape and black tape respectively, and a certain distance should be kept. If possible, try not to put two wire ends under the same metal sheet to avoid short circuit and danger.

d. During installation, the insulating shell of lamp cap should not be damaged to prevent electric leakage.

e. When installing large lamps such as ceiling lamps, operators should pay special attention to safety when working at heights, and special person should be assigned to assist in operation, as shown in Fig. 7-6.

Fig. 7-6　Installation of large lamps requires assistance

f. When installing various lamps on the flat roof of the decorative hanger, the installation should be carried out according to the requirements of the installation instructions for lamps. In addition, the concealed wires in the ceiling or wall panel must be protected by PVC flame retardant wire tubes. When the lamp mass is more than 3kg, the embedded hook or expansion bolt should be used to directly fix the support and hanger from the roof (the lamp cannot be installed with the keel support suspended from the ceiling).

The wire from the lamp head box should be protected to the lamp position with a hose to prevent the wire from being exposed to the flat roof.

g. When the steel pipe is used as the suspender of the lamp, the inner diameter of the steel pipe should not be less than 10mm; The thickness of steel pipe wall should not be less than 1.5mm. The lamp wire of the pendant chain lamp should not be subject to tension, and the lamp wire should be braided with the pendant chain. The two ends of the flexible cord wires of the pendant lamp should be provided with protective buckles, and the core wires at both ends should be tinned.

h. In flammable, explosive and damp places, lighting facilities should be explosion-proof and moisture-proof devices.

③ Height requirements

a. Lamps should not be installed directly above high-voltage and low-voltage power distribution equipment and buses in the substation. For lamps installed outdoors, the height from the ground should not be less than 3m; When installed on the wall, the height from the ground should not be less than 2.5m.

b. When there is no requirement in the design, the installation height of the lamp should not be less than the value specified in Table 7-3 (except when safe voltage is used). Workshop lighting, portable lights and partial lights of machine tools that are lower than the height specified in the table but without safety measures should be powered by safe voltage below 36V.

Table 7-3 Requirements for installation height of lamps

Places	Minimum installation height/m	Places	Minimum installation height/m
Outdoor (outdoor wall mounted)	2.5	Indoor	2
workshop	2.5	Lamps with soft suspension wire and lifter, after the suspension wire is deployed	0.8
Metal halide lamp	5		

④ Special requirements. The emergency lights and evacuation indicators used in public places should be equipped with obvious signs. The lighting in public places without special management should be equipped with automatic energy-saving switches.

In dangerous and severe dangerous places, when the height of the lamp from the ground is less than 2.4m, lighting lamps with rated voltage of 36V and below should be used, or special protective measures should be taken. When the height of the lamp from the ground is less than 2.4m, the accessible bare conductor of the lamp must be grounded (PE) or connected to neutral (PEN), and there should be a special grounding bolt with identification.

【Special reminder】

The basic requirements for the installation of lighting: safe and firm. At the same time, beauty should also be considered.

7.2.2 Lighting circuit trouble shooting

(1) Lighting circuit fault diagnosis

The main common faults of the lighting circuit are short circuit, open circuit and electrical leakage, and the fault diagnosis is shown in Table 7-4.

Table 7-4 Lighting circuit fault diagnosis

Fault type	Fault phenomenon	Cause of fault	Inspection method
Short circuit	The short circuit fault often causes the fusant of the fuse to melt, and there are obvious burn marks and insulation carbonization at the short circuit point. In serious cases, the insulation layer of the wire would be burnt and even cause a fire	(1) The installation is not in conformity with the specification, the multi strands wires are not tightly twisted, tin coated, crimped and burred (2) The crimping of phase wire and zero wire is loose, the distance between two wires is too close, and some external forces cause them to collide, causing relative zero short circuit or interphase short circuit (3) Water ingress into lamp holders, circuit breakers and other electrical appliances due to unexpected reasons (4) There is a lot of conductive dust in the environment where the electrical equipment is located (5) Human factors	Figure out the cause of short circuit first, then find out the fault point of short circuit, replace the fuse after treatment, and resume power transmission
Open circuit	In case of open circuit fault of phase wire and neutral wire, the load would not work normally. When single-phase circuit breaks, the load does not work; In case of phase failure of three-phase electrical power supply, adverse consequences would be caused; The three-phase-four-wire power supply line is unbalanced. If the neutral wire breaks, the three-phase voltage will be unbalanced. The phase with large load would have low voltage, and the phase with small load would have high voltage. If the load is an incandescent lamp, the light on one phase would be dim, and the light on the other phase would become very bright. Meanwhile, the load side of the neutral wire break would have voltage to ground	(1) Fusant is melted due to excessive load (2) The switch contacts are loose and in poor contact (3) The conducting wire is broken, the joint is severely corroded (especially the copper and aluminum conductors are directly connected without copper aluminum transition joint) (4) During installation, the crimping at the connection is not solid, and the contact resistance is too large, which would overheat the contact for a long time and cause oxidation at the contact of the wire and terminal (5) The wire is disconnected due to severe environments, such as strong wind, earthquake, etc. (6) Human factors, such as breaking the electric wire by carrying too high articles, and human destruction	A test pencil with neon tube could be used to measure whether the two poles of the lamp holder (lamp holder) are energized: if both poles are not lit, the phase wire is open circuit; If both poles are on (with bulb test), the neutral wire (zero wire) is open circuit; If one pole is on and the other is not, the filament is not connected. The digital display test pencil body is equipped with LED display screen, which can visually read the test voltage values. When measuring the lighting circuit, there is a voltage $U=220V$ between the live wire and the ground. Digital display test pencil has breakpoint detection function, which is very convenient for detecting open circuit faults. Press and hold the breakpoint detection key, and when moving along the wire longitudinally, the place without display in the display window is the breakpoint

Fault type	Fault phenomenon	Cause of fault	Inspection method
Electric leakage	(1) In case of electric leakage, the power consumption would increase; Sometimes it trips for no reason. (2) People would feel numb when touching the electric leakage. (3) When measuring the insulation resistance of the line, the resistance value would decrease	(1) The insulated conducting wire is damp or polluted (2) Wires and electrical equipment have been used for a long time, and the insulation layer has been aged (3) The insulation between the phase wire and the neutral wire is damaged by external forces, resulting in leakage between the phase wire and the ground	(1) Judge whether there is electric leakage (2) Judge whether the leakage is between the live wire and the neutral wire, or between the phase wire and the ground, or both (3) Determine the leakage range (4) Find out the leakage point

【Special reminder】

The open circuit fault of lighting circuit could be divided into three situations: total open circuit, partial open circuit and individual open circuit, which should be dealt differently during maintenance. The essence of electric leakage and short circuit is the same, but the degree of accident development is different. Serious electric leakage might cause short circuit.

(2) Maintenance method of lighting circuit fault

Methods for checking the failure of lighting circuit
- Fault investigation method
- Visual inspection method
- Testing method
- Branch and section inspection method

① Fault investigation method. Before handling the fault, the fault inspection should be taken to understand the situation before and after the fault from the person or operator on the site at the time of the accident, so as to preliminarily judge the fault type and the fault location.

② Visual inspection method. After fault investigation, further visual inspection should be carried out by sensory organs, namely: smell, listen and look.

Smell: if there is any smell caused by insulation burning due to high temperature.

Listen: whether there is an abnormal sound such as discharge.

Look: for the open laying line, patrol along the line to check whether there are obvious problems on the line, such as: broken wire sheath, collision, broken wire, damaged bulb, fuse burnt, fuse overheating, circuit breaker tripping, water penetration and burning of lamp holder, etc. Then check the key parts.

③ Testing method. In addition to visual inspection of lines and electrical equipment, the test pencil, multimeter, test lamp, etc., should be fully used for testing.

For example, when there is a phase failure, it is not enough to use a test pencil to check whether there is electricity. When the phase wire is indirectly loaded, the test pen would light up and mistakenly indicate that the phase is not failed, as shown in Fig. 7-7. So, the voltmeter or multimeter should be used for AC voltage test to accurately determine whether the phase is failed.

④ Branch and section inspection method. For the circuit to be checked, it could be checked by circuit, branch or "bisection method" to narrow the fault range and gradually approach the fault point.

(3) Safety measures for power cut-off maintenance

Generally, the maintenance of lighting lines should be carried out with power cut-off. Power cut-off maintenance could not only eliminate the electric shock hazard to the maintenance personnel, but also eliminate their concerns during work, which is conducive to improving maintenance quality and work efficiency.

① In case of power cut-off, all power supplies that might be input to the line or equipment to be repaired should be cut off, and clear breaking points should be provided. Hang a warning sign board on the breaking point, marking "No switching on, someone is operating", as shown in Fig. 7-8. If the breaking point is the fuse, it is better to take it off.

Fig. 7-7 Inspection of phase failure of line   Fig. 7-8 Hanging warning sign board at eye-catching position

② Before maintenance, the circuit to be repaired must be rechecked with a test pencil, and the maintenance could only be started when it is proved that the electricity is off.

③ If the lines to be repaired are pretty complex, temporary grounding wires should be installed near the maintenance point to short circuit all phase lines to each other before grounding, so as to artificially cause interphase short circuit or short circuit to ground, as shown in Fig. 7-9. In this way, in case of power transmission during maintenance, the main switch would trip or the fuse would be melted to avoid electric shock to the operators.

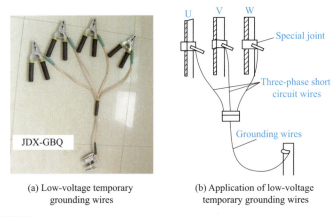

(a) Low-voltage temporary grounding wires

(b) Application of low-voltage temporary grounding wires

Fig. 7-9　Low-voltage temporary grounding wires and their application

④ After the line or equipment maintenance is completed, total checks, such as omissions and unqualified maintenance, is required, which includes whether the removed wires, components, fault points that should be eliminated, and the insulation layer that should be recovered have all been handled correctly. Whether there are tools, components, etc., left on the line and equipment, and whether all the staffs are evacuated from the site.

⑤ Remove the temporary grounding device installed for security before maintenance, and the temporary short circuit to ground or interphase short circuit of each phase. Finally take down the warning sign board of the power supply disconnection point.

⑥ Supply power to the repaired circuit or equipment.

Power capacitor

Low Voltage Electrician Certification Training Course

8.1 Brief introduction to power capacitor

Power capacitors are mainly used in power systems, but they are also widely used in industrial production equipment and high-voltage tests. It could be divided into high-voltage power capacitor and low-voltage power capacitor according to the applied voltage, bounded by the rated voltage of 1000V. High-voltage power capacitors are generally oil-immersed capacitors, while low-voltage power capacitors are mostly self-healing capacitors, also known as metallized capacitors.

8.1.1 Classification and application of power capacitors

(1) Shunt capacitor

Shunt capacitor is the abbreviation of parallel compensation capacitor. It is connected in parallel with the equipment to be compensated in the 50Hz or 60Hz AC power system to compensate the inductive reactive power, improve the power factor and voltage quality, reduce the line loss, and increase the output power of the system or transformer. Because shunt capacitors reduce the transmission of inductive reactive power on the line, reduce voltage and power losses, and thus improve the transmission capacity of the line.

Shunt capacitors could be divided into the following types.

① High voltage shunt capacitor. The rated voltage is above 1.0kV, and most of them are oil-immersed capacitors.

② Low voltage shunt capacitor. The rated voltage is 1.0kV and below, and most of them are self-healing capacitors.

③ Self-healing low-voltage shunt capacitor. Its rated voltage is 1.0kV and below.

④ Assembled shunt capacitor (also called dense capacitor). To be accurate, it should be called parallel capacitor bank, with rated voltage of 3.5~66kV.

⑤ Tank capacitor. The rated voltage is usually 3.5~35kV. The difference with assembled shunt capacitor is that the assembled shunt capacitor is composed of capacitor units (a single capacitor is sometimes called a capacitor unit) in series and parallel, and is placed in a metal box. Tank capacitor is composed core of components connected in series and parallel, and is placed in a metal box.

(2) Series capacitor

It is connected in series to 50Hz or 60Hz AC power system lines, and its rated voltage is mostly below 2.0kV. The functions of the series capacitor are as follows.

① Increase the line terminal voltage. Generally, the terminal voltage of the line could be increased by 10%~20% at most.

② Reduce the voltage fluctuation at the receiving end. When there are impact loads (such as electric arc furnace, electric welding machine, electrical rail, etc.) with great changes at the line receiving end, the series capacitor could eliminate the violent fluctuation of voltage.

③ Improve the transmission capacity of the line.

④ Improve the stability of the system.

(3) AC filter capacitor

It is connected with reactors and resistors to form AC filter capacitors, which are connected to 50Hz or 60Hz AC power systems to provide low-impedance channels for one or more harmonic currents. It could reduce the harmonic level of the network, and improve the power factor of the system. Its rated voltage is 15kV and below.

(4) Coupling capacitor

The coupling capacitor is mainly used in the carrier communication system of high-voltage and extra-high-voltage transmission lines, and could also be used as a component in the measurement, control and protection devices.

(5) DC filter capacitor

DC filter capacitors are used in high-voltage rectifier filter devices and high-voltage DC transmission. It could filter out residual AC elements, reduce ripple in DC, and improve the quality of DC transmission. The rated voltage is mostly about 12kV.

8.1.2 Structure and compensation principles of power capacitor

(1) Structure of power capacitor

The structures of various capacitors vary greatly according to their types, which are mainly composed of shell, core, lead and bushing.

The shell of the capacitor is generally welded with thin steel plate, the surface is coated with flame retardant paint, the shell cover is equipped with outlet bushing, and the side of the box wall is welded with hanging plate, grounding bolt, etc. Oil conservator or metal expander and pressure relief valve are also installed on the cover of large capacity assembled shunt capacitor, and sheet radiator and pressure type temperature control device are installed on the side of the wall.

The core of the capacitor is composed of components, insulating parts, etc. Fig. 8-1 shows the structure of power capacitor.

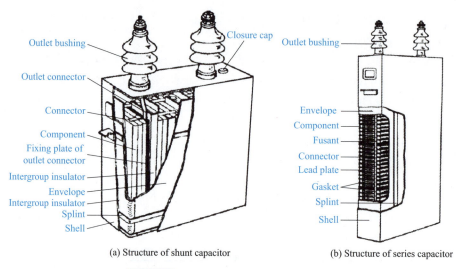

Fig. 8-1　Structure of power capacitor

(2) Compensation principle of power capacitor

Capacitance compensation is reactive power compensation or power factor compensation. The electrical equipment of the power system generates reactive power when it is used, and it is usually inductive, which would reduce the capacity use efficiency of the power supply. It could be improved by appropriately adding capacitance in the system. Power capacitance compensation is also called power factor compensation.

The principle of reactive power compensation of power capacitor is to connect the device with capacitive power load and inductive power load in parallel on the same capacitor, and the energy is converted between the two loads. In this way, the load of transformers and transmission lines in the power grid is reduced, thus increasing the output active power capacity. The loss of the power supply system is reduced when a certain amount of active power is output.

Capacitor is a simple and economical method to reduce the load of transformer, power supply system and industrial power distribution system. Therefore, it is very common to use shunt capacitors as reactive power compensation devices.

8.2　Installation requirements and wiring of power capacitor

8.2.1　Installation requirements of power capacitor

(1) Installation environment requirements

The ambient temperature of the power capacitor should not exceed ± 40 ℃ , the relative humidity of the surrounding air should not exceed 80%, and the altitude should not exceed 1000m. There should be no corrosive gas or steam, no large amount of dust or fiber, and the installation environment should be free of flammable, explosive hazards or strong vibration.

The power capacitor room should be a fire-resistant building, and the fire resistance rate should not be lower than Grade II; Good ventilation is required in the capacitor room. High-voltage capacitors with a total oil volume of more than 300kg should be installed in separate explosion-proof rooms; High-voltage capacitors and low-voltage capacitors with a total oil volume of less than 300kg should be installed in the interval with explosion-proof wall or partition depending on their oil volume.

The power capacitor should be protected from direct sunlight and the window glass should be painted white.

When the power capacitors are installed in layers, they generally should not be more than three layers. There should be no partition between layers to avoid obstructing ventilation. The distance between adjacent capacitors should not be less than 5m; The clear distance between upper and lower layers should not be less than 20cm; The height of the bottom of the lower capacitor to the ground should not be less than 30cm. The nameplate of capacitor should face the passageway.

(2) Installation conditions

① The construction drawings and technical data should be complete.

② The civil works are basically completed, and the ground and wall are completed. The elevation, size, structure and embedded parts meet the design requirements.

③ The roof should be free of water leakage. Doors, windows and glasses should be installed. The doors should be locked. The site should be cleaned and the road should be clear.

④ Before the installation of the complete set of capacitor frame bank, the section steel foundation should be prepared according to the design requirements. The capacitor ground frame should be made of non-combustible materials.

(3) Installation of capacitor

① The rated voltage of capacitor should be consistent with the grid voltage, and the Δ-configuration should be adopted generally. The capacitor bank should maintain three phase balance, and the three phase unbalanced current should not be greater than 5%.

② During capacitor installation, the nameplate should face the passageway; The capacitor must have a discharge process to ensure that the stored electrical energy could be discharged quickly after power failure; The metal shell of capacitors must be reliably grounded.

(4) Precautions for capacitor installation

① Any bad contact in the capacitor circuit may cause high-frequency oscillating arc, which would increase the working electric field strength of the capacitor and generate heat and cause accelerated damage. Therefore, good contact between the electrical circuit and the grounding part must be maintained during installation.

② When capacitors of lower-voltage level are operated in the network of higher-voltage level after being connected in series, the shell of each capacitor should be reliably insulated from the ground by installing insulators equivalent to the operating voltage level.

③ After Y-configuration, the capacitors are used for higher-rated voltage. When the neutral point is not grounded, the shell of the capacitors should be insulated to the ground.

④ Before capacitors installation, the primary capacitance should be distributed to balance its phases, and the deviation should not exceed 5% of the total capacity. When the relay protection device is installed, it should also meet the requirement that the balance current error during operation should not exceed the action current of relay protection.

⑤ The low-voltage capacitors for group compensation should be connected outside the low-voltage group bus power switch, to prevent self-excitation when the group bus switch is disconnected.

⑥ The low-voltage capacitor bank for centralized compensation should be equipped with a special switch and installed outside the main switch of the line, not on the low-voltage bus.

8.2.2 Wiring of power capacitor

(1) Capacitor wiring mode

The three phase capacitors are internally configured in Δ; The wiring mode of single-phase capacitor should be determined according to its rated voltage and the rated voltage of the line: when the rated voltage of the capacitor is consistent with the line voltage of the power lines, Δ-configuration should be adopted; Y-configuration should be adopted when the rated voltage of capacitor is consistent with the phase voltage of the power lines.

In order to obtain good compensation effect, the capacitor should be divided into several groups and connected to the capacitor bus respectively. Each group of capacitors should be able to control, protect and discharge separately. Several basic connection modes of capacitor are shown in Fig. 8-2.

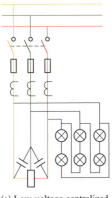 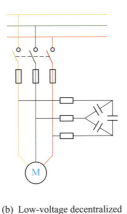 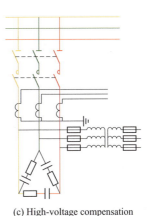

(a) Low-voltage centralized compensation   (b) Low-voltage decentralized compensation   (c) High-voltage compensation

Fig. 8-2 Basic wiring mode of capacitor

(2) Precautions for capacitor wiring

① The connecting wire of capacitors should be flexible wire, and the wiring should be symmetrical and consistent, neat and beautiful. The wire end should be added with a nose, and the crimping should be firm and reliable.

② When the capacitor bank is connected by buses, the capacitor bushings (connecting terminals) should not be subject to mechanical stress, the crimping should be tight and reliable. The buses should be arranged orderly, and the phase color should be painted.

③ The wiring of each capacitor should be connected to the buses with separate flexible wires. Buses should not be hard, to prevent the capacitor bushing from being damaged due to assembly stress and oil leakage caused by sealing damage.

④ For the wiring of individual compensation capacitors, it should be ensured that: in the case of induction motors that are directly started or started by rheostats, their capacitors that improve the power factor could be connected directly to the motor's outlet terminals, and no switch nor fuse should be installed between them; For the induction motor started by Y-Δ start, three single-phase capacitors should be used. Each capacitor is connected directly in parallel on two terminals of each phase winding, so that the connection of the capacitor is always consistent with that of the winding.

 【Special reminder】

> Insulation measurement should be carried out before powering on to capacitors. Capacitors below 1kV should be measured with 1000V megger, and 3~10kV capacitors should be measured with 2500V megger, and records should be documented.

8.3 Safe operation of power capacitor

8.3.1 Operation and maintenance of capacitor

(1) Protection measures for power capacitors

Proper protective measures should be taken for capacitor banks, such as balance or differential relay protection or instantaneous action overcurrent relay protection. For 3.15kV and above capacitors, separate fuses must be installed on each capacitor. The rated current of the fuse should be selected according to the characteristics of the fuse and the inrush current when it is connected. Generally, 1.5 times the rated current of the capacitor is appropriate to prevent the explosion of the capacitor tank.

If the capacitors are connected to the overhead line, an appropriate lightning arrester could be used for atmospheric overvoltage protection. In the high-voltage network, when the short circuit current exceeds 20A, and the protective device or fuse of the short circuit current could not reliably protect the short circuit to ground, the single-phase short circuit protective device should be used.

Capacitors are not allowed to be equipped with automatic reclosing devices, but non-pressure release automatic tripping devices should be installed.

(2) Daily maintenance of power capacitor

The daily maintenance of capacitors in operation could extend the service life of capacitors to a certain extent.

① The appearance of the capacitor bank in operation should be inspected every day according to the regulations. If the tank shells are found to be expanded, they should be stopped from use to avoid failure. Check the load of each phase of the capacitor bank with an ammeter.

② The surface of capacitor bushings and support insulators should be clean, and free from damage and discharge trace. The capacitor shells should be clean, and free from deformation and oil leakage, and the capacitors and iron frames should be free from dust and other dirt. Pay attention to checking the reliability of all contacts (energized busbar, grounding wire, circuit breaker, fuse, switch, etc.) on the electrical circuit connected with the capacitor bank. Because a fault occurs at a contact point on the line, even if a nut is not fastened tightly, the capacitors might be damaged early and the whole equipment might suffer accidents.

③ The capacitance and fuse of capacitors should be checked at least once a month. If withstand voltage test is required for capacitors in operation, the test should be carried out according to the specified value.

④ In case of the circuit breaker of the capacitor bank trips due to the action of the relay, it should not be closed again until the cause of the trip is found.

⑤ If the capacitor shell leaks oil during operation or transportation, it could be repaired by soldering with tin-lead solder.

(3) Switching operation of power capacitor bank

① Under normal conditions, when the whole substation is powered off, the capacitor bank circuit breakers should be disconnected first, and then disconnect the outlet circuit breakers of each circuit. The power transmission should be restored in reverse order.

② In case of an accident, the circuit breaker of the capacitor bank must be disconnected after the whole substation has no power.

③ The circuit breaker of capacitor bank should not be forced to power on after tripping. After the protective fuse is blown, it is not allowed to replace the fuse for power transmission without finding out the cause.

④ The capacitor bank is prohibited from being closed with charge. When the capacitor bank is closed again, it could only be done after the circuit breaker is disconnected for 3min.

(4) Capacitor operation monitoring

① Temperature monitoring. The ambient temperature for capacitor operation is generally $-40 \sim 40$ ℃. Temperature wax chips could be pasted on the shell of the capacitor for detection.

The temperature requirements for capacitor bank operation are: the temperature rise in 1h should not exceed 40 ℃, the temperature rise in 2h should not exceed 30 ℃, and the annual average temperature rise should not exceed 20 ℃. In case of exceeding, manual cooling (fan installation) should be adopted or the capacitor bank should be disconnected from the power grid.

② Voltage and current monitoring. The working voltage and current of capacitor should not exceed 1.1 times of rated voltage and 1.3 times of rated current during operation. The capacitor should not operate for more than 4h at 1.1 times of rated voltage. The difference of three-phase current of capacitor should not exceed $\pm 5\%$.

8.3.2 Trouble shooting of capacitors

(1) Trouble shooting during capacitor running

① In case of oil spout, explosion and fire of capacitor, the power supply must be cut off immediately, and sand or dry fire extinguisher should be used to extinguish the fire. Such accidents are mostly caused by overvoltage inside and outside the system, and serious internal faults of capacitors. In order to prevent such accidents, it is required that the fusant specifications of a single fuse must be matched. After the fusant is blown, the cause should be carefully found. The capacitor bank should not use reclosing, and after tripping, it should not be forced to power on, so as to avoid further accidents.

② In case of the circuit breaker of the capacitor trips, while the fusant of the shunt fuse is not blown. After the capacitor is discharged for 3min, check the circuit breaker, current mutual inductor, power cable and the capacitor external. If no abnormality is found, it might be caused by external fault or bus voltage fluctuation, and could be put into trial operation after it is checked to be normal, otherwise, further comprehensive power on test should be conducted for protection. No trial operation is allowed before the cause is found out.

③ When the fusant of the capacitor is blown, the circuit breaker of the capacitor should be disconnected with the consent of the dispatcher on duty. After cutting off the power supply and discharging the capacitor, conduct external inspection first, and then measure the insulation resistance between poles, and between poles and ground with a megger. If no fault is found, replace the fusant and continue to operate. If the fussant of the fuse is still melted after power transmission, the faulted capacitor should be taken out and the power transmission operation to the rest should be resumed.

(2) Safety precautions for handling faulty capacitors

The fault capacitor should be handled after the circuit breaker of the capacitor is disconnected, the isolating switches on both sides of the circuit breaker are opened, and the capacitor bank is discharged through the discharge resistance. After the capacitor bank is discharged by the discharge resistance (discharge transformer or discharge voltage mutual inductors), as part of the residual charge could be discharged completely for a while, a manual discharge should still be carried out. When discharging, first connect the grounding terminal of the grounding wire, and then discharge the capacitor for several times with the grounding rod until there is no discharge spark and discharge sound, and then fix the grounding terminal.

Due to the poor contact of the wire, internal disconnection or fusant fusing of the faulty capacitor, part of the charge might not be discharged completely, so the maintenance personnel should wear insulating gloves before contacting the faulty capacitor. First, short the two poles of the faulty capacitor with a short circuit wire, and then disassemble and replace it.

The neutral wire of the capacitor bank with dual Y-configuration and the serial connection wire of multiple capacitors should also be discharged separately.

(3) Capacitor repair

① The oil leakage on the bushing and shell could be repaired with tin-lead solder, but it should be noted that the soldering iron should not be overheated to prevent the silver layer from desoldering.

② In case of insulation breakdown to ground, loss tangent value of the capacitor increases, shell would expand, open circuit and other faults would occur, the capacitor needs to be repaired in a special repair shop.

Use of safety tools (K1)

Practical test subject 1 "the use of safety tools", with the test code K1, which mainly includes three sub-subjects of practical test: the safe use of electrical instruments, the use of electrical safety tools, and the identification of electrical safety signs. During the formal assessment, the system randomly selects the test questions of a sub-subject.

9.1 Safe use of electrical instruments (K11)

The electrical instruments involved in K11 include multimeter, clamp ammeter, megger, grounding resistance tester, etc.

Examination method: Practical operation and oral presentation.

Examination time: 10 minutes.

Safe operation steps:

① Select appropriate electrical instruments according to the given measurement tasks;

② Inspection of the selected instruments;

③ Use instruments correctly;

④ Read correctly and analyze the measured data.

Scoring standards: see Table 9-1.

Table 9-1 Scoring table for safe use of electrical instruments

Exam Items	Exam Contents	Score	Scoring Criteria
Safe use of electrical instruments	Selecting the appropriate electrical instruments	20	Oral description of the role of various electrical instruments, 3~10 points would be deducted for incorrect answer. For the measurement task assigned by the assessor, correctly select the appropriate electrical instruments (multimeter, clamp ammeter, megger, grounding resistance tester), 10 points would be deducted for incorrect selection of instruments
	Instrument Check	20	Correctly check the appearance of the instrument, 5 points would be deducted for not checking the appearance. Failure to check the certificate, deducts 5 points. Failure to check the integrity, deducts 10 points
	Proper use of meters	50	Follow the safety operation procedures and use the instrument correctly according to the operation steps. Zero points for violation of safety procedures and 5~50 points deduction for incomplete operation steps, depending on the situation
	Analysis of measurement results	10	Failure to analyze and judge the results of the measurement, deducts 10 points

Chapter 9 Use of safety tools (K1)

Exam Items	Exam Contents	Score	Scoring Criteria
Negative items	Negative item description	Deduct the score of this item	For the given measurement task, unable to select the appropriate instruments correctly, violating safety operation standards and making themselves or instruments in an unsafe condition, etc., the score is zero, and the examination should be terminated
Total		100	

[Training 9-1] Using a pointer multimeter

(1) Examination requirements

Be able to master the method of measuring resistance, voltage and current with a pointer multimeter.

(2) Training contents

1) Pre-use inspection

① Check the appearance: the meter should be intact, and the pointer of the meter should be free of jamming.

② The range switch should be flexible and the indicated range should be accurate.

③ Put the instrument flat, and check whether the pointer is pointing to the mechanical zero position, otherwise, mechanical zero adjustment should be carried out.

④ Before measuring resistance, the ohmic zero adjustment (electrical zero adjustment) should be carried out. Check the battery voltage, the voltage should be replaced when it is low.

⑤ The insulation of probes wire should be good, the black probe is inserted into the negative "−" or common end "*", and the red probe is inserted into the positive "+" or the corresponding measurement hole.

⑥ Use the ohmic range to check whether the probe wire is intact.

2) Measurement

① DC resistance measurement

a. Disconnect the power supply and connection of the circuit or component under test.

b. Choose the appropriate range according to the measured value, if the measured value cannot be estimated, the middle range should be chosen.

c. Ohmic zero adjustment should be done at every range conversion.

d. During measurement, the probes should be in good contact, the hands should not touch the metal part of the probes.

e. The measured value equals the indicating value multiplies the range multiplier. The indicating value should be in the range of 20% to 80% of the scale.

【Special reminder】

The resistance should not be measured with electricity, and the resistance under test should be disconnected from the circuit. Every time when changing the range, ohmic zero re-adjustment should be carried out. During measurement, the hands could not touch the two pins of the resistance at the same time, otherwise it would cause reading errors, use one hand to fix one end of the resistance, and the other end is pressed on the desktop.

② Measurement of AC/DC voltage

a. Choose the appropriate range according to the measured value, when the measured value could not be estimated, the maximum range should be selected.

b. The probes should be connected in parallel with the circuit under test. When measuring AC voltage, the probes are polarity free. When Measuring DC voltage, the red and black probes are connected in parallel to the two ends of the line under test; The red probe is connected to the high potential end of the circuit power "+", and the black probe is connected to the low potential end of the circuit power "−", which should be connected in parallel with the circuit under test.

c. The measured value could be got by multiplying the indicating value and the range multiplier. The indicating value should be at least in the 1/2 scale and above, preferably in the 2/3 scale and above.

【Special reminder】

When measuring voltage, pay attention to safety, hands cannot touch the charged parts. The lowest AC voltage range of some multimeters has a special line of the scale, when using this range, read according to the special scale. When switching measuring range, the probes should be disconnected from the power supply.

Wire voltage is 380V, which refers to the voltage between any two phases of the power supply. There are 3 wire voltages, they should be measured for 3 times. Phase voltage is 220V, which refers to the voltage of any one phase of the power supply to the zero wire. There are 3 phase voltages, they should be measured for 3 times, a diagram for illustration could be drawn.

There are some differences between measuring DC voltage and AC voltage: the measuring ranges are different; AC voltage measurement is polarity free, while the positive and negative polarities of DC voltage should be distinguished.

③ AC and DC current measurement

a. The probes must be connected in series with the circuit under test.

b. When measuring DC current, make sure to distinguish the polarities.

c. Choose the appropriate range according to the measured value, if the measured value could not be estimated, the maximum range should be selected.

d. The measured value could be got by multiplying the indicating value and the range multiplier. The indicating value should be at least in the 1/2 scale and above, preferably in the 2/3 scale and above.

【Special reminder】

After measurement, cut off the power first, and then disconnect the probes.

(3) Precaution

After measurement, the range should be at the maximum AC voltage or neutral.

【Special reminder】

The measurement method of the digital multimeter is similar to that of the pointer multimeter. Readers could refer to the relevant contents in Chapter 3 of this book.

[Training 9-2] Use of clamp ammeter

(1) Examination requirements

Be able to use a clamp ammeter correctly to measure the current in an AC circuit.

(2) Training content

1) Pre-use inspection

① Appearance inspection: all parts should be intact; clamp handle should be flexible; core of the jaw should be rust-free, and its closure should be tight; sheath of core insulation should be intact; pointer should be able to swing freely; range switch should be flexible, and the hand feeling of it should be normal.

② Adjustment: put the multimeter flat, the pointer should be pointed at zero, zero adjustment should be carried out.

2) Measurement

① Put the multimeter flat, and mechanical zero adjustment should be carried out if necessary.

② Choose the appropriate range according to the measured value, choose the largest range if the measured value could not be estimated. When measuring small current (20% or less of the full scale at the lowest range), the measured wire should be wound for several turns in the jaw. The actual measured value could be got by the indicating value dividing the number of turns.

③ The measured wire should be located in the center of the jaw, and the jaw should be tightly closed.

④ After measuring the high current, open and close the jaw several times to demagnetize before measuring the small current.

⑤ Read the indicating values correctly.

The current reading range of multimeter: the pointer should be in the range of 1~5A; The calculation is:

measured current value = indicating value of the multimeter ÷ number of turns,

or

the indicating value of the multimeter = measured current value × number of turns.

(3) Precautions

① Insulated gloves or dry thread gloves should be worn when measuring.

② When measuring, keep a safe distance from the charged body (0.1m) to prevent electric shock or short circuit.

③ When measuring, range shifting with electricity is not allowed, and measuring bare wire is not allowed.

④ It is not allowed to measure at the dynamic contact of the switch or fuse.

⑤ If the clamp ammeter is not used, the range switch should be shifted to the maximum range of AC voltage or current. It's not qualified to operate not in accordance with the provisions, requirements, or steps.

[Training 9-3] Use of grounding resistance tester

(1) Examination requirements

Be able to properly use a grounding resistance tester to measure the grounding resistance of lines and equipment.

(2) Training contents

1) Preparation

① Disconnect the power supply of the grounding pole under test, dismantle the reserved test point of grounding pole and polish it clean.

② Check the grounding resistance tester

a. Intact appearance without damage;

b. Adjust the pointer until it coincides to the central scale line;

c. Do short-circuit test to check the accuracy of the tester.

③ Select the appropriate multiplier, and generally choose the range of × 1.

④ Correct wiring, wiring diagram is as shown in Fig. 9-1.

a. 5m test wire, connected to the terminal E (or C2, P,) and the measured grounding pole.

b. 20m test wire, connected to the terminal P (or P1) and voltage auxiliary grounding pole.

c. 40m test wire, connected to the terminal C (or C1) and the current auxiliary grounding pole.

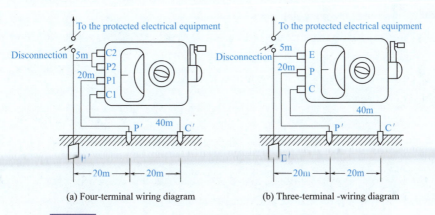

(a) Four-terminal wiring diagram   (b) Three-terminal -wiring diagram

Fig. 9-1　Wiring diagram of grounding resistance measurement

2) Measurement

After the correct wiring, the tester is placed horizontally and the handle is rotated at a constant speed of 120r/min, adjust the scale knob while rotating, and adjust until the pointer coincides with the central scale line. The value that the pointer indicated on the dial multiplied by the multiplier is the measured value.

3) Precautions

① The test wires should not be parallel to the overhead lines or metal pipes.

② During thunderstorm season and rainy days, it's not allowed to measure the grounding resistance of the lightning protection device.

③ Open circuit test is generally not done to the tester.

④ The measurement should be carried out by two people.

[Training 9-4] Measuring three-phase currents with ammeter and current mutual inductor

(1) Examination requirements

Master the wiring method of ammeter with current mutual inductor to measure three-phase current.

(2) Training contents

The wiring electrical diagram for measuring three-phase current with three ammeters connected to three mutual inductors, as shown in Fig. 9-2.

1) Selection of current mutual inductor, ammeter and wire

① The selection of ammeter should base on the load current, and the load current should often indicate at 1/3 ~ 3/4 of the full scale of the ammeter.

② The transformation ratio of current mutual inductor and ammeter should be the same, the polarities should not be reversed, and K2 terminal should be grounded or connected to neutral.

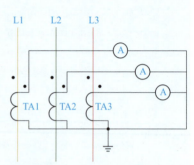

Fig. 9-2 Measuring three-phase current with current meter and mutual inductors

③ The connecting wire should be insulated copper wire with cross-section of not less than 2.5mm^2, the joint is not allowed, the end of the wire should be crimped clockwise firmly.

④ The secondary wires should be neatly arranged, the end of the wire should be covered to marker head with circuit marker and number.

For example: For a line with a calculated current of 510A, select an ammeter, a current mutual inductor, and a secondary wire.

Solution: ① Select the ammeter: 510 × 1.5 = 765 (A). Ammeter with a range of 750A could be used (such as square ammeter of 59L23- 750A).

② Select current mutual inductor: current mutual inductor of 750/5 could be selected (such as LMZJ1-0.5, 750/5 current mutual inductor).

③ Select the secondary wire: insulated copper wire of BV-2.5mm^2 could be used.

2) Actual wiring according to the diagram

K1 of three current mutual inductors are connected to any terminal of the three ammeters, the other end of the three ammeters are connected and then connected to the K2 terminal of the three current mutual inductors, and then connected to ground or zero.

3) Precautions

The polarities connection of the current mutual inductor must be correct. The secondary winding of the current mutual inductor is not allowed to be open-circuited during operation.

[Training 9-5] Phase check with voltmeter

(1) Examination requirements

Be able to correctly use the voltmeter to check the phase of two power supplies.

(2) Training contents

1) Occasions where phase checking is required

Among the six-phase wires of the two three-phase power supplies, there must be three corresponding groups of phase wires with the same phase. The phase verification is to check the phase of the corresponding two power supplies.

① Two or more power supplies are used as backup for each other or in parallel operation, before putting into operation the phase should be verified.

② After power supply system and equipment are changed, phases should be re-checked.

③ After power lines are repaired, phases should be re-checked.

2) Phase check operation

The same phase means that the two AC quantities reach their maximum or zero values ($\varphi = 0$) at the same time, referred to as the same phase. Verify the phase with a voltmeter as shown in Fig. 9-3.

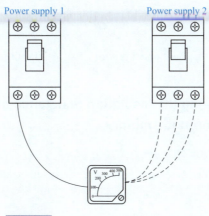

Fig. 9-3 Phase check with voltmeter

① Voltmeter of 0 ~ 450V or multimeter at a range of AC voltage 500V should be selected for phase check.

② Contrasting with any one of the known phases, measuring the three unknown phases once respectively, mark when meeting the same phase.

③ Turn to the other two phases of the known phases, measuring the three unknown phases once respectively, mark when meeting the same phase.

④ The total measurements should be 9 times, it's the same phase if the voltage is about 0V, it's different phases if the voltage difference is about 380V.

 【Special reminder】

When measuring, the voltage between the two phase wires with the same phase is not necessarily exactly equal to zero, as long as the approximate zero could be identified as the same phase. And the voltage with other phases must be approximately the line voltage.

During operation, the word "check" should be focused. That is, the corresponding relationship between the two power sources must be finally confirmed after repeatedly and carefully checked and make sure it's the error-free. Subjective assumptions or logical reasoning is prohibited to be used as the basis for judging the same phase.

3) Precautions

① Phase check should be carried out by two people, who need to wear insulated gloves.

② During phase check, working staff should keep a safe distance from the live body, hands should not directly touch the metal part of the probes, and it's prohibited to hold voltmeter by hand.

③ The length of the test wire should be moderate, the metal part of the probes should not be too long.

④ To prevent phase-to-phase short circuit or short circuit to the ground (if necessary, add shielding).

[Training 9-6] Measuring the insulation resistance of three-phase asynchronous motors

(1) Examination requirements

Be able to use a megger correctly to measure the insulation resistance of a three-phase asynchronous motor that is operating abnormally.

(2) Training content

1) Preparation

① Select meters correctly

For the newly installed motor, according to the regulation, 1000V megger should be used; For the running motor, 500V megger should be used.

② Measurement items

The insulation resistance value of stator winding (phase to phase and phase to ground) of cage motor is measured by megger.

③ Time of measurement

a. Before putting the newly installed motor into operation.

b. Before using it again when it has been out of use for more than three months.

c. When the motor is under major or minor repair.

d. When the motor has abnormal phenomenon or failure in operation.

2) Measurement and quality assessment

① Disconnect the power supply of the motor under test and remove power wires of the motor.

② Check the megger: the appearance is intact and undamaged, the pointer should point to the side of infinity when the megger is placed flat; In the open circuit test the pointer should point to infinity; In the short circuit test the pointer should instantly point to zero; Insulation of the test wires should be good, and the use of double twisted wire or parallel wire is prohibited.

③ Large motors should be discharged before and after measurement (large motors refer to J02 series with more than #16 base, and Y series with center height of more than 630mm).

④ When using megger to measure insulation resistance to the ground, terminal "E" of megger is connected the housing, and terminal "L" is connected to the three-phase windings, that is, the three phases to ground is completed testing once.

⑤ When using megger to measure the insulation resistance between phases, the joint piece should be removed from the motor terminals, terminal "E" of the megger is connected to one phase winding, terminal "L" is connected to another phase winding, that is, to measure phase U ~ phase V, phase V ~ phase W, phase W ~ phase U, respectively.

⑥ The instrument is placed flat during measurement, measured at a constant rotating speed of 120 rounds per minute, and read after the pointer becomes stable for one minute.

According to the minimum qualified value of insulation resistance, determine whether the tested three-phase asynchronous motor could be used.

① The newly installed motor should be measured with a megger of 1000V and less than 1MΩ.

② The running voltage of the motor should not be less than 1000Ω per volt, that is, the resistance of phase to the ground is not less than 0.22MΩ. It's not less than 0.38MΩ between phases, generally as long as it is greater than 0.5MΩ.

(3) Precautions

① Select the megger correctly and make a full inspection.

② Discharge the large motors after they are out of operation and measure them with megger according to the method of measuring capacitors. Also discharge after each test.

③ During measurement, keep a sufficient safety distance to the live body nearby (if necessary, a guardian should be assigned); human body should not touch the measured end, nor should it touch the exposed terminals on the megger.

④ To prevent unauthorized personnel from approaching the test site.

[Training 9-7]　Measurement of insulation resistance of low-voltage capacitors

(1) Examination requirements

Be able to use a megger correctly to measure the insulation resistance of low-voltage power capacitors.

(2) Training contents

1) Preparation

① Capacitor

The model number of the capacitor is BWO.4-12-3, where "B" indicates a shunt capacitor; "W" indicates that the liquid medium (impregnation) is dodecylbenzene; "0.4" indicates the rated voltage in kV; "12" indicates the number of nominal capacity in kvar; "3" indicates the number of phases. Here, according to the naming rules of capacitors, the code of solid medium for capacitor paper and the code for indoor use are omitted.

Therefore, this type of capacitor is a shunt capacitor with rated voltage of 0.4kV, with capacitor paper as solid medium and with dodecylbenzene as liquid medium; its nominal capacity is 12kvar, three-phase, indoor type.

② Selection of megger

Select megger of 500V or 1000V.

③ The method and requirements of checking megger are the same as training 4.

2) Measurement and quality assessment

① Measurement method

Do not measure insulation resistance between the poles of capacitors, only measure insulation resistance of three poles to the housing (pole to ground).

a. Before measurement, first connect terminal E of megger to the housing of capacitor, pick up the L wire, rotate the megger handle up to 120 rounds per minute, and then put the L wire on the three poles.

b. When measuring, the megger is placed flat, measured at a constant speed of 120r/min, and read the value after the pointer becomes stable for a minute. First remove the L wire and then stop the megger. The provisions of qualified value are as follows: newly installed (handover test), insulation resistance value should not be less than 2000MΩ; running (preventive test), insulation resistance value should not be less than 1000MΩ; according to the new national standard. During the above two tests, insulation resistance value should not be less than 3000MΩ.

c. After the measurement, discharge three poles to the shell (pole to ground) otherwise it is easy to cause electric shock accidents.

The method of manual discharge of power capacitors is as follows:

If the capacitor is not discharged with the discharge device for some reason, it should be discharged by the resistance buffer first, and then carry out short-circuit discharge. During manual discharge, pole to ground (housing) discharge should be done first, that is, phase U to the housing, phase V to the housing, phase W to the housing. It should be repeated several times until there are no spark or no sound. Then inter-pole discharge, that is, phase U to phase V, phase V to the phase W, phase W to phase U, it should be repeated several times, until there are no sparks or no sound, and then, short circuit the three poles with bare copper wire(phase U, phase V, phase W).

② Quality assessment. The capacitor would be short-circuited and discharged, which could be judged according to the following three results:

a. If the pointer of megger starts from zero, gradually increases to a certain value and tends to be smooth, there are crisp sounds and sparks of discharge when the capacitor is short-circuited after measurement, that means the charging and discharging performance of the capacitor is good, as long as the insulation is not lower than the specified value, it could be judged the capacitor is qualified, it is qualified to be put into operation.

b. If the megger has some readings, but there is no discharge spark when short-circuiting, it means that the connecting wire between the electrode plate and the terminal has been broken, and it must be withdrawn from operation or replaced by new products.

c. If the megger stops at zero, it indicates that the capacitor has been broken down and damaged, and should not be used again.

(3) Precautions

① The measurement should be carried out by two people, who need to wear insulated gloves.

② During discharging, do not touch the screw part of the terminal.

③ During measurement, people should keep a safe distance from the live parts.

④ During measurement, the megger should not be slowed down or stopped rotating.

⑤ For low-voltage shunt capacitor, only measure the pole-to-ground insulation, not inter-pole insulation.

⑥ To prevent unauthorized personnel from approaching the measurement site.

[Training 9-8] Measurement of the insulation resistance of low-voltage power cables

(1) Examination requirements

Be able to use a megger correctly to measure the insulation resistance of low-voltage four-core power cables.

(2) Training contents

1) Preparation

Selection of megger: choose 1000V megger, and check before use, the method and requirements are similar to the training 4.

2) Measurement by using megger and quality assessment

Measurement of Low-voltage four-core power cables are as follows.

Phase U to phase V, phase W, wire N and sheath; phase V to phase U, phase W, wire N and sheath; phase W to phase U, V phase, wire N and sheath; wire N to phase U, phase V, phase W and sheath. Before and after measurement discharging must be done fully to the sheath and among phases.

① The wiring of megger measurement of phase U to phase V, phase wire N and the sheath is shown in Fig. 9-4. The megger E is connected to the phase V, phase W, N-wire core and sheath, and terminal G is connected to the insulation layer of phase U (3-5 turns), pick up the wire L, rotate the megger handle up to 120r/min, and then put the wire L on the phase U core.

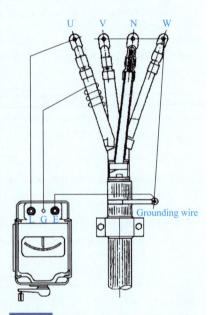

Fig. 9-4　Wiring diagram of insulation measurement of four-core cable

② When using megger to measure, the megger is placed flat, measured at a rotation speed of 120r / min, read the value after the pointer has been stable for a minute, and then withdraw the wire L and then stop rotating the megger.

③ Qualified value: before low-voltage power cable is used, use 1000V megger to measure the insulation resistance. The insulation resistance value should not be less than 10MΩ at 20 ℃.

(3) Precautions

① After the power of the tested cable is cut off, relevant safety measures should be taken, the other end of the tested cable should be guarded or installed with temporary fence, and hanging warning signs.

② Measurement with megger should be carried out by two people, who need to wear insulated gloves.

③ Before measurement with megger, the switch or equipment connected to both ends of the cable under test should be disconnected, and the original wiring should be restored after the measurement.

④ During measurement with megger, megger should not be slowed down or stopped. The rotating handle should be rotated at the rated speed, not fast or slow from time to time.

⑤ Before measurement, the cable should be discharged and connected to ground, to ensure safety. Measurement is completed or needs to repeat the measurement, the cable must be discharged, grounding, and the grounding time is generally not less than 1min.

⑥ The sheath of cable terminal should be wiped clean, and the "shielding layer" should be connected.

⑦ In the measurement report, the insulation resistance value, temperature of the cable and the relative humidity during measurement, etc., should be recorded.

9.2 Use of electrical safety appliances (K12)

During the K12 test for the use of electrical safety appliances, three kinds of appliances are selected for the test, including low-voltage electroscope (test pencil), insulating gloves, insulating shoes (boots), safety helmet, protective glasses, insulating clamps, insulating pads, portable grounding wires, foot buckles, safety belts, climbing pedals, etc.

Examination method: Practical operation, oral presentation.

Exam time: 10 minutes.

Safe operating procedures.

① Be familiar with the use and structure of various personal protective equipment for low-voltage electricians;

② Be able to check all kinds of personal protective equipment for low-voltage electricians;

③ Be familiar with the maintenance requirements of personal protective equipment for low-voltage electricians.

Scoring criteria: see Table 9-2.

Table 9-2 Use of electrical safety appliances (examination time: 10 minutes)

No.	Exam Items	Exam Content	Score	Scoring Criteria
1	The use of personal protective equipment of low-voltage electrician	The use and structure of personal protective equipment	30	Orally present the role and using occasions of low-voltage electrician's protective equipment, three of the following (low-voltage electroscope (test pencil), insulated gloves, insulated shoes (boots), helmets, protective glasses, insulated pliers, insulating mats, portable grounding wire, foot buckles, safety belts, climbing pedal) are selected to be tested. 3~15 points would be deducted for errors in the description. Orally describe the structure of various high-voltage electricians' protective equipment, with 3~15 points deducted for any errors.
		Inspection of personal protective equipment	15	Correctly check the appearance, 5 points would be deducted if it is not checked. Failure to check the certificate, deduct 5 points. Failure to check the usability, deduct 5 points
		Proper use of personal protective equipment	40	Follow the safety procedures and use personal protective equipment correctly according to the operation steps. If operation steps violate the safety regulations, zero point would be scored, if steps are incomplete, 5~40 points would be deducted depending on the situation.
		Maintenance of personal protective equipment	15	Failure to correctly present the maintenance points of the selected PPE, deduct 1~15 points
2	Total		100	

[Training 9-9] Inspection and use of electrician insulated gloves

(1) Examination requirements

Be able to master the method of checking whether the electrician's insulating gloves are leaking, and be able to use the electrician's insulating gloves correctly.

(2) Training contents

1) Inspection of insulating gloves

Inspection of insulating gloves is divided into ex-factory inspection, test inspection and pre-use inspection. Before using, the certificate of conformity and appearance should be inspected. The following is the pre-use inspection of insulating gloves.

① Check the label and certificate of conformity to see if it is within the expiration date.

② Appearance inspection, with or without damage, burn marks, burrs, cracks, holes, etc.

③ Inflation experiment, blow gas into the gloves, grasp the glove opening with one hand, the other hand squeeze the glove, to observe whether there is air leakage. No air leakage proves that the glove is intact, air leakage proves that the glove is damaged.

2) Use requirements of insulating gloves

① Insulating gloves should be selected according to the use of high and low voltage, different protective conditions. In accordance with the relevant requirements of the "safety regulations", for equipment testing, reverse switch operation, installation and removal of grounding wire and other work, insulating gloves should be worn.

② when wearing insulating gloves, cuffs of work uniform should be tied, and the cuffs should be put into the insulating glove.

3) Precautions

① After use, the inside and outside dirt should be scrubbed clean, after being dried, the gloves should be sprinkled with talcum powder and be placed flat to prevent damage by pressure, do not put the gloves on the ground.

② After using it for half a year, preventive tests must be taken.

[Training 9-10] Inspection and use of tools for pole-climbing operations

(1) Examination requirements

Be able to master the inspection and use of safety equipment for pole-climbing.

(2) Training contents

1) Inspection and use of safety belt and foot-buckles

① Regulations for using safety belt

Every six months the safety belt should be taken a tensile test to ensure it is qualified (big belt 225kg, small belt 150kg).

Before using, check whether there is decay, brittle crack, aging, broken strands and other phenomena, whether all the hooks and rings are firm, whether the hole eye on the safety belt is open.

The hook and ring on the safety belt should be intact and have reliable insurance device to prevent automatic decoupling.

The safety belt should be tied in the reliable place, forbidden to be tied in the pole tip, cross-arms, porcelain vase, bumping board, wire and the parts to be dismantled.

The belt (small belt) should be tied at the top of the hip, under the electrician tools, the tightness should be moderate. After the safety belt is fastened, the hook and loop should be fastened and the safety device should be installed before leaning or leaning back. "Hear the sound and lean" is prohibited.

The safety belt protection should not be lost when changing position on the pole.

The nylon safety belt should not come in to contact with high temperature above 120℃, the open fire, the acid material, and the hard object with sharp angles, etc.

② Regulations for using foot buckles

The form of the foot buckles should be adapted to the material of the electric pole, climbing cement pole with wooden buckles is prohibited.

The size of the foot buckles should be adapted to the diameter of the pole, climbing "small" pole with large foot buckles is prohibited.

Check whether the foot buckles have fallen damaged, whether the opening is too large or too small; Foot buckles with distortion and deformation should not be continued to be used.

The small claw of the foot buckles should be flexible, and the bolt should not be loose, and the rubber should not be worn.

The rubber layer on the foot buckles should be free from aging, smoothness, shedding, wear, fracture and the phenomenon of strand damage.

The limit of belt hole on the foot buckles should be no cracking, serious wear or fracture.

The welding of the foot buckles pedal and the iron pipe should be no welding failure and fracture phenomenon.

The static tension test of the foot buckles should not be less than 100kg, and the test cycle is six months.

2) Pole climbing

① Electrician insulated shoes should be worn, and the tightness of the belt on foot buckles should be appropriate for pole climbing.

② Foot buckles should be moved by foot, after foot buckles touch tightly with the pole, make an effort to step down for pole climbing.

③ Two hands should hold both sides of the pole, the chest and the pole should be parallel and maintain a distance of about 20cm for pole climbing.

④ After arriving at the work position, the safety belt should be fastened, the foot buckles should not be crossed and stacked.

⑤ Use "small rope" and tool bag to transfer tools and equipment upward and downward of the pole. It is prohibited to throw up and down.

⑥ Having a guardian when necessary.

⑦ Don't be on the pole when hearing the sound of thunder; In case of on the pole, get down immediately.

【Special reminder】

After the work is completed, use the safety belt to have the pole inside (not too tight to move up and down), climb downward with left and right feet moving alternately (action essentials are opposite with the method of climbing the pole, move downward).

(3) Precautions

① The model of foot buckles is adapted to the diameter of the rod on site. Before pole climbing, climbing test with force at the root of the rod should be done, to determine whether the foot buckles have deformation and damage.

② It is strictly prohibited to throw foot buckles from high down.

③ When using foot buckles for pole climbing, use the safety belt for the full process protection.

④ When climbing down the pole and near the ground, before feet cannot touch the ground, do not jump directly from the foot buckles down.

⑤ During the training of climbing pole with foot buckle, a guardian should be assigned.

【Special reminder】

The use of low-voltage electroscope (test pencil), insulating shoes (boots), helmets, protective glasses, insulating clamps, insulating mats, portable grounding wire, etc., is relatively simple, please refer to the relevant content of Chapter 3 of this book to get trained, here it's not to be introduced.

9.3 Identify electrical safety signs (K13)

The examination content of electrician safety sign identification is to identify 5 safety signs randomly selected by the computer in the question bank.

Examination method: Oral.

Examination time: 10 minutes.

Safe operation steps.

① Be familiar with safety signs commonly used in low-voltage electrical operations;

② Be able to give a written description of the prepared safety signs;

③ Be able to arrange the relevant safety signs reasonably for the designated operation sites.

Scoring criteria, see Table 9-3.

Table 9-3　Identification of electrical safety signs

No.	Exam items	Exam content	Score	Scoring Criteria
1	Identification of commonly used safety signs	Familiar with commonly used safety signs	20	Identify the safety signs listed on the picture (5), 20 points for all correct, 4 points for each wrong
		Explanation of the use of commonly used safety signs	20	Could explain the use of the prepared safety signs (5) and explain their use, deduct 4 points for each wrong answer
		Correctly arrange safety signs	60	Correctly arrange the relevant safety signs (2) according to the prepared operation sites. Select the wrong sign, deduct 20 points, the wrong placement would be deducted 10 points.
2		Total	100	

[Training 9-11] Identification and hanging of electrical safety signs

(1) Examination requirements

Be able to identify the type of safety signs and master their use.

(2) Training content

1) Hang the signs

① Maintenance signs. When checking the lines, the safety warning sign "No switching on, someone is working on the line" should be selected and hung on the handle of line switch or knife switch.

When checking the equipment, the safety warning sign of "No switching on, someone is working" should be selected and hung on operating handle of the switch and knife switch that could be sent to the construction equipment once the switch is on.

Memorizing tips: Switching on the corresponding switch; Indicating the corresponding line, as shown in Fig. 9-5.

Fig. 9-5　Signs during maintenance

1—Switching on the corresponding switch; 2—Indicating the corresponding line

② Warning signs

The sign "Stop, high-voltage danger" should be hung in the following locations: on the aisle where passage is prohibited; on the high-voltage test site; on the shelter of the workplace adjacent to live equipment; on the fence of the outdoor workplace; on the beam of the workplace adjacent to live equipment; on the outdoor structure.

The sign "No climbing, high-voltage danger" should be hung in the following locations: on ladders of running transformer, pole frame adjacent area and less than 2.5m climbing ladders; another iron frame near the one that staffs go up and down to the work site; the line tower/pole in city or residential areas.

The two signs are shown in Fig. 9-6.

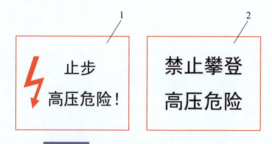

Fig. 9-6 Hanging of warning signs
1—Stop, high-voltage danger; 2—No climbing, high voltage danger

③ Permit signs

Sign "Work Here" is hung on outdoor and indoor operation sites of construction equipment.

Sign "Upwards and downwards from here" is hung on the iron frame or ladder where the staff is up and down. As shown in Fig. 9-7.

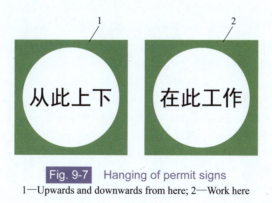

Fig. 9-7 Hanging of permit signs
1—Upwards and downwards from here; 2—Work here

④ Reminding signs

The sign "Grounded" is hung on the operating handle of the disconnecting switch that is already grounded; or on equipment where the grounding wire is invisible.

2) Precautions

① The installation location and height of the same type of equipment signs should be consistent, size and letterform of the signs should also be consistent.

② The permanent safety signs (including barriers and signs) must be firmly installed, reasonably located, and the safety distance from the charged part should be in accordance with the "Safety Regulations", and not affect the inspection and maintenance of the equipment. Warning signs should be eye-catching, and could really play a warning role.

Chapter 10

Safe operating technology (K2)

The subject two of practical test is "Safe operating technology", the test code is K2. Due to the actual differences in different provinces and cities, in some places the simulation equipment is currently used for the practical test for subject two, while there are places where actual equipment is used for the practical test. Therefore, readers who are preparing to register for the test should know the equipment configuration of the test center in advance so that they could train for the test targeted.

10.1 Practical examination platform and examination instructions

10.1.1 Cognition of practical examination platform

Low-voltage electrician practical intelligent network assessment system is divided into two specifications of A and B practical simulation boards. According to the control principle of motor start-up, starting self-locking, forward and reverse dual self-locking; the principle of daily single switch and two two-way switches to control lighting respectively; the principle of single switch to control single-phase five-hole 10A socket, to assess the students' correct wiring installation, the ability to find and troubleshoot the line, to achieve the purpose of simulation of actual operation.

Each test position is equipped with a multimeter, and a number of conducting wires with special socket and plug on both ends.

(1) Practical platform boards

The practical platform boards are shown in Fig. 10-1, which are divided into board A and B. The code of board A is odd number, and the code of board B is even number, each platform is connected quickly to the server of safe production assessment center of the province (municipalities directly under the central government, autonomous regions) through the network of the production to download test questions, and upload test results.

During examinations board A or B is randomly selected by the system. However, in the training, we should learn both of them.

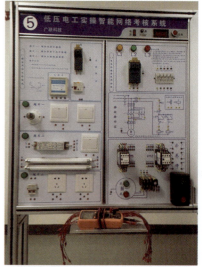

Fig. 10-1　Boards of practical platform

(2) Board A

The test content of board A is the simulation of problem analysis and troubleshooting of lighting wiring and motor drive circuit. Lighting wiring is divided into three topics.

Topic 1: Simulating two two-way switches to control a single incandescent lighting wiring, the circuit principle diagram is shown in Fig. 10-2.

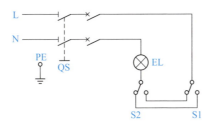

Fig. 10-2 Lighting wiring, circuit principle diagram of topic 1

Topic 2: Simulating a single switch to control fluorescent lighting wiring, the circuit principle diagram is shown in Fig. 10-3.

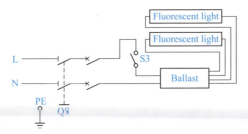

Fig. 10-3 Lighting wiring, circuit principle diagram of topic 2

Topic 3: Simulating the wiring of two single switches to control two 10A five-hole sockets respectively, circuit principle diagram as shown in Fig. 10-4.

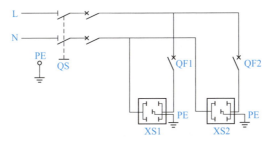

Fig. 10-4 Lighting wiring, circuit principle diagram of topic 3

Simulate the motor drive circuit to find the faults and troubleshoot, as shown in Fig. 10-5.

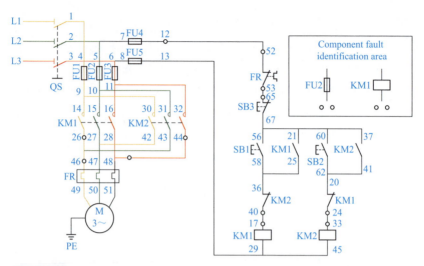

Fig. 10-5　Fault finding and troubleshooting schematic of motor drive circuit

(3) Practical platform board B

The test content of board B is three questions about motor drive circuit wiring, fault finding and troubleshooting of lighting line.

Topic 1: Simulate the wiring of the motor drive, the circuit principle diagram is shown in Figure 10-6.

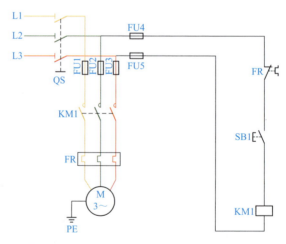

Fig. 10-6　Motor drive circuit schematic diagram

Topic 2: Simulate the starting the motor (self-locking), stop control wiring, as shown in Fig. 10-7.

Topic 3: Simulate the dual interlock of motor, forward and reverse starting and stop control wiring, as shown in Fig. 10-8.

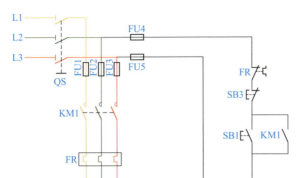

Fig. 10-7 Motor starting (self-locking), stop control wiring diagram

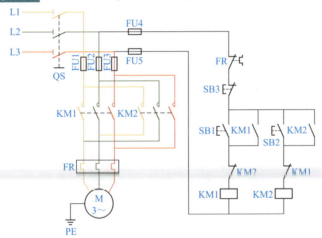

Fig. 10-8 Wiring diagram of dual motor interlock, forward and reverse start and stop control circuit

Simulate fault finding and troubleshooting of lighting line, as shown in Fig. 10-9.

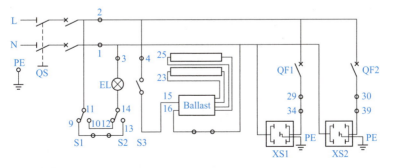

Fig. 10-9 Fault finding and troubleshooting diagram of lighting line

10.1.2 Instruction to practical examination

(1) Examination topics

The examination is divided into board A and board B; examiners are arranged randomly!

Board A1: Simulation of lighting circuit wiring (choose one from 3 topics)
Board A2: Troubleshooting of motor drive circuit
Board B1: Simulation of motor drive circuit wiring (choose one from 3 topics)
Board B2: Troubleshooting of lighting circuit.

(2) Examination process

1) Examination process of board A

① Cut off the circuit breaker → cut off the leakage switch → cut off the knife switch → wiring → switch on the leakage switch and circuit breaker → press the switch (start button) to take the power-on test.

② Cut off the knife switch → troubleshooting → switch on the knife switch → press the switch (start button) take the power-on test.

③ Click the "Submit " button.

④ Reset: Cut off the circuit breaker, leakage switch and the knife switch → unplug and sort out the connecting wire → switch on the knife switch, leakage switch and the circuit breaker → reset the multimeter.

2) Examination process of board B

① Cut off the knife switch → wiring → switch on the knife switch → press the switch (start button) take the power-on test.

② Cut off the circuit breaker → cut off the leakage switch → cut off the knife switch → troubleshooting → switch on the leakage switch and circuit breaker → press the switch (start button) take the power-on test.

③ Click the "Submit " button.

④ Reset: cut off the circuit breaker, leakage switch and the knife switch → unplug and sort out the connecting wire → switch on the knife switch, leakage switch and the circuit breaker → Reset the multimeter.

【Special reminder】

During Operation, power should be off, whether it is board A or board B, four switches disconnect together (order of power-off: circuit breaker → leakage switch → knife switch→ knife switch), and then start wiring and troubleshooting. After wiring and troubleshooting, supply power together(power-supply sequence: knife switch → knife switch → leakage switch → circuit breaker), and then click on "Submit", then reset.

(3) Precautions

① All wiring and measurement operations, must be carried out in the case of power off.

② In the measurement process of wiring and troubleshooting, "short-circuit operation" is strictly prohibited, which would burn the fuse. Fuse damage is considered unqualified.

③ Answer the question according to instructions for indicator light.

④ Strictly follow the correct order of cutting and supplying of power, " disconnect from the bottom to the top ", " supply power from top to bottom ".

⑤ During lighting troubleshooting, before measurement S3 switch should be opened for the pathway first.

⑥ During troubleshooting, measurement must be at the red jack, plug at the yellow jack for troubleshooting.

⑦ When wiring or troubleshooting is completed, all buttons (green, black, red) must be started and all switches must be switched on for power-on testing.

⑧ During the operation of lighting wiring and lighting troubleshooting, the circuit breaker of topic 3 must be cut off.

⑨ Do not touch the main power switch (red switch) above the test equipment.

⑩ When testing fault 26-46 on the electric drive circuit, please measure point 46 on the thermal relay directly (screw head on the thermal overload).

⑪ In the wiring and troubleshooting of sub-control socket, multimeter could be used to measure the state of power-on of the jack, voltage is 24V (This step could be omittted).

⑫ After the 2 topics are completed, the examiner should carefully double-check himself, confirm that there is no error, keep the power on, and click "Submit". The examination is all over and no operation is allowed on the drawing board of test bench.

【Special reminder】

Operate with electricity, the score would be zero. Before operation, a multimeter should be used to select to get at least 20 good wires.
When submitting, it should be kept that the wiring and troubleshooting are in the energized state.

The operation test for low-voltage electrical operation is mainly wiring and troubleshooting, consisting of six basic circuits. The score of the actual wiring operation and troubleshooting operation is 30 points for each, that is, only 48 points and above could pass the examination. The test results are automatically scored by the test software and instantly typed out by the printer.

During the test, please do not panic, pay attention to the operation steps. Since it is computerized scoring, every step of the operation is recorded. Don't do extra actions and try to finish the test in the way the trainer did. After the operation, please return the tools and wires to the right places.

10.2 Simulation equipment examination training project

Subject two, the simulation equipment examination includes four sub-projects: simulation of installation, wiring and troubleshooting of single-phase energy meter with lighting, simulation of installation, wiring, troubleshooting and other of control circuit of three-phase asynchronous motor, during the examination one question is selected randomly from these four sub-projects.

10.2.1 Wiring and safe operation of electric drive circuit

[Training 10-1] Wiring and safe operation of three-phase asynchronous motor one-way continuous running (with inching) (K21)

Examination method: Simulation operation, oral presentation.

Examination time: Operating simulation equipment 25 minutes.

Examination board configuration: The electrical components on the board should be configured to match the specifications, and their capacity should be able to meet the reasonable and normal use of the motor; the layout of the electrical components should be reasonable, firmly installed, and easy to wire.

(1) Training (examination) requirements

① Master the safety measures for the entire operation process, be skilled and standardized to use electrical tools for safe technical operation.

② Correctly wired in accordance with the circuit diagram, each control function could be realized; Correctly use a multimeter to measure the voltage in the circuit, and be able to read correctly.

③ Prepare for the test and start only after the examiner agrees.

④ Scoring criteria (see Table 10-1)

Table 10-1 Scoring table of motor one-way continuous running wiring (with inching control)

Exam Items	Examination content and requirements	Score	Scoring Criteria
Preparation before operation	The correct wearing of protective equipment	2	(1) 1 point would be deducted if the work uniform is not worn correctly (2) 1 point would be deducted if insulated shoes are not worn
Safety before operation	Inspection of safety hazards	4	(1) 2 points would be deducted for not checking whether there are safety hazards on the operating stations and platforms (2) 1 point would be deducted if the safety hazards on the operating platform were not dealt with (3) 2 points would be deducted if the damaged insulated wires or components on the operating platform are not pointed out
Safety of Operation process	Procedures of safe operation	11	(1) 5 points would be deducted if the electricity is turned on and off without the permission of the assessor (2) Deduct 5 points if the operation sequence of power on and off violates the safe operating procedures (3) If the operation of knife switch (or circuit breaker) is not standardized, deduct 3 points (4) If the candidate has unsafe behavior in the operation process, deduct 3 points
	Safe operation techniques	16	(1) If a component (or contact) is not connected, deduct 5 points (2) Inlet and outlet wiring of the fuses, circuit breakers, thermal relays are not standardized, deduct 3 points for each (3) If the color of start-stop control buttons is not standardized, deduct 5 points (4) If the color of insulated wire is not standardized, deduct 5 points (5) If the PE wire is not properly connected, deduct 3 points (6) If the wiring of the circuit board is not reasonable and standardized, deduct 2 points (7) If the three-phase load is not correctly connected, deduct 3 points (8) If the use of tools is not skilled or not standardized, deduct 2 points
Safety of post-operation	Safety inspection of the work site after operation	3	(1) 2 points would be deducted if the operating station is not cleaned or tidy (2) If the tools and instruments are not placed in a standardized manner, deduct 1 point (3) If the components is damaged, deduct 1 point

Exam Items	Examination content and requirements	Score	Scoring Criteria
The use of instrument	Measuring voltage with a pointer multimeter	4	(1) 4 points would be deducted if the candidate cannot use the multimeter or if it is used incorrectly (2) 2 points would be deducted for each time that do not know how to read the values
Time limit of examination	25 minutes	Items of points deduction	Every 1 minute overtime deduct 2 points, until 10 minutes overtime, terminate the entire practical project examination
Negative items	Explanation of negative item	Deduct the score of this question	The question will be scored as zero if one of the following occurs (1) The wiring principle is wrong (2) The circuit is faulty such as short circuit or damaged equipment (3) The function is not fully realized (4) Safety accidents appear in the process of operation
Total		40	

(2) Training (examination) operation steps

① Check if there are any safety hazards artificially set on the operating platform and the console, if there are, the safety hazards should be eliminated first.

② According to the sequence number marked as shown in Fig. 10-10, cut off the power in 4 steps in sequence. Note that incorrect order of power off would lead to deduction of points or failure.

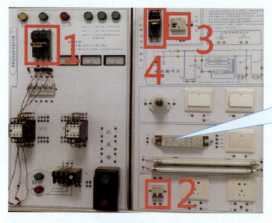

1. Turn off the "motor wiring" switch.
2. Turn off the circuit breaker of "sub control socket".
3. Turn off the circuit breaker of "lighting power".
4. Turn off the "lighting power" switch.
(Must operate in sequence)

Fig. 10-10　Sequence to cut off the power supply

③ Select the required electrical components and determine the wiring scheme on the already installed circuit board according to the given electrical schematic diagram as shown in Fig. 10-11.

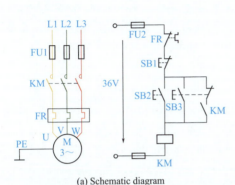

(a) Schematic diagram (b) Wiring diagram

Fig. 10-11 Wiring of motor one-way continuous running with inching

④ Complete the wiring the motor one-way continuous running (with inching control) as required, as shown in Fig. 10-12.

⑤ Use a meter to check the circuit before powering on to ensure that there are no safety hazards before power on.

⑥ Measure the voltage in the circuit with a multimeter.

⑦ Check whether the motor can achieve inching, continuous running and stopping.

⑧ Safety check of the work site after operation.

(3) Precautions

① Be sure to wire in the case of power off, and then power on to commission after wiring.

The recommended wiring order is: power supply → circuit breaker → contactor → thermal relay → motor.

② Don't miss the grounding wire PE at the motor case.

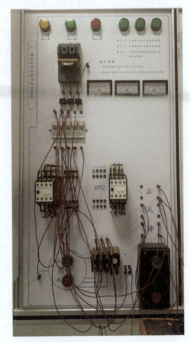

Fig. 10-12 Wiring of three-phase motor one-way continuous running with inching

[Training 10-2] Wiring and safe operation of three-phase asynchronous motor forward and reverse control circuit (K22)

Examination method: Simulation operation and oral presentation.

Examination time: 30 minutes of simulation equipment operation.

Examination board configuration: The electrical components on the board should be configured to match the specifications, and their capacity should be able to meet the reasonable and normal use of the motor; the layout of the electrical components should be reasonable, firmly installed, and easy to wire.

(1) Training (examination) requirements

① Master the safety measures for the entire operation process, be skilled and standardized to use electrical tools for safe technical operation.

② Correctly wire in accordance with the circuit diagram, the normal realization of each control function; Correctly use a multimeter to measure the voltage in the circuit, and be able to read correctly.

③ Prepare for the test and start only after the examiner agrees.

④ Scoring criteria (see Table 10-2)

Table 10-2 Scoring table of wiring for forward and reverse operation control of three-phase asynchronous motor

Exam Items	Examination content and requirements	Score	Scoring Criteria
Preparation before operation	The correct wearing of protective equipment	2	(1) 1 point would be deducted if the work uniform is not worn correctly (2) 1 point would be deducted if insulated shoes are not worn
Safety before operation	Safety inspection of operating stations and platforms	4	(1) 2 points would be deducted for not checking whether there were safety hazards on the operating stations and platforms (2) 1 point would be deducted if the safe hazards on the operating platform are not disposed of (3) 2 points would be deducted if the damaged insulated wires or components on the operating platform are not pointed out
Operation process safety	Safety operating procedures	11	(1) 5 points would be deducted if the power is turned on without the consent of the assessor (2) The operation sequence of power on and off violates the safety operating procedures, deduct 5 points (3) Operation of knife switch (or circuit breaker) is not standardized, deduct 3 points (4) If there is unsafe behavior in the operation process, deduct 3 points

Exam Items	Examination content and requirements	Score	Scoring Criteria
Operation process safety	Safe operating techniques	16	(1) If a component (or contact) is not connected, deduct 5 points (2) Inlet and outlet wiring of fuses, circuit breakers, thermal relays are not standardized, deduct 3 points each (3) If the color of start-stop control buttons is not standardized, deduct 5 points (4) If the color of insulated wire is not standardized, deduct 5 points (5) If the PE wire is not properly connected, deduct 3 points (6) If the wiring of the circuit board is not reasonable and standardized, deduct 2 points (7) If the three-phase load is not correctly connected, deduct 3 points (8) If the use of tools is not skilled or not standardized, deduct 2 points
Safety of post-operation	Safety inspection of the work site after operation	3	(1) 2 points would be deducted if the operating station is not cleaned or tidy (2) If the tools and instruments are not placed in a standardized manner, deduct 1 point (3) If the components is damaged, deduct 1 point
Use of instrument	Measuring the current in a three-phase motor with a pointer clamp meter	4	(1) 4 points would be deducted if the candidate cannot use the multimeter or if it is used incorrectly (2) 2 points would be deducted for each time that do not know how to read the values
Time limit of examination	30 minutes	Points deducted	Every 1 minute overtime deduct 2 points, until 10 minutes overtime, terminate the entire practical project examination
Negative items	Negative items description	Deduct the score of the question	The question would be scored as zero if one of the following occurs (1) The wiring principle is wrong (2) The circuit is faulty such as short circuit or damaged equipment (3) The function is not fully realized (4) Safety accidents appear in the process of operation
	Total	40	

(2) Training (examination) operation steps

① Check whether there are any artificial safety hazards on the operating station and the console, if there are, they should be eliminated first.

② Cut off the power in 4 steps according to the order marked as shown in Fig. 10-10. Note that incorrect order of power off would result in score deduction or failure.

③ In accordance with the given electrical schematic diagram as shown in Fig. 10-13, select the required electrical components on the already installed circuit board and determine the wiring scheme.

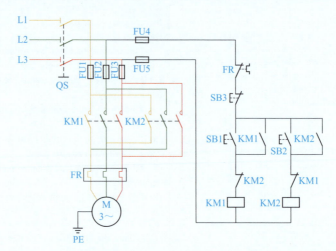

Fig. 10-13 Forward and reverse control circuit schematic of three-phase asynchronous motor

④ Complete the wiring of forward and reverse control circuit of the motor contactor interlock as required, as shown in Fig. 10-14.

⑤ Use the meter to check the circuit before powering on to ensure that there are no safety hazards before power on.

⑥ Check whether the motor can achieve forward and reverse rotation and stopping.

⑦ Use a clamp ammeter to detect the current of the motor in operation, and know how to read it correctly.

⑧ Check the safety of the work site after operation.

 【Special reminder】

When checking, the switch should be pressed in place. Whether it is forward to reverse or reverse to forward, press the stop switch and wait for the motor to stop running before reversing. Otherwise, it is easy to damage the motor.

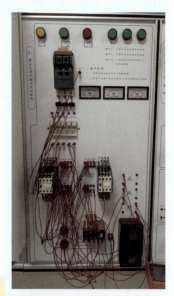

Fig. 10-14 Wiring of forward and reverse control of three-phase asynchronous motor

[Training 10-3] Wiring and safe operation of the control circuit for the operation of three-phase asynchronous motors (K23)

Examination method: Simulation operation and oral presentation.

Examination time: 25 minutes of simulation equipment operation.

Examination board configuration: The electrical components on the board should be configured to match the specifications, and their capacity should be able to meet the reasonable and normal use of the motor; the layout of the electrical components should be reasonable, firmly installed, and easy to wire.

(1) Training (examination) requirements

① Master the safety measures for the entire operation process, be skilled and standardized to use electrical tools for safe technical operation.

② Correctly wire in accordance with the circuit diagram, normal realization of each control function; Correctly use a multimeter to measure the voltage in the circuit, and be able to read correctly.

③ Be prepared to start the test only after the consent of the examiner.

④ Scoring criteria, see Table 10-1 for reference.

(2) Training (examination) operation steps

① Check whether there are any artificial safety hazards on the operation station and the console, if there are, the safety hazards should be eliminated first.

② According to the sequence number marked as shown in Fig. 10-10, cut off the power in 4 steps in sequence. Note that incorrect order of power off would result in score deduction or failure.

③ In accordance with the given electrical schematic diagram as shown in Fig. 10-15, select the required electrical components on the already installed circuit board and determine the wiring scheme.

④ Complete the start (self-locking) and stop control circuit for the motor as required, as shown in Fig. 10-16.

⑤ Use a meter to check the circuit before powering on to ensure that there are no safety hazards before power on.

⑥ Measure the voltage in the circuit with a multimeter.

⑦ Check whether the motor can achieve pointing, continuous running and stopping.

⑧ Safety inspection of the work site after operation.

(3) Precautions

① Be sure to wire in the case of power off, and then power on to commission after connecting the line.

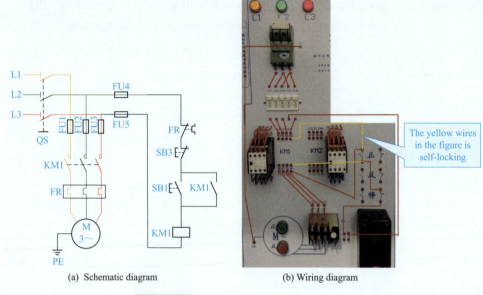

(a) Schematic diagram   (b) Wiring diagram

Fig. 10-15　Motor operation control circuit

The recommended wiring order is: power supply → circuit breaker → contactor → thermal relay → motor.

② Don't miss the grounding wire PE to the motor shell.

【Special reminder】

Both the motor inching and self-locking operation control circuits use buttons and contactors to control the motor to operate in a single direction. These two circuits are basically the same, the difference is that the self-locking operation control circuit uses the normal-open contact of the contactor itself to keep the contactor coil energized, which is called self-locking or self-protection, and the auxiliary normal-open contact is called self-locking (or self-protection) contact. It is because of the self-locking contact that the motor continues to run when SB1 is released, rather than running on point.

Fig. 10-16　Wiring of motor operation control

[Training 10-4] Troubleshooting of electric drive circuit (K22-1)

Examination method: Simulation operation, oral description.

Examination time: 25 minutes of simulation equipment operation.

(1) Training (examination) requirements

Troubleshoot the 2 preset faults in the forward and reverse control circuit of a three-phase asynchronous motor.

Use the multimeter at resistance range to find between the corresponding two numbers on the wiring diagram with the actual components, and then find the corresponding two numbers with wire connections between them (i.e., yellow terminal area) in the troubleshooting operation area.

Component search and troubleshooting is performed in the "Component Fault Identification Area".

(2) Training (examination) procedures

Disconnect the power supply, according to the fault phenomenon, combined with the circuit diagram of "troubleshooting operation area" shown in Fig. 10-17(b) to analyze the general scope of the fault, according to the eight possible points of failure provided in Table 10-5, detect with a multimeter whether the line or component is open (broken) circuit.

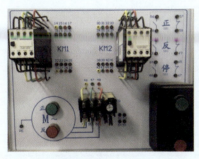

(a) Troubleshooting measurement area   (b) Troubleshooting operation area

Fig. 10-17 Troubleshooting and operation area of electric drive circuit

The test is to generate 2 random faults out of the 8 fault ranges. First, a multimeter is used to measure in the measurement area to find the fault point; then the two fault points are connected with test wires in the troubleshooting operation area. Priority is given to measure component faults, followed by measuring circuit faults.

The digital multimeter would be placed in the beeper range (pointer multimeter is according to the required range in Table 10-3), measure in the two terminals (red) of the fault point, if the buzzer does not beep, it means that the line is not working. At this moment, the corresponding number of yellow terminal in the troubleshooting operation area is connected with short circuit wires, as shown in Fig. 10-18. After the exclusion of the two faults, the power test circuit is normal, the result could be submitted.

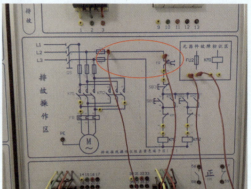

Fig. 10-18 Example of troubleshooting in short circuit with wire

Table 10-3 Possible fault points in the electric drive circuit

Fault point	Pointer multimeter range	Test points	Fault type	Description
Fault 1	R×1	26#—46#	Circuit fault	Directly measuring point 46 on the thermal relay (screw head on the thermal overload)
Fault 2	R×1	48#—44#	Circuit fault	—
Fault 3	R×1	12#—52#	Circuit fault	—
Fault 4	R×1	53#—65#	Circuit fault	—
Fault 5	R×1	40#—17#	Circuit fault	—
Fault 6	R×1	24#—33#	Circuit fault	—
Fault 7	R×1	5#—10#	Component fault	Measuring fuse FU2
Fault 8	R×10	17#—29#	Component fault	Measurement of KM1 coil

(3) Precautions

① Troubleshooting must be operated after the power supply is off.

② After the completion of troubleshooting, power on to commission whether the line is normal before submitting the results.

③ Fault detection is in the red terminal area, repair the fault in the yellow terminal area.

10.2.2 Wiring and safe operation of lighting circuit

On the test platform of the simulation equipment, the wiring of lighting circuits is divided into the following three sub-projects (K24-1 to k24-3), the computer randomly selects 1 out of them.

Examination method: Simulation operation, oral presentation.

Examination time: 20 minutes of simulation equipment operation; 30 minutes of physical equipment operation.

Examination circuit board configuration: configuration specifications of the electrical components on the board should be matched, its capacity should be able to meet the reasonable normal use of the motor; the layout of the electrical components should be reasonable, installed firmly, and easy to wire.

[Training 10-5] Wiring and safe operation of single-phase electric energy meter with dual control light (K24-1)

(1) Training (examination) requirements

① Master the safety measures in the whole operation process, be skilled and standardized to use electrical tools for safe technical operation.

② Correctly wiring in accordance with the circuit diagram, the lighting is lit, and the energy meter is operating properly; Use electrical instruments correctly, and read values correctly.

③ Make preparations for the test, start only after the consent of the assessor.

④ Scoring criteria (see Table 10-4)

Table 10-4 Scoring table of installation and wiring of single-phase energy meter with lighting

Exam Items	Examination content and requirements	Score	Scoring criteria
Preparation before operation	The correct wearing of protective equipment	2	(1) 1 point would be deducted if the work uniforms are not worn correctly. (2) 1 point would be deducted if insulated shoes are not worn
Safety before operation	Inspection of safety hazards	4	(1) 2 points would be deducted for not checking whether there are safety hazards on the operating stations and platforms. (2) 1 point would be deducted if the safety hazards on the operating platform were not disposed of. (3) 2 points would be deducted if the damaged insulated wires or components on the operating platform are not pointed out.
Operating safety	Safety operating procedures	11	(1) 5 points would be deducted if the electricity is turned on without the consent of the assessor (2) The operation sequence of power on and off violates the safe operating procedures, deduct 5 points (3) Operation of knife switch (or circuit breaker) is not standardized, deduct 3 points (4) Candidates have unsafe behavior in the operation process, deduct 3 points

Exam Items	Examination content and requirements	Score	Scoring criteria
Operating safety	Safe operating techniques	16	(1) 3 points would be deducted if in and out line of the power meter is incorrectly wired (2) 5 points would be deducted if the leakage circuit breaker is incorrectly wired (3) Incorrect position of the control switch installation, deduct 5 points (4) Socket wiring is not standardized, deduct 5 points (5) incorrect position of various types of load catch fire, deduct 3 points (6) The color of insulated line is not standardized, deduct 5 points (7) Working zero wire and protection zero wire are mixed, deduct 5 points (8) Wiring in circuit board is not reasonable or standardized, deduct 2 points (9) The PE wire is not correctly connected, deduct 3 points (10) It is not skilled or not standardized to use tools, deduct 2 points
Post-operation safety	Safety inspection of the work site after operation	3	(1) 2 points would be deducted if the operating station is not cleaned or tidy (2) Tools and instruments are not placed in a standardized manner, deduct 1 point (3) Damaging components, deduct 1 point
The use of instruments	Measurement of insulation resistance of motor with megger	4	(1) Unable to use megger or using incorrect methods, deduct 4 points (2) Deduct 2 points for each time that do not know how to read the values
Examination time limit	20minutes	Items of deducting points	Every 1 minute overtime deduct 2 points, until 10 minutes overtime, terminate the entire practical project examination
Negative items	Description of Negative items	Deduct the score of the question	The question would be scored as zero if one of the following occurs (1) The wiring principle is wrong (2) The circuit is faulty such as short circuit or damaged equipment (3) The function is not fully realized (4) Safety accidents appear in the process of operation
	Total	40	

(2) Training (examination) operation steps

① Check whether there are safety hazards (artificially set) on the operating station and the console, if there are, the safety hazards should be eliminated first.

② Cut off the power in accordance with the sequence number marked as shown in Fig. 10-19 in four steps. Note that incorrect order of power off would result in score deduction or failure.

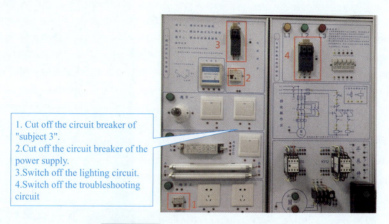

1. Cut off the circuit breaker of "subject 3".
2. Cut off the circuit breaker of the power supply.
3. Switch off the lighting circuit.
4. Switch off the troubleshooting circuit

Fig. 10-19 Steps of power cut off

③ According to the given electrical schematic diagram as shown in Fig. 10-20, select the required electrical components on the already installed circuit board and determine the wiring scheme. Insulating wires are available in two colors, and the color of the wires could be disregarded (the same for each of the following training requirements).

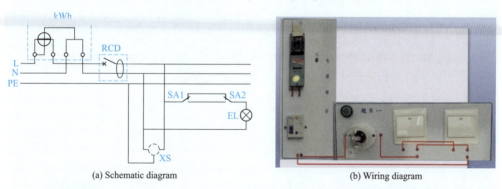

(a) Schematic diagram   (b) Wiring diagram

Fig. 10-20 Wiring diagram of single-phase energy meter with dual control light

④ Complete the installation and wiring of single-phase energy meter with lighting as required, as shown in Fig. 10-21.

a. The energy meter has four terminals, from left to right, the wiring rules are: 1 for live wire inlet, 2 for live wire outlet, 3 for neutral wire inlet, 4 for neutral wire outlet.

b. The switch is connected to the live wire in series to control the light and safety.

⑤ Use the meter to check the circuit before power on, to ensure that there are no safety hazards before powering on.

⑥ The lighting is lit and the energy meter is running normally. Check whether the single-phase leakage circuit breaker could play the role of leakage protection, and whether incandescent lamps could realize the role of dual control.

(3) Precautions

① Be sure to complete the wiring in the case of power off. Before power supplying, use the meter to check the wiring to ensure that there are no safety hazards before power supplying.

② Sequence of switching on: switch on "power" switch → switch on the "power" circuit breaker → switch on the circuit breaker in "topic 3" → switch on the switch in "topic 1" → light on.

After the above switching-on operation, if the light does not light up, please check the circuit after power off in accordance with the sequence.

③ The order of disconnection: circuit breaker → knife switch.

④ After the operation, the work site should be checked for safety.

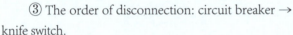

Fig. 10-21 Installation and wiring of single-phase energy meter with lighting

[Training 10-6] Wiring and safe operation of energy meter with single control fluorescent lighting (K24-2)

(1) Training (examination) requirements

① Master the safety measures in the whole operation process, skilled and standardized use of electrical tools for safe technical operation.

② Correctly wire in accordance with the circuit diagram, fluorescent lights are lit and the energy meter is operating normally; Correctly use the electrical instrument and be able to read the values correctly.

③ Be prepared, start the test only after the examiner agrees.

④ Scoring criteria, refer to Table 10-4.

(2) Operation steps of training (examination)

① Check whether there are safety hazards (artificially set) on the operating station and the console, if so, safety hazards should be eliminated first.

② Cut off the power in accordance with the sequence number shown in Fig. 10-19 in four steps. Note that incorrect order of power off would result in score deduction or failure.

③ In accordance with the given electrical schematic diagram as shown in Fig. 10-22, select the required electrical components on the already installed circuit board and determine the wiring scheme.

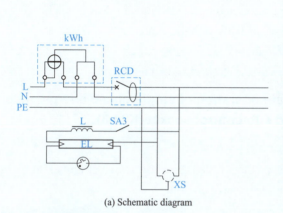
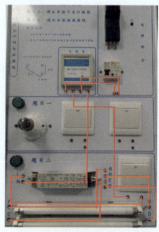

(a) Schematic diagram   (b) Wiring schematic

Fig. 10-22   Wiring of single control fluorescent light with energy meter

④ The installation and wiring of single control fluorescent light with single-phase energy meter should be carried out as required, as shown in Fig. 10-23.

a. The energy meter has four terminals, from left to right, the wiring rules are: 1 for live wire inlet, 2 for live wire outlet, 3 for neutral wire inlet, 4 for neutral wire outlet.

b. The switch is connected to the live wire in series to control the light and safety.

⑤ Use the meter to check the circuit before powering on to ensure that there are no safety hazards before power on.

⑥ The fluorescent lights are lit and the energy meter is running normally. Check whether the single-phase leakage circuit breaker could play the role of leakage protection.

Fig. 10-23   Installation and wiring of single control fluorescent light with single-phase energy meter

(3) Precautions

① Be sure to complete the wiring in the case of power off. Before power supplying, use the meter to check the wiring to ensure that there are no safety hazards before power supplying.

② The sequence of switching on and power supplying: switch on the "power" knife switch → switch on the "power" circuit breaker → switch on the "topic 3" circuit breaker → switch on the "topic 2" fluorescent light switch → fluorescent light is on.

After the above switching-on operation, if the fluorescent light does not light up, please check the wiring after powering off in accordance with the sequence.

③ The order of disconnection: circuit breaker → knife switch.

④ After the operation, the work site should be inspected for safety.

[Training 10-7] Wiring and safe operation of single-phase energy meter with switched sub-control socket (K24-3)

(1) Training (examination) requirements

① Master the safety measures in the whole operation process, and skillfully and standardly use electrician tools for safe technical operation.

② Correctly wire in accordance with the circuit diagram, the normal realization of each control function; Correctly use the electrical instrument and be able to read the values correctly.

③ Prepare for the test, start only after the consent of the examiner.

④ Scoring criteria, refer to Table 10-4.

(2) Operation steps of training (examination)

① Check whether there are safety hazards (artificially set) on the operation station and the console, if there are, safety hazards should be eliminated first.

② Cut off the power in 4 steps according to the sequence number as shown in Fig. 10-19. Note that incorrect order of power off would result in score deduction or failure.

③ In accordance with the given electrical schematic diagram as shown in Fig. 10-24, select the required electrical components on the already installed circuit board and determine the wiring scheme.

④ Complete the installation and wiring of the single-phase energy meter with switch to control the socket respectively as required, as shown in Fig. 10-25.

a. The energy meter has four terminals, from left to right, the wiring rules are: 1 for live wire inlet, 2 for live wire outlet, 3 for neutral wire inlet, 4 for neutral wire outlet.

b. The switch is connected to the live wire in series to control the light and safety.

c. Socket wiring for single-phase two-hole is left zero wire and right live wire, socket wiring for single-phase three-hole is left zero wire, live wire, and the upper grounding wire.

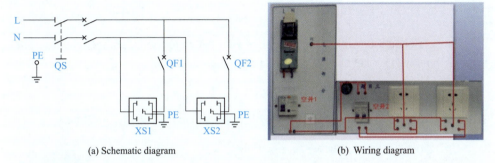

(a) Schematic diagram (b) Wiring diagram

Fig. 10-24 Wiring of single-phase energy meter with switches to control the socket respectively

⑤ Use the meter to check the circuit before power supplying to ensure there are no safety hazards before power on.

⑥ Use a test pencil to check if the socket is wired correctly; Use a multimeter to measure the voltage value of the socket. Check if the switch can control the power of the socket.

⑦ Check whether the single-phase leakage circuit breaker could play the role of leakage protection.

(3) Precautions

① Be sure to complete the wiring in the case of power off. Before power supplying, use the meter to check the wiring to ensure that there are no safety hazards before power supplying.

② The sequence of switching on and power supplying: switch on the "power" switch → switch on the "power" circuit breaker → switch on the "topic 3" circuit breaker.

After the above switching-on operation, if there is no electricity in the socket, please check the wiring after powering off in sequence.

Fig. 10-25 Installation and wiring of single-phase energy meter with switch to control the socket respectively

③ The order of disconnection: circuit breaker → knife switch.

④ After operation is completed, the work site should be checked for safety.

【Special reminder】

Sockets should be wired in accordance with the regulations, polarity could not be wrong. Three-hole socket must be connected to the grounding wire.

[Training 10-8]　Troubleshooting of lighting control circuit (K24-4)

(1) Training (examination) requirements

Examination, the energy meter with dual control incandescent circuit, energy meter with fluorescent light circuit, and two single-switch respectively control socket circuit, in each of the three circuits there are randomly 2 pre-determined faults.

Use the resistance range of multimeter to search between the corresponding two numbers in the wiring diagram with the actual components. After finding them, connect the corresponding two numbers with wire in the troubleshooting operation area (that is, the yellow terminal area).

Component search and troubleshooting is performed in the "Component fault identification area".

(2) Training (examination) procedures

Disconnect the power supply, according to the fault phenomenon, combined with the circuit diagram as shown in Fig. 10-26(b) of "troubleshooting operation area" to analyze the general scope of the fault. Referring to the 8 possible fault points provided in Fig. 10-26(a) and Fig. 10-26(c), use a multimeter to detect whether the circuit or component is open (broken), if open, then use the test wire to shortly circuit.

During the examination, 2 faults are randomly selected from these 8 faults. First, use the multimeter to measure the measurement area and find the fault points; then connect the two fault points with test wires in the troubleshooting operation area.

Put the digital multimeter in the beeping range (pointer multimeter according to the required range in Table 10-5), and measure at the two terminals (red) in the fault point, if the buzzer does not beep, it means the line is not working. At this moment, the corresponding yellow terminal in troubleshooting operation area is shortly connected with a test wire, as shown in Fig. 10-27(b). After the two faults, power on test circuit is normal, the results could be submitted.

Chapter 10 Safe operating technology (K2) 253

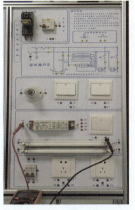

(a) Troubleshooting measurement area

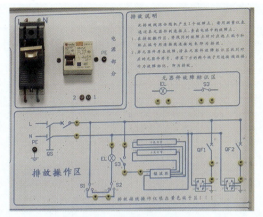

(b) Troubleshooting operation area

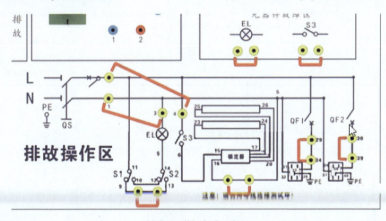

(c) 8 possible fault points

Fig. 10-26 Measurement area and operation area of lighting circuit troubleshooting

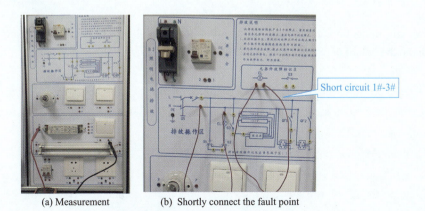

(a) Measurement (b) Shortly connect the fault point

Fig. 10-27 Troubleshooting method

Table 10-5 Possible fault points in the lighting circuit

Fault point	Range of pointer multimeter	Test points	Fault type	Description
Fault 1	$R\times 1$	2#—4#	Circuit fault	
Fault 2	$R\times 1$	1#—3#	Circuit fault	—
Fault 3	$R\times 1$	9#—13#	Circuit fault	—
Fault 4	$R\times 1$	5#—16#	Circuit fault	—
Fault 5	$R\times 1$	29#—34#	Circuit fault	—
Fault 6	$R\times 1$	30#—39#	Component fault	—
Fault 7	$R\times 10$	3#—5#	Component fault	Measurement of lighting EL
Fault 8	$R\times 1$	4#—6#	Circuit fault	Measure the switch S3 of the fluorescent light, S3 must be pressed when measuring

(3) Precautions

① Before troubleshooting, power supply must be off.

② After the troubleshooting is completed, the power is turned on to commission whether the circuit is normal before submitting the results.

③ Fault detection is in the red terminal area, and repairing the fault is in the yellow terminal area.

【Special reminder】

When inspecting, the priority is to measure the failure of components, and then measure the failure of circuit.

After submitting the results, the test equipment should be restored to the original state before the test, which is called the reset operation, as shown in Fig. 10-28. The steps and methods of reset operation are as follows:

① Power off in order.
② Unplug the test wire and place it neatly.
③ Reset the circuit breaker and the knife switch in the sequence of power on.
④ Reset the multimeter.
⑤ End of the test.

Fig. 10-28 The examination is over and the examination device is reset

10.3 Equipment examination training project

Equipment examination of subject 2, including five sub-projects: wiring and safety operation of three-phase asynchronous motor one-way continuous running (with inching control) (K21); forward and reverse control circuit wiring and safety operation of three-phase asynchronous motor(K22); wiring and safety operation of motor control circuit with fuse (circuit breaker), instrumentation, transformer(K23); wiring and safety operation of single-phase energy meter with lighting (K24); wiring and safety operation of indirect three-phase-four-wire active energy meter(K25). During examination, a question would be selected from them randomly.

10.3.1 Wiring and safe operation of electric drive circuit

[Training 10-9] Wiring and safe operation of one-way continuous running of three-phase asynchronous motor (with inching) (K21)

Examination method: Operation method of practical equipment.

Examination time: 35 minutes.

The configuration of the examination circuit board: specification configuration of the electrical components on the board should be matched, its capacity should be able to meet the reasonable normal use of the motor; the layout of the electrical components should be reasonable, installation should be fixed and reliable; the wiring length of the main and auxiliary circuits should be the same, and easy to wire.

(1) Training (examination) requirements

① Master the safety measures of the electrician before, during and after the operation.
② Skilled and standardized to use electrical tools for safe technical operation.
③ Correctly use common electrical instruments, and read the values.
④ Before the start of the practical test, the test site should be equipped with intact circuit boards, insulated wires and loads of various colors other test equipment and measuring instruments and tools in place, to ensure there aren't any safety hazards. After the consent of the examiner, the test could be started. If the equipment in the process of the test has a safety hazard or the fault could not be immediately eliminated, the test of this practical project is terminated, and the test site should be responsible for that.
⑤ Scoring criteria, see Table 10-6.

Table 10-6 Scoring table for wiring and safe operation of three-phase asynchronous motor of one-way continuous running (with inching) circuit

Exam items	Examination content and requirements	Score	Scoring criteria
Preparation before operation	The correct wearing of protective equipment	2	(1) 1 point would be deducted if the work uniforms are not worn correctly (2) 1 point would be deducted if insulated shoes are not worn
Safety before operation	Inspection of safety hazards	4	(1) 2 points would be deducted for not checking whether there are safety hazards on the operating stations and platforms (2) 1 point would be deducted if the safety hazards on the operating platform were not eliminated (3) 2 points would be deducted if the insulated wires or components on the operating platform are not damaged
Operating safety	Safe operating procedures	11	(1) 5 points would be deducted if the electricity is turned on and off without the consent of the assessor (2) The operation sequence of power on and off violates the safe operating procedures, deduct 5 points (3) Operation of knife switch (or circuit breaker) is not standardized, deduct 3 points (4) Candidates have unsafe behaviors in the operation process, deduct 3 points
	Safe operating techniques	16	(1) 1 point would be deducted for each wiring where the exposed copper exceeds the standard provisions (2) 2 points would be deducted for each loose crimp joint (3) The PE wire is not properly connected, deduct 3 points (4) The color of insulating wire is not standardized, deduct 5 points (5) Inlet and outlet wiring of fuses, circuit breakers, thermal relays is not standardized, deduct 2 points for each

Exam items	Examination content and requirements	Score	Scoring criteria
Operating safety	Safe operating techniques	16	(6) The wiring of circuit board is unreasonable, or not standard, deduct 2 points (7) The color of start-stop control buttons is not standardized, deduct 3 points (8) The three-phase load is not correctly connected, deduct 3 points (9) Irregular arrangement of terminals, deduct 1 point for each place (10) The use of tools is not skilled or not standardized, deduct 2 points
Post-operation safety	Safety inspection of the work site after operation	3	(1) 2 points would be deducted if the operating station is not cleaned or tidy (2) Tools and instruments are not placed in a standardized manner, deduct 1 point (3) Damaging the components, deduct 1 point
The use of Instrument	Measuring voltage with a pointer multimeter	4	(1) 4 points would be deducted if the multimeter cannot be used or if it is used incorrectly (2) Deduction of 2 points for each time that does not know how to read the values
Examination time limit	35minutes	Points deducted	Every 1 minute overtime deduct 2 points, until 10 minutes overtime, terminate the entire practical project examination
Negative items	Description of negative items	Deduct the score of the question	The question would be scored as zero if one of the following occurs (1) The wiring principle is wrong (2) The circuit is faulty such as short circuit or damaged equipment (3) The function is not fully realized (4) Safety accidents appear in the process of operation
	Total	40	

(2) Operation steps of the training (examination)

① Check whether there are safety hazards (artificial settings) on the operating station and the console, and be able to eliminate the safety hazards that exist.

② According to the electrical schematic diagram given as shown in Fig. 10-29, select the required electrical components on the installed circuit board and determine the wiring scheme.

③ Match the connection wires with different colors according to the given conditions.

④ Wire the three-phase motor for one-way continuous running (with inching control) circuit as required. After installation, it must be carefully checked to prevent wiring errors or omission of wiring from causing abnormal circuit operation or even short circuit accidents.

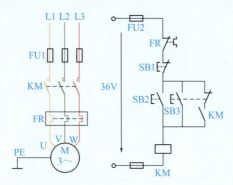

Fig. 10-29　Schematic of one-way continuous running (with inching) of three-phase asynchronous motor

a. Check the wiring. According to the electrical schematic diagram or electrical wiring diagram from the power end, check the circuit number of wiring and wiring terminals section by section, focusing on checking whether the main circuit has omissions, misconnections and easily wrong-connected line number in control circuit, also check whether the circuit number of both ends of the same wire is consistent.

b. Check whether the terminal wiring is solid. Check whether all the wiring crimps on the terminals are solid, whether the contact is good. Loose, or dropping is not allowed, to avoid failure caused by wire false connection when powering on and testing.

⑤ Use a meter to check the wiring before power supplying to ensure that there are no safety hazards before power supplying. When the control circuit is not energized, manually simulate the operation action of the appliance and use a multimeter to measure the on/off condition of the circuit. The inspection steps and contents should be determined according to the action of the control circuit, and the measurement points should be selected according to the schematic diagram and wiring diagram.

⑥ Check whether the motor can be operated, run continuously and stop. The steps of power-on are as follows.

a. Introduce power into the distribution board (note that it is not allowed to be introduced with electricity).

b. Switch on and test whether the power supply has power (test with a test pencil).

c. Operate the circuit according to the working principle. Check the function of the control circuit without the motor. Check the function of the main circuit, and check whether the motor runs normally, with the motor connected.

⑦ Use a pointer multimeter to test the voltage in the circuit and read the values correctly.

⑧ Safety inspection of the work site after operation.

(3) Circuit installation

① When installing the electrical components, do not install the components to the side of the control board, leaving a distance of at least 50mm. The leaving distance among components should be at least 50mm, not only for safety distance, but also convenient with routing.

② Before installing the electrical components, check whether the device is complete, whether there is damage, and whether the voltage and current of the contacts meet the requirements. Use resistance range of a multimeter to check whether the dynamic switching-on, dynamic breaking contacts and coil resistance of each component meet the requirements.

③ Wiring order is: Control circuit first, then main circuit, in the principle of not interfering with the subsequent wiring. Wiring is strictly prohibited to damage the core and wire insulation layer.

(4) Process requirements of wiring

① The routing of the wires is as close as possible to the components; Try to separate the phases by wire colors. The routing must be straight, neat and reasonable.

② the exposed wire is required to be horizontal and vertical, directly crossing between the wires should be avoided. The turning of wire should be 90° with a right angle with arc. Wiring could be bent with the help of the rod of a screwdriver, avoid directly bending with sharp pliers, etc., to avoid damage to the wire insulation.

③ Control circuit should be routed close to the control board, the main return wire could be "overhead routing" between the adjacent components of the short distance.

④ During open wiring on the board, the wiring channels should be as few as possible, parallel wires concentrate by classification according to the main, control circuit.

⑤ The connection wire of removable control button must be used with soft wire, and must be connected with components of power distribution board through the terminal, and numbered.

⑥ Routing of all wires must be continuous from one terminal to another terminal, and there must be no joints in the middle.

⑦ All the wires must be firmly connected, crimping with plastic layer is not allowed, and copper exposure should not exceed 3mm. For the wiring of wire and terminal: generally a terminal is connected to only one wire, not more than two.

⑧ The label of wire number should be consistent with the schematic diagram and wiring diagram. Each end of the wire must be covered with sleeves marked with a wire number. The location should be close to the terminal. The inverted number such as 6 and 9 or 16 and 91, must make a mark to distinguish, so as not to cause confusion with the wire number. The method of numbering wire should be in accordance with the relevant national standards.

⑨ The order of the installation of the wires is generally centered on the contactor from inside to outside, from low to high, first control circuit then the main circuit, in the principle of not interfering with subsequent wiring. For inlet and outlet wires of electrical components, they must be wired in the following principle: the above for the inlet wire, the below for the outlet wire, the left for the inlet wire, and the right for the outlet wire.

⑩ The center piece of spiral fuse should be connected to the incoming end, and the screw shell should be connected to the load side; The spare screws on the electrical appliances should be tightened.

⑪ If there are gaskets on terminals, flat gaskets should be placed on top of the connecting coil, and spring gaskets should be placed on top of the flat gaskets.

(5) Precautions

① Correctly select the button, green for starting, red for stopping, black for inching.

② Wear formal work uniform and insulated shoes, and be supervised when the circuit is energized.

③ After installation of the control circuit board, it must be carefully checked before powering on to test drive, so as to prevent abnormal line operation, or even short-circuit accidents caused by wiring errors or leakage of wiring.

[Training 10-10] Wiring and safe operation of forward and reverse control circuits of three-phase asynchronous motors (K22)

Examination method: operation method of practical equipment.

Examination time: 35 minutes.

Configuration of examination board: specification configuration of the electrical components on the board should be matched, and its capacity should be able to meet the reasonable normal use of the motor. The layout of the electrical components should be reasonable, and the installation should be fixed and reliable. The wiring length of the main and auxiliary circuits should be the same and easy to wire.

(1) Training (examination) requirements

① Master the safety measures of electricians before, during and after operation.

② Skilled and standardized to use electrical tools for safe technical operation.

③ Correctly use common electrical instruments, and could read the values.

④ Before the start of the practical test, the test site should put intact circuit boards, insulated wires and loads of various colors and other test equipment, measuring instruments and tools in place to ensure no existence of any safety hazards. After the consent of the examiner, the test could be started. If in the process of the examination, there are safety hazards of the examination equipment or the fault could not immediately be eliminated, the practical operation of the project examination is terminated, the consequences of which are the responsibility of the test center.

⑤ Scoring criteria, see Table 10-7.

Table 10-7 Scoring table of wiring and safety operation of forward and reverse control circuit of three phase asynchronous motor

Exam items	Examination content and requirements	Score	Scoring criteria
Preparation before operation	The correct wearing of protective equipment	2	(1) 1 point would be deducted if the work uniforms are not worn correctly (2) 1 point would be deducted if insulated shoes are not worn
Safety of pre-operation	Inspection of safety hazards	4	(1) 2 points would be deducted for not checking whether there are safety hazards on the operating stations and platforms (2) 1 point would be deducted if the safety hazards on the operating platform were not disposed of (3) 2 points would be deducted if the insulated wires or components on the operating platform are not damaged
Operating safety	Safe operating procedures	11	(1) 5 points would be deducted if the electricity is turned on and off without the consent of the assessor (2) The operation sequence of power on and off violates the safe operating procedures, deduct 5 points (3) Knife switch (or circuit breaker) operation is not standardized, deduct 3 points (4) Candidates have unsafe behavior in the process of operation, deduct 3 points
	Safe operating techniques	16	(1) 1 point would be deducted for each wiring where the exposed copper exceeds the standard provisions (2) 2 points would be deducted for each loose crimp joint (3) The PE wire is not properly connected, deduct 3 points (4) The color of insulating wire is not standardized, deduct 5 points (5) The wiring of inlet and outlet wires of fuses, circuit breakers, and thermal relays is not standardized, deduct 2 points each (6) The wiring of circuit board is unreasonable, or not standard, deduct 2 points

Exam items	Examination content and requirements	Score	Scoring criteria
Operating safety	Safe operating techniques	16	(7) The color of start-stop control buttons is not standardized, deduct 3 points (8) The three-phase load is not correctly connect, deduct 3 points (9) Irregular arrangement of terminals, deduct 1 point for each place (10) The use of tools is not skilled or not standardized, deduct 2 points
The use of Instrumentation	Safety inspection of the work site after operation	3	(1) 2 points would be deducted if the operating station is not cleaned or tidy (2) Tools and instruments are not placed in a standardized manner, deduct 1 point (3) Damaging the components, deduct 1 point
Time limit of examination	Measuring the current in a three-phase motor with a pointer clamp meter	4	(1) Unable to use clamp meter or the methods of using is incorrect, deduct 4 points (2) Unable to read the number, deduct 2 points for each
Negative items	40minutes	Points deducted	Every 1 minute overtime deduct 2 points, until 10 minutes overtime, terminate the entire practical project examination
The use of Instrument	Description of negative item	Deduct the score of the question	The question would be scored as zero if one of the following occurs: (1) The wiring principle is wrong (2) The circuit is faulty such as short circuit or damaged equipment (3) The function is not fully realized (4) Safety accidents appear in the process of operation
	Total	40	

(2) Operation steps of training (examination)

① Check whether there are safety hazards (artificially set) on the operating station and the console, and could eliminate the safety hazards that exist.

② According to the electrical schematic diagram shown in Fig. 10-30, select the required electrical components on the installed circuit board, and determine the wiring scheme.

③ Select the connecting wires of different colors according to the given conditions.

④ Wiring forward and reverse control wires of the three-phase asynchronous motor as required.

⑤ Use the meter to check the wiring before power supplying to ensure that there are no safety hazards before power supplying.

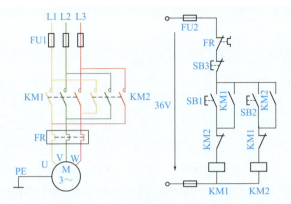

Fig. 10-30 Schematic of forward and reverse control circuit of three-phase asynchronous motor

When the control circuit is not energized, use manual to simulate the operation action of the appliance and use a multimeter to measure on/off condition of the circuit. The inspection steps and contents should be determined according to the action of the control circuit, and the measurement points should be selected according to the schematic diagram and wiring diagram.

⑥ Check whether the motor could achieve forward and reverse operation and stop.

⑦ Use a pointer clamp ammeter to detect the current of motor in operation and know how to read values correctly.

⑧ Safety inspection of the operating site after operation.

(3) Circuit installation

① When arranging electrical components, do not install the components to the side of the control board, the leaving distance should be at least 50mm; The leaving distance among components should be at least 50mm, not only for safety distance, but also for routing.

② Before installing the electrical components, check whether the device is complete, whether there is damage, and whether the voltage and current of the contacts meet the requirements; Check whether the dynamic switching-on, dynamic breaking contacts and coil resistance of each component meet the requirements with resistance range of a multimeter.

③ The wiring order is control circuit first, then main circuit, in the principle of not interfering with the subsequent wiring. The connection wire of the main circuit is generally 2.5mm^2 single-stranded plastic wire with copper core; The control circuit is generally 1mm^2 single-stranded plastic wire with copper core, and wires of different colors should be used to distinguish the main circuit, control circuit and grounding wire. Wiring installation is characterized by neat and beautiful lines, clear direction of the wire, easy to find faults.

(4) Precautions

① The process requirements of wiring is the same as training 10.9.

② Simple electrical control circuits could be wired directly; As for more complex electrical control circuits, it is recommended to draw electrical wiring diagrams before layout.

③ Wear formal work uniforms and insulated shoes, and be supervised when it is energized.

[Training 10-11] Wiring and safe operation of motor control circuits with fuses (circuit breakers), meters, and mutual inductor (K23)

Examination method: Equipment operation

Examination time: 30 minutes.

Examination board configuration: Specifications of the electrical components on the board should match the configuration, and their capacity should be able to meet the reasonable and normal use of the motor. The layout of the electrical components should be reasonable, and the installation should be fixed and reliable. The wiring length of the main and auxiliary circuits should be the same, and it should be easy to be wired.

(1) Training (examination) requirements

① Master the safety measures of electricians before, during and after operation.

② Skilled and standardized to use electrical tools for safe technical operation.

③ Correctly use common electrical instruments, and could read the values correctly.

④ Before the practical test, the test site should be equipped with intact circuit boards, insulated wires and loads of various colors and other test equipment and measuring instruments and tools in place, to ensure no existence of any safety hazards. After the consent of the examiner, the test could be started. If the test equipment in the process of the test has a safety hazard or a fault which could not immediately be eliminated, the test of this practical project is terminated, and the consequences of which are the responsibility of the test site.

⑤ Scoring criteria see Table 10-8.

Table 10-8 Scoring table of wiring of motor control circuit with fuses (circuit breakers), meters, and mutual inductor

Exam items	Examination content and requirements	Score	Scoring criteria
Preparation before operation	The correct wearing of protective equipment	2	(1) 1 point would be deducted if the work uniforms are not worn correctly (2) 1 point would be deducted if insulated shoes are not worn

Exam items	Examination content and requirements	Score	Scoring criteria
Safety before operation	Inspection of safety hazards	4	(1) 2 points would be deducted for not checking whether there are safety hazards on the operating stations and platforms (2) 1 point would be deducted if the safety hazards on the operating platform were not disposed of (3) 2 points would be deducted if the damaged insulated wires or components on the operating platform are not pointed out
Operating safety	Safe operatingty procedures	11	(1) 5 points would be deducted if the electricity is turned on and off without the consent of the assessor (2) The operation sequence of power on and off violates the safe operating procedures, deduct 5 points (3) The operation of knife switch (or circuit breaker) is not standardized, deduct 3 points (4) The candidates have unsafe behavior in the operation process, deduct 3 points
Operating safety	Safe operating techniques	16	(1) 1 point would be deducted for each wiring where the exposed copper exceeds the standard provisions (2) 2 points would be deducted for each loose crimp joint (3) The PE wire is not properly connected, deduct 3 points per place (4) The color of insulating wire is not standardized, deduct 5 points (5) The wiring of inlet and out let of the fuses, circuit breakers, thermal relays is not standardized, deduct 2 points per place (6) The wiring of circuit board is not reasonable or not standardized, deduct 2 points (7) The color of start-stop control buttons is not standardized, deduct 3 points (8) The safety installation of the mutual inductor is not correct, deduct 1 point (9) The wiring of mutual inductor, ammeter is incorrect, deduct 2 points per location (10) not correctly connect the three phase load, deduct 3 points (11) The arrangement of terminal is not standardized, deduct 1 point per place (12) The use of tools is not skilled or not standardized, deduct 2 points
Post-operation safety	Safety inspection of the work site after operation	3	(1) 2 points would be deducted if the operating station is not cleaned and tidy (2) Tools and instruments are not placed in a standardized manner, Deduct 1 point (3) Damaging the components, deduct 1 point

Exam items	Examination content and requirements	Score	Scoring criteria
The use of Instrument	Measuring voltage with a pointer multimeter	4	(1) 4 points would be deducted if candidates cannot use the multimeter or if it is used incorrectly (2) Deduct 2 points for each time that do not know how to read the numbers
Time limit of examination	30minutes	Points deducted	Every 1 minute overtime deduct 2 points, until 10 minutes overtime, terminate the entire practical project examination
Negative items	Description of negative item	Deduct the score of the question	The question would be scored as zero if one of the following occurs (1) The wiring principle is wrong (2) The circuit is faulty such as short circuit or damaged equipment (3) The function is not fully realized (4) Safety accidents appear in the process of operation
	Total	40	

(2) Operation steps of Training (examination)

① Check whether there are safety hazards (artificially set) on the operating station and the console, and could eliminate the safety hazards that exist.

② According to the electrical schematic diagram as shown in Fig. 10-31, select the required electrical components on the installed circuit board and determine the wiring scheme.

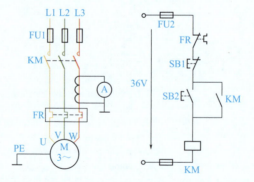

Fig. 10-31 Schematic diagram of motor control circuit with fuse (circuit breaker), meter and mutual inductor

③ Match the connection wires with different colors according to the given conditions.

④ Wire the motor control circuit with fuse (circuit breaker), meter and current mutual inductor as required. K1 and K2 on the secondary side of the mutual inductor enter the ammeter through the current terminal.

⑤ Use the meter to check the circuit before power supplying to ensure there are no safety hazards.

⑥ Check whether the motor could start and stop, and whether the ammeter could indicate during continuous running.

⑦ Use a pointer multimeter to detect the voltage in the circuit and could read the values correctly.

⑧ Safety check of the operation site after operation.

(3) Precautions

① Requirements of wiring process are the same as training 10-9.

② When connecting the current mutual inductor, attention should be paid to the polarities of the primary side, secondary side, and the terminals with the same number should correspond, a wrong connection is not permitted.

③ During installation, if the pointer of current meter does not point to the "0" position, the mechanical zero knob should be adjusted, so that the pointer is in the zero position.

10.3.2 Wiring and safe operation of lighting circuit

[Training 10-12] Wiring and safe operation of single-phase energy meters with lighting (K24)

Examination method: Practical operation.

Examination time: 30 minutes.

(1) Requirements of training (examination)

① Master the safety measures of electrician before, during and after the operation.

② Skilled and standardized to use electrical tools for safe technical operation.

③ Correctly use common electrical instruments, and could correctly read the values.

④ Before the practical test, the test site should put intact circuit boards, insulated wires and loads of various colors and other test equipment and measuring instruments and tools in place, to ensure the existence of any safety hazards, only after the consent of the examiner, the test could be started. If the test equipment in the process of the test has a safety hazard or a fault that could not immediately be eliminated, the test of this practical project is terminated, the consequences of which are the responsibility of the test site.

⑤ Scoring criteria, see Table 10-9.

Table 10-9 Scoring table of single-phase energy meter with lighting wiring

Exam items	Examination content and requirements	Score	Scoring criteria
Preparation before operation	The correct wearing of protective equipment	2	(1) 1 point would be deducted if the work uniforms are not worn correctly (2) 1 point would be deducted if insulated shoes are not worn
Safety before operation	Inspection of safety hazards	4	(1) 2 points would be deducted for not checking whether there are safety hazards on the operating stations and platforms (2) 1 point would be deducted if the safety hazards on the operating platform were not disposed of (3) 2 points would be deducted if the damaged insulated wires or components on the operating platform are not pointed out
Operating safety	Safe operating procedures	11	(1) 5 points would be deducted if the electricity is turned on and off without the consent of the assessor (2) The operation sequence of power on and off violates the safe operating procedures, deduct 5 points (3) Operation of knife switch (or circuit breaker) is not standardized, deduct 3 points (4) Candidates have unsafe behavior in the operation process, deduct 3 points
Operating safety	Safe operating techniques	16	(1) 3 points would be deducted for incorrect inlet and outlet wires of the energy meter (2) 2 points would be deducted for each incorrect meter crimp connector (3) Incorrect position of installation of the control switch, deduct 5 points (4) Wiring errors of leakage circuit breaker, deduct 5 points (5) Wiring of the socket is not standard, deduct 5 points (6) The PE wire is not correctly connected, deduct 3 points (7) The working zero wire and protection zero wire are mixed, deduct 5 points (8) Exposed copper of wiring is beyond the standard provisions, deduct 1 point per location (9) For loose crimping, deduct 2 points per place (10) Wiring of circuit board is unreasonable or non-standard, deduct 2 points (11) The color of insulating wire is not standardized, deduct 5 points (12) The arrangement of the terminals is not standardized, deduct 1 point per place (13) The use of tools is not skilled or not standardized, deduct 2 points

Exam items	Examination content and requirements	Score	Scoring criteria
Post-operation safety	Safety inspection of the work site after operation	3	(1) 2 points would be deducted if the operating station is not cleaned and tidy (2) Tools and instruments are not placed in a standardized manner. Deduct 1 point (3) Damaging the components, deduct 1 point
The use of instrumentation	Measuring insulation resistance of motor with megohmmeter	4	(1) Unable to use megger or use incorrect methods, deduct 4 points (2) 2 points of deduction for each if you do not know how to read the numbers
Time limit of the examination	30minutes	Points deducted	Every 1 minute overtime deduct 2 points, until 10 minutes overtime, terminate the entire practical project examination
Negative items	Description of negative item	Deduct the score of the question	The question would be scored as zero if one of the following occurs (1) The wiring principle is wrong (2) The circuit is faulty such as short circuit or damaged equipment (3) The function is not fully realized (4) Safety accidents appear in the process of operation
	Total	40	

(2) Operation steps of the training (examination)

① Check whether there are safety hazards (artificially set) on the operating station and the console, and could eliminate the safety hazards that exist.

② According to the electrical schematic diagram shown in Fig. 10-32, select the required electrical components on the installed circuit board, and determine the wiring scheme.

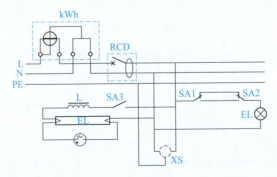

Fig. 10-32 Wiring schematic diagram of single-phase energy meter with lighting

③ Match the connection wires with different colors according to the given conditions.

④ Wiring the circuit of single-phase energy meter with lighting as required.

a. The energy meter has 4 terminals, from left to right, the wiring rules are: 1 for the inlet live wire, 2 for the outlet live wire, 3 for the inlet zero wire, 4 for the outlet zero wire.

b. Connect the switch in series with the live wire, in order to control the light and safety.

c. The live wire is connected with the center contact of the screw lamp holder.

d. Wiring in accordance with the power and load signs on the leakage circuit breaker, do not reverse the two connections.

⑤ Use the meter to check the wiring before power supplying to ensure that there are no safety hazards.

⑥ Check whether the single-phase leakage circuit breaker could play a role in leakage protection, incandescent lamps could achieve dual-control role, and fluorescent light (or LED lamps) could achieve single-control role, etc.

⑦ Use megger to test the insulation of the three-phase motor, and could read the values correctly.

⑧ The safety inspection of the work site after operation should be completed.

(3) Precautions

① Requirements of wiring process are the same as training 10-9.

② After the energy meter is installed, close the isolation switch and turn on the electrical equipment, the dial would rotate from left to right (the dot indicates that there is usually a turning indication icon). After turning off the electrical equipment, the dial would sometimes have a slight rotation, but no more than one turn is normal.

③ During installation, the neutral wire must be strictly distinguished from protective grounding wire. The protective grounding wire must not be connected to the leakage circuit breaker.

④ Operate the test button to check whether the leakage circuit breaker could operate reliably. In general, it should be tested more than 3 times, to ensure all could act normally.

[Training 10-13] Wiring and safe operation of indirect three-phase-four-wire active energy meter (K25)

Examination method: Equipment operation

Examination time: 30 minutes.

(1) Requirements of the training (examination)

① master the safety measures for the entire operation process, skilled and standardized to use electrical tools for safe technical operation.

② Wiring correctly according to the schematic diagram, power on for normal operation. Correctly use a multimeter to measure the voltage in the circuit, and be able to read the values correctly.

③ Three-phase load could be replaced by a three-phase asynchronous motor or by a combination of three light bulbs.

④ Be prepared to start the test only after the examiner agrees.

⑤ Scoring criteria, see Table 10-10.

Table 10-10 Scoring table of wiring of indirect three-phase-four-wire active energy meter

Exam items	Examination content and requirements	Score	Scoring criteria
Preparation before operation	The correct wearing of protective equipment	2	(1) 1 point would be deducted if the work uniforms are not worn correctly (2) 1 point would be deducted if insulated shoes are not worn
Pre-operation Safety	Inspection of safety hazards	4	(1) 2 points would be deducted for not checking whether there are safety hazards on the operating stations and platforms (2) 1 point would be deducted if the safety hazards on the operating platform were not disposed of (3) 2 points would be deducted if the damaged insulated wires or components on the operating platform are not pointed out
Operation process Safety	Safe operating procedures	11	(1) 5 points would be deducted if the electricity is turned on and off without the consent of the assessor (2) The operation sequence of power on and off violates the safe operating procedures, deduct 5 points (3) Operation of knife switch (or circuit breaker) is not standardized, deduct 3 points (4) Candidates have unsafe behavior in the operation process, , deduct 3 points
	Safe operating techniques	16	(1) Wiring error of the inlet and outlet wires of the three-phase power meter, deduct 3 points (2) Deduct 2 points for each crimping connection error of three-phase energy meter (3) Primary wiring and secondary wiring of mutual inductor is not standardized, deduct 2 points per each error (4) Wiring error of inlet and outlet wires of the circuit breaker, deduct 2 points (5) Primary wiring and secondary wiring are mixed, deduct 5 points (6) The three-phase load is not correctly connected, deduct 4 points

Exam items	Examination content and requirements	Score	Scoring criteria
Operation process Safety	Safe operating techniques	16	(7) The PE wire is not correctly connected, deduct 3 points (8) The working zero wire and protective zero wire are mixed, deduct 5 points (9) Wiring of circuit board is not reasonable or not standardized, deduct 2 points (10) Arrangement of wiring terminals is not standardized, deduct 1 point per place (11) Exposed copper of wiring is beyond the standard provisions, deduct 1 point for each (12) Wiring is loose, deduct 2 points each (13) The color of insulating wire is not standardized, deduct 5 points for each (14) The use of tools is not skilled or not standardized, deduct 2 points
Post-operation safety	Safety inspection of the work site after operation	3	(1) 2 points would be deducted if the operating station is not cleaned and tidy (2) Tools and instruments are not placed in a standardized manner. Deduct 1 point (3) Damaging the components, deduct 1 point
The use of instrument	Measurement of insulation resistance of motor with megger	4	(1) Unable to use The megger or use in incorrect methods, deduct 4 points (2) Deduct 2 points for each time that does not know how to read the numbers
Examination time limit	30minutes	Points deducted	Every 1 minute overtime deduct 2 points, until 10 minutes overtime, terminate the entire practical project examination
Negative items	Negative item description	Deduct the score of the question	The question would be scored as zero if one of the following occurs (1) The wiring principle is wrong (2) The circuit is faulty such as short circuit or damaged equipment (3) The function is not fully realized (4) Safety accidents appear in the process of operation
Total		40	

(2) Operation steps of the training (examination)

① According to the wiring schematic diagram of the three-phase active meter via current mutual inductor as shown in Fig. 10-33, select the appropriate components and insulated wires.

② The components should be reasonably installed on the distribution board as required, and the indirect three-phase-four-wire active energy meter should be correctly wired.

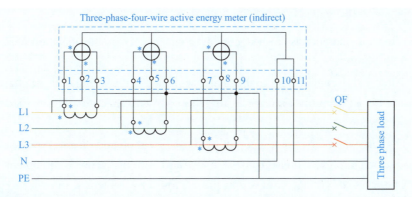

Fig. 10-33 Wiring diagram of three phase active energy meter through current mutual inductor

1, 4, 7 of the energy meter are connected to the S1 end of the secondary side of the current mutual inductor, that is, the current inlet end; 3, 6, 9 are connected to the S2 end of the secondary side of the current mutual inductor, that is, the current outlet end; 2, 5, 8 are connected to the three-phase power supply; 10, 11 are connected to the zero end, as shown in Fig. 10-34. For safety, S2 end of the current mutual inductor should be connected and grounded.

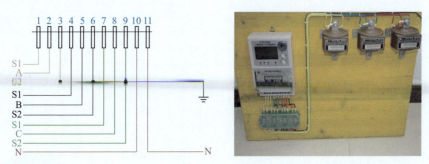

Fig. 10-34 Wiring diagram of indirect three-phase-four-wire active energy meter

Open circuit is not allowed on the secondary side of current mutual inductor. In the case of the secondary open circuit, serious consequences would occur: One is overheating of iron core, and even burning the mutual inductor; Secondly, due to the number of turns of the secondary winding, it would induce a dangerous high voltage, endangering the safety of people and equipment.

③ Three-phase load could be replaced by a three phase asynchronous motor or a combination of three light bulbs.

④ Check whether the wiring with the same name end of 3 current mutual inductors and three-phase-four-wire active energy meter is correct.

⑤ Use the meter to check the wiring before power supplying to ensure there are no safety hazards.

⑥ When the operation is finished, conduct a safety check on the operation site.

(3) Precautions

① Attention should be paid to the phases sequence of the wiring of three phase power meter. The wire of the mutual inductors wiring should be copper-core wire with diameter not be less than 2.5mm^2.

② The zero wire of the power supply could not be cut off and directly connected to the user's load switch to prevent breaking the zero wire and burning the user's equipment.

③ The screw of voltage connection piece should be tightened, in order to prevent loosening or to result in voltage loss.

④ Check the wiring to ensure it is correct. The joints should be solid, the contact should be good, and there should be no false connection. Make the power meter vertical to the ground when power is on.

Chapter 11

Elimination of potential safety hazards at the operation site (K3)

Low Voltage Electrician Certification Training Course

The subject three of practical test is "elimination of potential safety hazards at the operation site ", its test code is K3, and mainly includes the troubleshooting of potential safety hazards of power distribution cabinet, safety risk identification and hidden danger analysis of workshop distribution box, safety risk identification and hidden danger analysis of the electrical control room.

[Training 11-1]   Analysis of the safety risks and occupational hazards existing in the operation site (K31)

(1) Examination method
Oral presentation or written examination

(2) Safe operation steps
① Carefully understand the pictures or videos of the work site provided by the examiner.

② Point out the safety risks and occupational hazards that exist, which might involve the following:

　　a. Personal protective measures are not done properly when working on site.

　　b. Unsafe wire laying or power utilization at the operation site.

　　c. At operation site there are no appropriate safety signs: for example, during equipment maintenance, on switch operation handle the sign "No switching on, someone is working" is not hung.

　　d. No safety zone is planned for live equipment, and the sign "Stop, high voltage danger!" is not hung.

　　e. There are operation errors during switching operation.

　　f. Improper methods of handling emergency.

　　g. The tools are disorderly placed at the work site.

(3) Scoring criteria (see table 11-1)

Table 11-1   Scoring table of judging the safety risks and occupational hazards at the operation site

Examination items	Examination content	Score	Scoring criteria
Determine the safety risks and occupational hazards that exist at the operation site	Observe the operation site, pictures or videos to clarify the operation task or electricity environment	25	By observing the operation site, pictures or videos, orally describing the tasks or electricity environment. 25 points if correct, 5~25 points deduction if incorrect
	Judgment of safety risk and occupational hazard	75	Oral description of the existence of safety risks and occupational hazards, 15 points would be scored for each correct finding
Total		100	

(4) Training content

During the examination, any three pictures in Table 11-2 are usually combined to ask candidates to point out the potential safety hazards and occupational hazards in them.

Table 11-2　Examples of inspection of low-voltage electrician's hidden danger

No.	Pictures	Safety hazards (reference answer)
1		(1) No protective door, easy to cause accidents of electric shock (2) Some switches have more wires per phase (in principle, no more than 2), which could easily cause the contact to heat up and cause burnout accidents (3) The wire of no use is not completely removed or intact wrapping, the wire head is put towards the edge to prevent misuse (4) The wiring is messy
2		(1) Round plug is prohibited in China. Because it has the possibility of wrongly insert, and causing accidents of electric shock (2) The lead wire sheath is not into the plug, which makes it easy to break at the junction and causing electric shock (3) Three-phase equipment is not connected to the protection wire. Once the equipment has leakage, it would cause accidents of electric shock
3		(1) The wire pipe erection is not in place, and no fixed with a clamp. The wire protection is missing, once there are wear and leakage, it would cause electrocution accidents (2) The power distribution board is used as hanging shelves, miscellaneous shelves, which easily leads to electrocution and fire accidents (3) Improper selection of mobile switch, which would easily lead to damage and electrocution in frequent operation
4		(1) The wires are disordered, according to speculation it should be zero wire bar. According to the regulations, light blue should be used, but yellow and dark blue are in use, which is easy to cause wiring errors, and cause accidents (2) Copper nose is not used at the joint, and more than 2 wires on each joint, which is easy to cause the joints heat and lead to accidents
5		(1) The power distribution box is a workshop power box, and according to the regulations, it should not be set together with flammable materials, work stations and rest positions, which is prone to fire and personal safety accidents (2) The button board is suspended too high, once an accident occurs, it cannot be controlled in time

No.	Pictures	Safety hazards (reference answer)
6		(1) The wire diameter is too small, exposed copper could be found at the wire head. Copper nose is not used for wiring, the color of wire is messy, and the wires are disordered (2) The disconnected wires are not lined up to the edge (3) It is easy to cause misconnection, electric shock, leakage accidents
7		(1) The welders are too close to each other, poorly ventilated, it is easy to overheat (2) The spacing between the welders and the distribution box is less than 1m, so it is impossible to stop the machine in time in case of an accident
8		(1) Directly inserting the wire into the socket is easy to cause dislodgement, poor contact, heat, prone to electric shock, burn accidents (2) The lead wires do not meet the regulations. The sheath does not insert into the box, which is easy to cause wear and leakage, and endangers the safety of personnel
9		(1) The location of the distribution box is improper, near the pool, which is prone to electrocution accidents caused by leakage of electricity (2) The distribution box is not closed and locked, which does not conform to the management regulations
10		(1) The distribution box should be zero protected, after changing, the line should be sorted up, while it is not. The grounding protection contact on door is too long and easy to break, thus losing protection once the leakage of electricity appears, and would cause electric shock accidents (2) The color of the wire is confusing

No.	Pictures	Safety hazards (reference answer)
11		The inlet and outlet wires are not standardized, stuffs and equipment are piled up in front of power distribution cabinet, which is prone to fires and affects emergency handling
12		The protective tube of the inlet wire is not in place and there is no pipe clamp. The inlet wire of the socket should be three. There is no protective zero wire and the wire color is wrong. Once electric leakage occurs, electric shock accidents are likely to occur
13		Disconnect the power supply of the emergency light without permission and use it for other purposes. Once an accident occurs, it cannot play its due role
14		The equipment and connection points are oxidized and rusted without maintenance. It is easy to cause accidents due to poor contact. Line color confusion
15		This is a five strand sheathed wire, which is used to transmit three-phase-five-wire power supply. It uses two plugs (one is three holes, one is four holes) to transmit the above power supply. It might have the possibility of current mismatch, missed plug, incorrect plugging or no plugging; It also violates the stipulation that no equipment should be installed on the PE wire of the N wire. It is easy to cause burning of plugs and sockets. Once the equipment leaks electricity, it will cause electric shock if the grounding socket is not plugged
16		The residual current circuit breaker has the wrong wiring (no input and output live wires; the output terminal and N terminal wiring do not conform to the regulations, it should be connected to the terminal block of the output N wire, and the wire color is wrong), which does not play a protective role; There are too many contact wires and copper is exposed, and the wire color is confused. It is easy to cause accidents due to poor contact

No.	Pictures	Safety hazards (reference answer)
17		Temporary lines are not arranged as required. One plug is damaged and the other plug is connected in parallel, which is easy to cause electric shock
18		The three-phase current is seriously unbalanced. According to the regulations, the neutral wire current of the transformer should not exceed 25% of the wire current, otherwise the neutral point would shift and the voltage would change, which may easily cause equipment failure
19		Two wires are not firmly connected, which is easy to make the wire heat
20		In the laying cassette, the wires do not obviously distinguish the color of phase wire, neutral wire and grounding wire

No.	Pictures	Safety hazards (reference answer)
21		(1) There are too many wires through the pipe, the total cross-sectional area is more than 40% of the pipe (2) Concealed wire laying of wire through the pipe is extremely irregular, which must be rectified
22		(1) There are too many wires in the junction box through the pipe (2) The wire in the junction box is not marked, which is easy to cause wiring errors and safety accidents
23		(1) There is no rain-proof device on the temporary electrical control box, which is prone to leakage and electric shock accidents on rainy days (2) There are conductors exposed on the crimping terminals, which could be dangerous to be electrically shocked, and should be wrapped with insulation tape
24		(1) The exposed part of the lower outlet of the circuit breaker is too long, which is prone to electric shock accidents (2) Wires of the lower end of the circuit breaker do not lead from each outlet, if the insulation is aging or damaged, a short-circuit accident will occur

No.	Pictures	Safety hazards (reference answer)
25	Single-phase inlet device (The main fuse box, 2.4m)	(1) The height from household inlet point of low-voltage power supply to the ground is wrong (it should be greater than or equal to 2.9m) (2) The outdoor end of the inlet pipe does not take rainproof measures (waterproof elbow should be made on the outdoor end of the inlet pipe) (3) Inlet wires should be used with soft wires (inlet wires must use copper-core insulating wire, not less than 6mm^2)
26	Single-phase distribution device (The main fuse box, 15m, Joint)	(1) The main fuse box is too far away from the energy meter (the length should be less than or equal to 10m) (2) There is a joint in the power meter bus (no joint is allowed in the power meter bus, it should be replaced) (3) Wiring of the leakage protector is wrong (it should be changed to inlet at upper, outlet at lower)
27	N L1 L2 L3, Leakage protection device, Test button	(1) Power and lighting share one leakage protection device (leakage protectors should be installed separately) (2) Repeated grounding is connected at the outlet end of the leakage protection device (it should be connected at the inlet end) (3) There is no switch control on lighting (switch control should be added) (4) Test button of leakage protection device is damaged (replace the leakage protection device)
28	The sagging of wires is called "sag", 110mm, Span 35m	(1) The distance of outdoor open wire is too large (it could not exceed 25m) (2) Space between wires is too small (not less than 150mm)

Chapter 11 Elimination of potential safety hazards at the operation site (K3)

No.	Pictures	Safety hazards (reference answer)
29		(1) Vertical laying is too low from the ground (1.8m) (2) The height of the protection pipe is too low (1.5m) (3) Crossover does not trough protection pipe (trough protection pipe) (4) Wire color is not distinguished (the color should be distinguished)
30		(1) Connection of the steel pipe and button switch box in explosion-proof line is not sealed (should be sealed). (2) Steel pipe grounding and motor grounding are in series (to be grounded separately or in parallel) (3) The left steel pipe laying is not continuously grounded (add grounding)
31		(1) The flat iron of grounding device is joined by butt (flat iron should lap, at least three sides of welding) (2) The space of two grounding bodies is too close, only 2.4m (not less than 5m) (3) The length of grounding body is not enough, only 1.4m (not less than 2.5m)
32	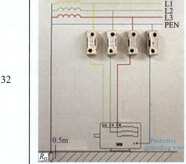	(1) The protective zero wire is connected in series with fuse, it's not permitted (should be removed) (2) The depth of buried grounding body is too shallow, distance from the top to the ground is only 0.5m (should be not less than 0.6m) (3) Motor has no switch control (add switch control, and overload protection device) (4) There is no repeated grounding in TN system of the PEN wire (There should be repeated grounding)

No.	Pictures	Safety hazards (reference answer)
33	Power device / PEN	(1)The case of iron case button box is not grounded (add grounding wire) (2)In the uppermost group of spiral fuse, the middle one is installed backwards (should be corrected) (3)A group of thermal relays is connected to normal open contacts (should be connected to normal closed contacts)

(5) Precautions

Carefully observe and analyze all the details of the picture and try to list the safety hazards and occupational hazards present in the scenes shown.

[Training 11-2] Eliminate the safety risks at the operation site (K32)

(1) Examination method

Practical operation, simulation operation, and oral presentation.

(2) Safe operation steps

Eliminate the safety risks and occupational hazards on the operation site in combination with the actual work tasks.

① Clarify the work tasks and do personal protection.

② Observe environment of the operation site.

③ Eliminate the safety risks at the operation site.

④ Conduct safe operations.

(3) Scoring criteria (see Table 11-3)

Table 11-3 Score table for eliminating safety risks on the operation site

Examination items	Examination content	Score	Scoring criteria
Eliminate the safety risks and occupational hazards on the operation site in combination with the actual work tasks	Awareness of personal security	20	Failure to clarify operational tasks and personal protection, deduct 5~20 points depending on the preparation
	Risk elimination	50	Observe the environment of the operation site and eliminate the safety risks at the operation site. 15 points would be deducted for each missing elimination. If the items not eliminated would affected the safety of personnel and equipment during operation, 50 points would be deducted

Examination items	Examination content	Score	Scoring criteria
Safe operation	Safe operation	30	Orally present the safety rules for this operation. 5 points would be deducted for each missing item
	Total	100	

(4) Training content

During the examination, any three pictures in Table 11-2 are usually combined to let candidates point out the potential safety hazards at the operation site. In fact, as long as we could point out the potential safety hazards in the figure, it is easy to eliminate the safety risks and occupational hazards existing in the operation site.

For example, if the potential safety hazard is "sundries stacked in front of the power distribution cabinet", the measure to eliminate the potential safety hazard is "remove the sundries stacked in front of the power distribution cabinet"; Another example is that the potential safety hazard is "the wire diameter is too small, and the wire end is exposed", so the measure to eliminate the potential hazard is "replace the wire with the appropriate wire diameter, and pay attention to the wire end not to be exposed during installation".

(5) Precautions

Carefully observe and analyze each detail of the picture, try to find out the potential safety hazards and occupational hazards in the scene in the picture, and eliminate the potential safety hazards one by one.

Chapter 12

Emergency response at operation site (K4)

The subject 4 of the examination of low-voltage electrician's operation qualification certificate is "emergency response at operation site", and the test code is K4. Randomly select any one of the following three sub items from the computer: Emergency response at the site of electric shock accident; Single-handed cardiopulmonary resuscitation(CPR) and fire extinguisher simulation test.

12.1 Emergency response and first aid for electric shock

[Exercise 12-1] Emergency response of electric shock at accident site (K41)

Examination method: Oral presentation.

Examination time: 10 minutes.

(1) Safe operation steps

① Methods and precautions for disconnecting from power supply in case of low-voltage electric shock

a. In case of low voltage electric shock, immediately find the nearest power switch for emergency power-off. If the switch could not be disconnected, cut off the power by insulation.

See Table 12-1 for the steps and methods to separate the person who gets an electric shock from the power supply when a large machine is shocked.

Table 12-1 Steps and methods of disconnecting from the power supply in the accident of electric shock of a large machine

Step	Operation methods
1	Turn off the load switch (stop button) of the leakage machine, pull the knife switch back and cut off the power as soon as possible
2	When the power supply could not be cut off immediately due to the site conditions, the rescue personnel could use non-conductive objects (such as dry wooden sticks, gloves, dry clothes, etc.) as tools to push (pull) the person who has received an electric shock away from the power supply
3	If the person's clothes are dry and not tightly wrapped around the body, the clothes could be pulled, but be careful not to touch his skin
4	The rescue personnel should pay attention to their own safety and try to stand on dry boards, insulating mats or wear insulating shoes for rescue. Generally, it should be operated by one hand

See Table 12-2 for the steps and methods to separate the person who gets an electric shock from the power supply when small equipment or electric tools get an electric shock.

Table 12-2 Steps and methods of disconnecting from the power supply in the accident of electric shock of a small machine

Step	Operation methods
1	If the plug or switch can be pulled off in time, the power supply should be cut off as soon as possible
2	When the power supply cannot be cut off immediately due to site conditions, the rescue personnel could use non-conductive objects (such as dry wooden sticks, gloves, dry clothes, etc.) as tools to push (pull) away the electric shock person or leakage equipment from the power supply. Generally, it should be operated by one hand
3	When the electric shock person has cramped and grasped the electrified body, it is dangerous to open his hand directly. Meanwhile, a dry wooden handle hoe or insulated rubber pliers and other insulated tools could be used to break the wire, but be careful to cut the wire one by one
4	If the power supply forms a circuit through the person who gets an electric shock entering the ground, a dry wooden board could be inserted under or under the feet of the person who gets an electric shock to cut off the current circuit

b. While the person who gets an electric shock is away from the power supply, the rescuer should prevent himself/herself from getting an electric shock, and should also prevent the person who gets an electric shock from secondary injury after getting away from the power supply.

c. Let the person who gets an electric shock lie down in a ventilated and warm place and make preparations for first aid according to the physical characteristics of the person who gets an electric shock.

d. If the person who has received the electric shock has suffered from trauma, the treatment of trauma should not affect the rescue work.

e. If someone gets an electric shock at night, the problem of temporary lighting should be solved during first aid.

② Methods and precautions for disconnecting from the power supply in case of high-voltage electric shock

a. If someone is found to have high-voltage electric shock, immediately notify the relevant power supply department at the higher level to cut off the power in an emergency. If the power could not be cut off, use the insulation method to pick up the wires and try to make them separated from the power supply as soon as possible.

b. While the person who gets an electric shock is away from the power supply, the rescuer should prevent himself/herself from getting an electric shock, and should also prevent the person who gets an electric shock from secondary injury after getting away from the power supply.

c. According to the physical characteristics of the person who got an electric shock, arrange for someone to closely observe to determine whether a doctor should be called from the hospital for diagnosis.

d. Let the person who gets an electric shock lie down in a ventilated and warm place and make preparations for first aid according to the physical characteristics of the person who gets an electric shock; If someone gets an electric shock at night, the temporary lighting should be solved during first aid.

e. If the person who has received the electric shock has suffered from trauma, the treatment of trauma should not affect the rescue work.

(2) Scoring criteria (Table 12-3)

Table 12-3 Scoring criteria for emergency response at the site of electric shock accident

Examination items	Examination content	Score	Scoring criteria
On-site emergency response of electric shock accident	Emergency procedures for power-off of low-voltage electric shock	50	In case of incomplete oral description on the method of separating from the power supply for low-voltage electric shock, deduct 5 to 25 points; And deduct 5 to 25 points if the oral description on precautions is inappropriate or incomplete
	Emergency procedures for power-off of high-voltage electric shock	50	In case of incomplete oral description on the method of separating from the power supply for high-voltage electric shock, deduct 5 to 25 points; And deduct 5 to 25 points if the oral description on precautions is inappropriate or incomplete
Negative items	Description	Deduct the score of this item	In case of incorrect oral description of the method of separating from the high and low-voltage electric shock, this test is terminated
Total		100	

(3) Precautions

① The rescuer should not use metal and other damp articles as rescue tools.

② Before taking insulation measures, the rescuer should not directly touch the skin and wet clothes of the person who gets an electric shock.

③ In the process of pulling the person who gets an electric shock away from the power supply, the rescuer should use one-hand operation, which is safer for the rescuer.

④ When the person is in a high position, measures should be taken to prevent the person from falling to the ground or dying after leaving the power supply.

[Exercise 12-2]　Single-handed cardiopulmonary resuscitation(CPR)(K42)

Examination method: Simulated operation of cardiopulmonary resuscitation with dummy.

Examination time: 3 minutes.

Examination method: The examination method of simulated operation of cardiopulmonary resuscitation is adopted.

Examination requirements:

① Candidates should implement single-handed cardiopulmonary resuscitation within the yellow and black warning lines;

② Master the operation skills of single-handed cardiopulmonary resuscitation, and be able to conduct the operations correctly;

③ The examination could be started only after the examiner agrees; When the dummy fails to be pressed in place and could also be counted and confirmed or could not be blown, this test would be terminated if the failure could not be eliminated immediately, and the test site would be responsible for the consequences.

(1) Safe operation steps

① Awareness checks: Pat the patient on the shoulder and call the patient loudly.

② Call for help: Look around, ask someone to help, unbutton the patient's clothes, loosen the belt, and lay the patient properly.

③ Check the carotid pulsation: the operation should be correct (touch one side for at least 5s).

④ Positioning: The following three methods could be used to determine the heart location.

Method 1: At the intersection of the sternum and ribs, commonly known as two fingers cross upward and one finger cross on the left of the "cardiac fossa".

Method 2: One finger cross left to the center of the two breast transverse lines.

Method 3: That is, the rescuer is facing the person who has received the electric shock, offset downwards with the distance of the extended right hand when the middle finger pointing at the intersection point of the collarbone under the neck of the person who has received the electric shock, and then press his palm.

⑤ External cardiac compression: the compression rate should be at least 100 times per minute and the compression amplitude should be at least 5cm (30 times per cycle for 15~18s). The compression posture is shown in Fig. 12-1.

⑥ Smooth the respiratory tract: Remove false teeth and clean the mouth.

⑦ Open the respiratory tract: Commonly used methods are head up chin lifting and jaw support. The standard is that the line between the mandibular angle and the earlobe is perpendicular to the ground.

⑧ Blowing: When blowing, the chest rise and fall could be seen. After blowing, leave the mouth immediately, loosen the nasal cavity, and blow again (twice per cycle) as the patient's chest falls.

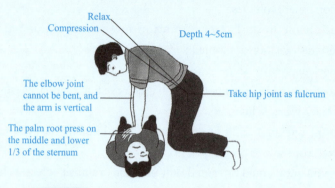

Fig. 12-1　External cardiac compression posture

⑨ After completing 5 cycles, check whether there is spontaneous respiration and heartbeat, and observe bilateral pupils. Oral statement after examination: the patient's pupil shrinks, carotid artery pulsates, spontaneous respiration recovers, and the face, lips, and nail bed (gently pressed) turn ruddy. Cardiopulmonary resuscitation was successful.

⑩ Sorting out: Place the person who gets an electric shock well, tidies up the clothes, puts the body in position, and tidies up the articles for use.

【Special reminder】

The three basic measures of cardiopulmonary resuscitation are to open the respiratory tract; Mouth to mouth (nose) artificial respiration; External cardiac compression (artificial circulation). The effective indicators for judging the overall quality of the single-handed CPR exam: effective blowing 10 times, effective pressing 150 times, and checking the effect (the total time from the beginning of checking the carotid pulse to the last blowing should not exceed 130s).

(2) Precautions

① After the electrically shocked person is disconnected from the power supply, if his/her breathing stops and his/her heart does not beat, if there is no other fatal injury, he/she could only be considered as suspended animation and must be rescued immediately on the site.

② The rescue work should be carried out continuously and should not be interrupted easily, even during the process of being sent to the hospital. If CPR is successful, the patient should be closely observed, waiting for rescue or receiving advanced life support.

③ The rescuer should dress neatly.

(3) Scoring Criteria (Table 12-4)

Table 12-4 Scoring criteria for single-handed cardiopulmonary resuscitation

Examination items	Examination content	Score	Scoring criteria
Awareness check	Pat the patient on the shoulder and call the patient loudly	1	0.5 points would be deducted if one item is not achieved
Call for help	Look around and evaluate the surrounding, and ask someone to help	1	0.5 points would be deducted if the on-site environmental safety is not assessed; 0.5 points would be deducted for those who fail to mention to call 120 first aid
Lay the patient properly	The patient should be laid on the hard board or on the ground, positioned correctly, unbutton and loosen the belt	2	0.5 points would be deducted if the patient is not placed on the hard board orally; 0.5 points would be deducted if the positioning place is not stated or is incorrect; Deduct 1 point if the button and belt are not loosened
Check the carotid pulsation	Correct operation (unilateral touch, 5~10s)	2	If no thyroid cartilage is found, 0.5 point would be deducted; If the position is incorrect, 0.5 point would be deducted; Deduct 1 point if the check time is less than 5s or more than 10s
Positioning	At the middle and lower 1/3 of the sternum, the palm root of one hand is placed on the pressing part, the other hand is parallel and overlapped on the back of the previous hand, the fingers are close together and bend upwards, the arms are directly above the patient's sternum, the elbows are straight, and the body weight is used to press vertically	2	If the pressing position is incorrect, 0.5 point would be deducted; 0.5 points would be deducted if one hand does not overlap the other hand in parallel; 0.5 points would be deducted if the chest is not pressed with the palm root and the fingers do not leave the chest; Deduct 0.5 points if the body is not vertical when pressing
External cardiac compression	The pressing frequency should be kept at 100~120 times/min, and the pressing depth should be 5~6cm (adults) (30 times per cycle, 15~18s)	3	If the pressing frequency is fast and slow, and it is not kept at 100~120 times/min, 1 point would be deducted; 0.5 points would be deducted if the pressing is of impact type; 0.5 points would be deducted for the palm away from the chest on each compression; 0.5 points would be deducted if pressing continues after each cycle (30 times); Deduct 1 point if the chest does not rebound due to the unequal proportion of pressing and releasing time

Examination items	Examination content	Score	Scoring criteria
Smooth the respiratory tract	Remove false teeth and clean the mouth	1	0.5 points would be deducted if the head does not turn to one side, 0.5 points would be deducted if the mouth is not cleaned and the denture is removed
Open the respiratory tract	Commonly used methods are head up chin lifting and jaw support. The standard is that the line between the mandibular angle and the earlobe is perpendicular to the ground	1	If the respiratory tract is not opened, 0.5 point would be deducted; 0.5 points would be deducted for excessive leaning back or insufficient degree
Blowing	When blowing, the chest rise and fall could be seen. After blowing, leave the mouth immediately, loosen the nasal cavity, and blow again (twice per cycle) as the patient's chest falls	3	If the nostrils are not pinched when blowing, 1 point would be deducted; 0.5 points would be deducted if the nostrils are not loosened between two blows; 0.5 points would be deducted if air blowing continues after each cycle (twice); Deduct 1 point if the proportion of blowing time to releasing time is different
Check	After completing 5 cycles, check whether there is spontaneous respiration and heartbeat, and observe bilateral pupils	1	0.5 points would be deducted if respiration and heartbeat are not observed, and 0.5 points would be deducted if bilateral pupils are not observed
The effective indicators for judging the overall quality	After completing 5 cycles(i.e., effective blowing 10 times, effective pressing 150 times), check the results(The total time from compression to the last blowing does not exceed 160s)	2	Deduct 1 point if the operation is unskilled and the technique is wrong; If the total time is exceeded, 1 point would be deducted
Sorting out	Place the patient, tidy up the clothes, position the body, and tidy up the articles for use	1	Deduct 0.5 points for each nonconformity
Negative item	Time limit: 3 minutes	Deduct the score of this item	If the dummy fails to be rescued within the specified time, the score is zero
Total		20	

12.2 Fire extinguishing simulation with fire extinguisher

[Exercise 12-3] Selection and use of fire extinguishers (K43)

Examination method: Use simulation equipment for operation.

Examination time: 2 minutes.

Examination requirements:

① Master the safety operation steps in the firefighting process;

② Be familiar with the operation skills of the fire extinguisher and be able to operate the fire extinguisher correctly;

③ Before the practical operation test, the test site should prepare complete simulation test equipment, and the test could only be started after the approval of the examiner; When the fire extinguishing assessment system fails to work normally or a fire extinguisher fails to work normally, the examination of this exercise would be terminated if the failure could not be eliminated immediately, and the examination site would be responsible for the consequences.

(1) Safe operation steps

① Preparations: Check the pressure, lead seal, and factory certificate, period of validity, bottle bottom and nozzle of the fire extinguisher.

② Fire situation check: According to the fire situation, select the appropriate fire extinguisher and rush to the fire site quickly, and correctly check the wind direction.

③ Extinguishing operation: Stand at the upper air vent of the fire source and pull down the safety ring quickly 3~5m away from the fire source; Hold the nozzle to aim at the ignition point, press down the handle, and aim at the root of the fire source to spray fire from near to far; Evacuate from the fire site immediately (3s) before the dry powder is ready to be sprayed. If the fire is not extinguished, replace the extinguisher and continue to extinguish the fire.

④ Check and confirm: Check the fire extinguishing results; Confirm that the fire source is extinguished; Put the used fire extinguisher at the designated position; Indicate that it has been used; Report the firefighting circumstance.

⑤ Check the tools and clean the site.

(2) Scoring Criteria (Table 12-5)

Table 12-5 Scoring criteria for the use of fire extinguishers

Examination items	Examination content	Score	Scoring criteria
Preparation	Check the pressure, lead seal, factory certificate, period of validity, bottle bottom and nozzle of the fire extinguisher	2	Deduct 2 points if the fire extinguisher is not checked; 0.5 points would be deducted if one item of the pressure, lead seal, bottle body, spray pipe, validity period and factory certificate is not stated
Fire situation check	Select the appropriate fire extinguisher according to the fire situation(The fire extinguisher could be re-selected before extinguishing) , and correctly check the wind direction	5	Deduct 5 points for each wrong fire extinguisher selection; Deduct 3 points for wrong wind direction check

Examination items	Examination content	Score	Scoring criteria
Extinguishing operation	Determine the location of the fire extinguisher according to the wind direction of the fire source; The safety ring could be quickly pulled down at a safe distance from the fire source	5	Deduct 4 points if the standing position is incorrect during firefighting; Deduct 3 points if the fire extinguishing distance is not correct; Deduct 2 points if the safety ring is not pulled down quickly
	Hold the nozzle to aim at the ignition point, press down the handle, and aim at the root of the fire source to spray fire from near to far. Continue to replace the fire extinguisher if the fire is not extinguished	4	When spraying at the root of the fire source, 2 points would be deducted if the examinee does not stand correctly; 2 points would be deducted if the fire is not extinguished from near to far; Deduct 4 points for stopping operation before the fire goes out
Check and confirm	Put the used fire extinguisher at the designated position; Indicate used	3	Deduct 1 point if the fire extinguisher is not restored; If it is not placed at the designated position, 1 point would be deducted; 1 point would be deducted for those not indicated as used
	Report the firefighting circumstance	1	1 point would be deducted if the firefighting circumstance is not reported
Negative item	Time limit: 2 minutes	Deduct the score of this item	If the fire is not extinguished according to the regulations or the fire is not successfully extinguished within the specified time, the score is zero
Total		20	

(3) Fire extinguishing simulating operation with fire extinguisher

① The examinee could enter the fire extinguisher simulation test system after entering the examination permit and ID card number and confirming. After entering the fire site, the system would pop up the prompt of fire inducement; Candidates should correctly choose fire extinguishers according to the fire inducements. See Table 12-6 for the corresponding fire type of fire extinguisher.

Table 12-6 Corresponding fire types of fire extinguishers

Fire extinguisher selection	Type of fire
Dry powder extinguisher, water-based foam extinguisher	Wood, cardboard boxes, curtains, trash cans, clothes, etc
Carbon dioxide (CO_2) extinguisher	Electrical motor
Carbon dioxide (CO_2) extinguisher, dry powder extinguisher, water-based foam extinguisher	Gasoline tank
Carbon dioxide (CO_2) extinguisher, dry powder extinguisher	Power distribution box

② Visual inspection of fire extinguisher, as shown in Fig. 12-2, to check whether the filler in the fire extinguisher is at the standard position [the area pointed by the pointer in (a)], and the inspection date and time of validity [the date in (b) is 2009, 10 years old, unqualified]; Whether the components of the fire extinguisher are damaged [(c) Nozzle is damaged]. Only one of the three fire extinguishers automatically provided by the system meets the requirements of the conditions (if the fire extinguisher is selected incorrectly, 0 point would be obtained in the exam). If the first two choices are both wrong, generally speaking, the third one is correct (optional).

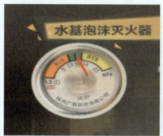
(a) Filler inspection

(b) Date inspection

(c) Component inspection

Fig. 12-2 Visual inspection of fire extinguisher

③ Select the fire extinguishing position. The effective distance between the fire extinguisher and the fire source is 3~5m. (As shown in Fig. 12-3, the distance is 3.86m, which is acceptable)

④ According to the prompts, take up the physical fire extinguisher to extinguish the fire, pull out the safety plug first, and control the fire extinguisher on the screen through the fire extinguisher in hand. Aim the collimator of the fire extinguisher at the root of the fire point, and aim the nozzle of the fire extinguisher at the "root of the fire point" to extinguish the fire.

To ensure that the sensor could sense the nozzle, move the nozzle position slowly until the red arrow turns green. Meanwhile, keep standing at the original position, press the nozzle, and the fire would go out after about 30s, as shown in Fig. 12-4.

Fig. 12-3 Selecting the fire extinguishing position

Fig. 12-4 Fire extinguishing process

⑤ After the fire is extinguished, insert the safety plug back into the original position, click to submit the score, and the exam is completed, as shown in Fig. 12-5.

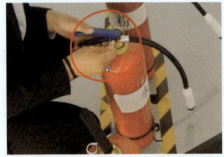

(a) Insert the safety plug back into the original position   (b) Submit the score

Fig. 12-5 End of examination

(4) Precautions

① The type of fire extinguisher must be selected correctly, and check whether the selected fire extinguisher could be used according to the steps.

② Observe the prompts and information on the large screen during firefighting.

Single choice question bank

1. Among the following materials, the one with the best conductivity is (　　).

　　A. aluminium　　B. copper　　C. iron

2. If a wire is evenly stretched to twice the original length, its resistance is (　　) times of the original resistance.

　　A. 1　　B. 2　　C. 4

3. The purpose of *The Production Safety Law of the People's Republic of China* is to strengthen production safety, prevent and reduce (　　), ensure the safety of people's lives and property, and promote economic development.

　　A. Production safety accidents

　　B. Fire and traffic accidents

　　C. Major and extraordinarily serious accidents

4. According to the provisions of *The Law on the Prevention and Control of Occupational Diseases the People's Republic of China*, in addition to meeting the establishment conditions stipulated by laws and administrative regulations, the establishment of the employer that causes occupational disease hazards should follow the following principles for the layout of its workplaces: (　　).

　　A. Separate production operation from storage operation

　　B. Separate processing operations from packaging operations

　　C. Separate hazardous operation from harmless operation

5. In the power control system, the (　　) type AC contactor is most widely used.

　　A. pneumatic　　B. electromagnetic　　C. hydraulic

6. The test button of the leakage protection device should be pressed once every (　　).

　　A. month　　B. half a year　　C. three months

7. *The Production Safety Law of the People's Republic of China* stipulates that any entity or (　　) has the right to report to the department responsible for the supervision and administration of production safety on potential accidents or violations of production safety.

　　A. employee　　B. individual　　C. administrator

8. The direction of the induced current always makes the magnetic field of the induced current block the change of the magnetic flux of the induced current, which is called (　　).

　　A. Faraday's Law　　B. Tesla's Law　　C. Lenz's Law

9. The ultra-low voltage limit refers to the (　　) voltage value between any two conductors under any conditions.

　　A. minimum　　B. maximum　　C. intermediate

10. Check the motor and its nameplate, mainly including (　　), frequency and

---

Solutions: 1.B; 2.C; 3.A; 4.C; 5.B; 6.A; 7.B; 8.C; 9.B; 10.A

configuration method of stator winding.

    A. power supply voltage

    B. power supply current

    C. working system

11. In flammable and explosive dangerous places, the power supply line should adopt the ( ) mode for power supply.

    A. Single-phase-three-wire system, three-phase-four-wire system

    B. Single-phase-three-wire system, three-phase-five-wire system

    C. Single-phase-two-wire system, three-phase-five-wire system

12. According to some data, about 80% of the patients could be saved if their heartbeat and respiration stop and they are rescued within ( ) minutes.

    A. 1   B. 2   C. 3

13. When using the wire stripper, the cutting edge with a diameter ( ) the wire diameter should be selected.

    A. the same with   B. bigger than   C. smaller than

14. Check the lubrication of motor bearing, ( ) the motor shaft, to check whether it rotates flexibly, and listen for abnormal noise.

    A. power on to rotate   B. manual rotate

    C. drive with other equipment

15. When checking the socket, the two holes of the test pencil in the socket are not lit, ( ) should be judged first.

    A. short circuit   B. phase wire disconnection

    C. neutral wire disconnection

16. ( ) is a necessary protective tool for climbing pole operation, which should be used together with pedal or foot buckles.

    A. Safety belt   B. Iadder   C. Gloves

17. The direction of the magnetic field generated by the energized coil is not only related to the current direction, but also related to the coil ( ).

    A. length   B. wound direction   C. volume

18. When the voltage relay is used, its suction coil is directly or through the voltage transformer ( ) in the controlled circuit.

    A. connected in parallel   B. connected in series

    C. connected in series or in parallel

19. When measuring the insulation resistance of the motor coil to the ground, the L and

---

Solutions: 11.B; 12.A; 13.B; 14.B; 15.B; 16.A; 17.B; 18.A; 19.B

E terminals of the megger should be ( ).

    A. "E" is connected to the terminal of the motor outgoing wire, and "L" is connected to the motor housing

    B. "L" is connected to the terminal of the motor outgoing wire, and "E" is connected to the motor housing

    C. random, no rules

20. The travel switch consists of ( ).

    A. Coil part   B. protective part   C. reaction system

21. The clamp ammeter should be used with a larger range first, and then the range should be changed depending on the magnitude of the measured current. When switching the measuring range, ( ) should be used.

    A. rotate the range switch directly

    B. retreat from the wire first, then rotate the range switch

    C. rotate the range shift while entering to the wire

22. Electric soldering iron is used for ( ) wire joints, etc.

    A. copper soldering

    B. tin soldering

    C. iron soldering

23. Check the insulation of each winding of the motor. If the insulation resistance is measured to be zero, drying could be carried out when no obvious burning is found. Meanwhile, ( ) power on for operation.

    A. it is allowed   B. it is not allowed   C. after drying, it is OK to

24. Step-down starting refers to reducing the voltage applied to the motor ( ) winding during starting, and then restoring the voltage to the rated voltage for normal operation after starting.

    A. stator   B. rotator   C. stator and rotator

25. After the air switch acts, touch its case with hand. If the switch case is found to be hot, the action might be ( ).

    A. short circuit   B. overload   C. undervoltage

26. The normal operating state of the zener diode is ( ).

    A. connecting state   B. cut-off state

    C. reverse breakdown state

27. Before climbing the pole, ( ) for the foot buckles should be conducted.

    A. human body static load test   B. human body load impact test

---

Solutions: 20.C; 21.B; 22.B; 23.B; 24.A; 25.B; 26.C; 27.B

C. human body load tensile test

28. In the Y-configuration of three-phase symmetrical AC power supply, the wire voltage is ( ) ° ahead of the corresponding phase voltage.
   A. 120   B. 30   C. 60

29. When positive voltage is applied to both ends of PN junction, its positive resistance ( ).
   A. becomes lower   B. becomes higher
   C. is the same

30. When the wire joint is wrapped with insulating tape, the last loop is pressed against the width of the previous loop ( ).
   A. 1/3   B. 1/2   C. 1

31. ( ) should be used in places with higher requirements for color difference.
   A. Coloured lights   B. Incandescent lights
   C. Purple lights

32. Since the ( ) of motor is the path of motor magnetic flux, the material of it should have good magnetic conductivity.
   A. base   B. end cover   C. stator core

33. The ( ) voltage and step voltage generated by lightning current could directly cause people to be electrocuted to death.
   A. induced   B. contact   C. direct strike

34. ( ) belongs to paramagnetic material.
   A. Water   B. Copper   C. Air

35. Electrostatic phenomenon is a very common electrical phenomenon, ( ) is its greatest harm.
   A. discharge electricity to human body and directly kill people
   B. high-voltage breaks the insulation down
   C. easy to cause fire

36. In the circuits shown in the figures, after the switches S1 and S2 are on, the touchable one is ( ).

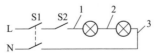

   A. the 2nd part   B. the 3rd part   C. nowhere

37. The direction of the electromotive force is ( ).

Solutions: 28.B; 29.A; 30.B; 31.B; 32.C; 33.B; 34.C; 35.C; 36.B; 37.A

A. from negative to positive B. from positive to negative
C. the same as voltage direction

38. The main cause of electric ophthalmia is ( ).
A. infrared B. visible light C. ultraviolet rays

39. Pure capacitor element ( ) electric energy in the circuit.
A. stores B. distributes C. consumes

40. The signboard "No climbing, high-voltage danger!" should be made as ( ).
A. red letters on white background
B. white letters with red background
C. black letters with red edge on white background

41. A relay is an automatic electrical appliance that controls the circuit "on" or "off" according to ( ).
A. external input signal (electrical signal or non-electrical signal)
B. electrical signal
C. non-electrical signal

42. There are many static electricity protection measures. The following common and effective method to eliminate static electricity in equipment shell is ( ).
A. grounding B. zero connection C. series connection

43. The electrical fire is caused by the existence of dangerous temperature, which is mainly caused by ( ).
A. light equipment load B. voltage fluctuation
C. excessive current

44. In the case of live extinguishing, the distance between the body and nozzle of carbon dioxide extinguisher and the high-voltage live body below 10kV should not be less than ( ) m.
A. 0.4 B. 0.7 C. 1

45. In thunderstorm weather, doors and windows should be closed to prevent ( ) intrusion into the house, causing fire, explosion or casualties.
A. spherical thunder B. induction thunder
C. direct strike thunder

46. ( ) apparatus is composed of fixed coil, rotatable coil, rotating shaft, hairspring, pointer, mechanical zero adjustment mechanism, etc.
A. Magnetoelectric B. Electromagnetic
C. Electric

Solutions: 38.C; 39.A; 40.C; 41.A; 42.A; 43.C; 44.A; 45.A; 46.C

47. If the main person in charge of a production and business operation entity escapes after a major production safety accident occurs in the entity, he should be detained by ( ) for not more than 15 days.

  A. police

  B. procuratorial

  C. production safety supervision and management department

48. The principle of using AC contactor for undervoltage protection is that when the voltage is insufficient, the coil would produce insufficient ( ) and the contact would break.

  A. magnetic force   B. vortex   C. heat

49. When the pointer multimeter measures resistance, the rightmost side of the scale is ( ).

  A. ∞   B. 0   C. uncertain

50. The current carrying conductor would be affected by ( ) in the magnetic field.

  A. electromagnetic force   B. flux

  C. electrodynamic force

51. The floor socket should have a firm and reliable ( ).

  A. sign board   B. protective cover   C. switch

52. The value of AC voltage and current indicated by general electrical appliances or instruments is ( ).

  A. maximum values   B. effective values

  C. average values

53. The fluorescent lamp belongs to the ( ) light source.

  A. gas discharge   B. thermal radiation

  C. biological discharge

54. The insulation resistance of newly installed and overhauled low-voltage lines and equipment should not be lower than ( ) MΩ.

  A. 1   B. 0.5   C. 1.5

55. The electrical diagram of the circuit breaker is ( ).

  A.   B.   C.

56. The electrical life of AC contactor is about ( ) times of the mechanical life.

  A. 10   B. 1   C. 1/20

57. The insulation resistance of lines or equipment is measured with ( ).

  A. resistance grade of multimeter   B. megger

  C. grounding resistance meter

Solutions: 47.A; 48.A; 49.B; 50.A; 51.B; 52.B; 53.A; 54.B; 55.A; 56.C; 57.B

58. The interior of the inductive fluorescent lamp ballast is ( )
    A. electronic circuit   B. coil
    C. oscillating circuit
59. The iodine tungsten lamp belongs to ( ) light source.
    A. gas discharge   B. electric arc   C. thermal radiation
60. It is stipulated in *The Electrical Safety Operation Regulations* that the equipment with a voltage of ( ) V and below to the ground is low-voltage equipment.
    A. 400   B. 380   C. 250
61. The rated current of the fusant of fuse in the lighting circuit is ( ) times the calculated current of the circuit.
    A. 0.9   B. 1.1   C. 1.5
62. The insulation strength of the conductor joint should be ( ) the insulation strength of the original conductor.
    A. higher   B. equal   C. lower
63. The height from the wall switch to the ground is ( ) m.
    A. 1.3   B. 1.5   C. 2
64. The basic structure of universal transfer switch includes ( ).
    A. reaction system   B. contact system
    C. coil section
65. The regrounding of other grounding points on PE wire or PEN wire except for working grounding is called ( ) grounding.
    A. indirect   B. direct   C. repeated
66. When the voltage is 5V, the resistance of the conductor is 5 Ω, and when the voltage at both ends of the resistance is 2V, the resistance of the conductor is ( ) Ω.
    A. 10   B. 5   C. 2
67. The sectional area of the neutral wire of three-phase-four-wire system is generally ( ) the sectional area of the phase wire.
    A. bigger than B. smaller than C. equal to
68. In the multi-level fuses protection, the rated current of the fusant of the latter level is higher than that of the former level, so as to ( ) in case of the fuse from fusing beyond the level.
    A. avoid difficulty in fault detection   B. reduce the scope of power failure
    C. expand the scope of power failure
69. The use of safety rope should be ( ).

Solutions: 58.B; 59.C; 60.C; 61.B; 62.B; 63.A; 64.B; 65.C; 66.B; 67.B; 68.C; 69.A

A. hanging on high and using on low

B. hanging on low and using in adjustment

C. ensuring safety

70. The two main components of the megger are hand rotated ( ) and magnetoelectric current ratio meter.

A. current mutual inductor   B. DC generator

C. alternator

71. In case of electrical fire in the workshop, the power supply should be cut off first. The method of cutting off the power supply is ( ).

A. disconnecting the knife switch

B. disconnecting the circuit breaker or the magnetic switch

C. reporting to the person in charge, and requesting to cut off the main power supply

72. According to the national standard, all motors above ( ) kW should be configured in Δ.

A. 3   B. 4   C. 7.5

73. In normal places, ( ) electric tools should be selected to ensure safe use.

A. catagory I   B. catagory II   C. catagory III

74. Low-voltage electrical apparatus could be classified as low-voltage power distribution apparatus and ( ) electrical apparatus.

A. low-voltage control   B. voltage control

C. low-voltage electric drive

75. The structure of the frequency sensitive rheostat is similar to the three-phase reactance, that is, it is composed of three iron cores and ( ) winding(s).

A. one   B. two   C. three

76. The configuration between three-phase windings of motor stator and AC power supply is called configuration method, where Y is ( ).

A. configuration in Δ   B. configuration in Y

C. configuration in extended triangle

77. In the AC circuit, the current is behind the voltage by 90°, which belongs to the ( ) circuit.

A. pure resistance   B. pure inductance

C. pure capacitance

78. When selecting the sensitivity of the leakage protection device, it should avoid

---

Solutions: 70.B; 71.B; 72.B; 73.B; 74.A; 75.C; 76.B; 77.B; 78.A

affecting the normal power supply due to unnecessary actions caused by normal ( ).
   A. leakage current   B. leakage voltage
   C. leakage power

79. From the perspective of manufacturing, low-voltage apparatus refers to electrical apparatus with AC 50Hz, rated voltage ( ) V or DC rated voltage of 1500V and below.
   A. 400   B. 800   C. 1000

80. When the motor runs under rated working condition, the mechanical power of ( ) is called rated power.
   A. allowed input   B. allowed output   C. driving motor

81. The unit of capacitance is ( ).
   A. F   B. Var   C. A·h

82. When soldering transistor and other weak current components, ( ) W electric soldering iron should be used.
   A. 25   B. 75   C. 100

83. The electrical machinery equipment used on the construction site ( ) be equipped with leakage protection device.
   A. should not   B. should   C. is not stipulated to

84. The grounding resistance meter is a device for measuring ( ).
   A. insulation resistance   B. DC resistance
   C. grounding resistance

85. When the tanker transporting liquefied gas, petroleum, etc. is running, metal chains or conductive rubber should be used at the bottom of the tanker to make it contact with the ground for the purpose of ( ).
   A. neutralizing the static charge generated during the running of the tanker
   B. discharging the static charge generated during the running of the tanker
   C. equaling the potential between the tanker with the ground

86. The following figure ( ) is the electrical figure of button.
   A. ∕   B. E∕   C. ㅋ\

87. Before capacitor measurement, ( ) is required.
   A. cleaning   B. fully charged   C. fully discharged

88. The special operation certificate should be reviewed once every ( ) year.
   A. 5   B. 4   C. 3

89. In flammable and explosive hazardous areas, electrical lines should be laid ( ) or

---

Solutions: 79.C; 80.B; 81.A; 82.A; 83.B; 84.C; 85.B; 86.A; 87.C; 88.C; 89.C

armored cables.

    A. in cable trench with sand after passing through metal snake skin pipe

    B. through water gas pipe

    C. through steel pipe

90. Which following statement about the domestic legal system of production safety is correct ( ).

    A. *The Production Safety Law of the People's Republic of China*, *The Fire Control Law of the People's Republic of China*, *The Road Traffic Safety Law of the People's Republic of China*, *The Law on Mine Safety of the People's Republic of China* are separate laws on production safety in the domestic production safety legal system

    B. *The Production Safety Law of the People's Republic of China* is a common law in the field of production safety, which is widely used in all fields of production and business activities

    C. *The Fire Control Law of the People's Republic of China*, *The Road Traffic Safety Law of the People's Republic of China* are different from *The Production Safety Law of the People's Republic of China*, they should be used together with *The Production Safety Law of the People's Republic of China*.

91. Three phase asynchronous motor could be divided into open type, protective type and closed type according to its ( ).

    A. power supply mode   B. housing protection mode

    C. structure

92. The current from the left hand to both feet causes ventricular fibrillation effect. It is generally believed that when the product of power on time and current is greater than ( ) mA · s, life is in danger.

    A. 16   B. 30   C. 50

93. The rotation direction of the rotating magnetic field is determined by the phase sequence of the three phase AC power supply connected to the stator windings. As long as the phase sequence of the AC power supply connected to ( ) of the motor is arbitrarily changed, the rotating magnetic field will reverse.

    A. one phase winding   B. two phase windings

    C. three phase windings

94. When the electrical equipment has a grounding fault, the grounding current flows to the ground through the grounding body. If a person walks around the grounding short-

Solutions: 90.C; 91.B; 92.C; 93.B; 94.B

circuit point, the electric shock caused by the potential difference between his two feet is called ( ) electric shock.

    A. single-phase   B. step voltage   C. induced electricity

95. If the person who gets an electric shock stops heartbeat and breathes, the ( ) first aid should be given to the person who gets an electric shock immediately.

    A. supine chest compression   B. external cardiac compression

    C. prone back compression

96. The grounding resistance meter is mainly composed of hand generator, ( ), potentiometer and galvanometer.

    A. current mutual inductor   B. voltage mutual inductor

    C. transformer

97. Insulation safety appliances are divided into ( ) safety appliances and auxiliary safety appliances.

    A. direct   B. indirect   C. basic

98. In semiconductor circuit, fast fuse is mainly used for ( ) protection.

    A. short circuit   B. overvoltage   C. overheated

99. At the starting moment of asynchronous motor, the induced current in the rotor windings is very large, so the starting current flowing through the stator is also very large, which is about ( ) times of the rated current.

    A. 2   B. 4~7   C. 9~10

100. ( ) V safety extra low voltage should be adopted in particularly humid places.

    A. 42   B. 24   C. 12

101. The protection characteristics of the thermal relay are close to the overload characteristics of the motor, which is to give full play to the ( ) capability of the motor.

    A. overload   B. controlling   C. throttling

102. The single-phase electric energy meter is mainly composed of a rotatable aluminum disk, a ( ) wound on different iron cores respectively, and a current coil.

    A. voltage coil   B. voltage mutual inductor

    C. resistance

103. A maximum of ( ) conductor joints are allowed in the conducting wire through pipe.

    A. 2   B. 1   C. 0

104. The step-down starting of cage asynchronous motor could reduce the starting current, but because the motor torque is ( ) to the square of voltage, the torque is reduced

---

Solutions: 95.B; 96.A; 97.C; 98.A; 99.B; 100.C; 101.A; 102.A; 103.C; 104.B

more during step-down starting.

  A. inversely proportional   B. proportional

  C. corresponding

 105. Turn on the power switch, the fusant would be burnt out immediately, then the circuit is ( ).

  A. short circuit   B. in electric leakage

  C. voltage too high

 106. ⊣ is the ( ) contact.

  A. delayed closing and dynamic closing

  B. delayed opening and dynamic closing

  C. delayed opening and dynamic opening

 107. There are many master switches, including ( ).

  A. contactor   B. travel switch   C. thermal relay

 108. The thread of screw lamp cap should be connected to ( ).

  A. neutral wire   B. phase wire   C. grounding wire

 109. The resistance of the conducting wire joint should be small enough, and the resistance ratio to the conducting wire of the same length and section should not be greater than ( ).

  A. 1   B. 1.5   C. 2

 110. The unit of power of capacitor is ( ).

  A. Var   B. Watt   C. V·A

 111. When the motor is running, the motor should be monitored in time by ( ), looking, smelling and other methods.

  A. recording   B. listening   C. wind blowing

 112. In order to avoid direct lightning strike on high-voltage power transformation and distribution station, which may cause large-scale power outage, ( ) is generally used for lightning protection.

  A. lightning rod   B. valve type arrester

  C. lightning net

 113. In an ungrounded system, if a single-phase ground fault occurs, the voltage of other phase wires to ground would be ( ).

  A. increased   B. decreased   C. unchanged

 114. When the three-phase symmetrical load is configured in Y, the total three-phase current ( ).

Solutions: 105.A; 106.B; 107.B; 108.A; 109.A; 110.A; 111.B; 112.A; 113.A; 114.A

A. equal to zero   B. equal to three times of one phase current
C. equal to one phase current

115. There are two kinds of starting methods for three-phase cage asynchronous motor, namely direct starting under rated voltage and (　) starting.
   A. rotor with resistance in series
   B. rotor with frequency sensitivity transformer in series
   C. reduce starting voltage

116. The multimeter is essentially a (　) instrument with a rectifier.
   A. magnetoelectric   B. electromagnetic
   C. electric

117. Conducting wire joints should be in close contact and (　), etc.
   A. undamaged by tension   B. firm and reliable
   C. not heat generated

118. Although there are many kinds of three-phase asynchronous motors, their basic structures are composed of (　) and rotor.
   A. housing   B. stator   C. housing and base

119. The ventricular fibrillation current in humans is about (　) mA.
   A. 16   B. 30   C. 50

120. When the cage asynchronous motor adopts resistance step-down starting, the starting times (　).
   A. should not be too few
   B. should be no more than 3 times per hour
   C. should not be too frequent

121. Low-voltage circuit breaker is also called (　).
   A. knife switch   B. main switch   C. automatic air switch

122. The spray water gun could be used for live extinguishing, but for safety reasons, the firefighters should wear insulating gloves and insulating boots, and also require the head of the water gun (　).
   A. to be grounded   B. must be made of plastic
   C. could not be made of metal

123. The setting current of the thermal relay is (　)% of the rated current of the motor.
   A. 100   B. 120   C. 130

124. (　) could be used to operate high-voltage drop-out fuses, single pole disconnectors and install temporary grounding wires.

Solutions: 115.C; 116.A; 117.B; 118.B; 119.C; 120.C; 121.C; 122.A; 123.A; 124.C

A. insulating gloves B. insulated shoes
C. insulating rod

125. The ampere rule is also called ( ).
A. left-hand rule B. right-hand rule   C. right-hand spiral rule

126. ( ) and economy are the two basic principles for correct selection of electrical appliances.
A. Safety   B. Performance   C. Price

127. When the speed of a four-pole motor is 1440r/min, the slip of this motor is ( )%.
A. 2   B. 4   C. 6

128. When the human body directly contacts a live equipment or a phase in the line, the current flows into the earth through the human body. This electric shock phenomenon is called ( ) electric shock.
A. single-phase   B. two-phase   C. three phase

129. The voltage range 2.5V of the multimeter means the voltage when the pointer is at ( ).
A. 1/2 of the scale   B. full scale
C. 2/3 of the scale

130. The upper hole of single-phase three holes socket should be connected to ( ).
A. neutral wire   B. phase wire   C. grounding wire

131. When the conducting wire of 6~10kV overhead line passes through the residential area, the minimum distance between the line and the ground is ( ) m.
A. 6   B. 5   C. 6.5

132. The lighting voltage used is 220V, which is the ( ) of AC.
A. effective value   B. maximum value
C. constant value

133. The total resistance of three resistors with equal resistance in series is ( ) times of that in parallel.
A. 6   B. 9   C. 3

134. The symbol of the ammeter is ( ).
A. Ω   B. A   C. V

135. The magnitude of the electromagnetic force is ( ) to the effective length of the conductor.
A. proportional   B. inversely proportional

Solutions: 125.C; 126.A; 127.B; 128.A; 129.B; 130.C; 131.C; 132.A; 133.B; 134.B; 135.A

C. unchanged

136. In principle, the specifications of the new fusant and the old fusant should be (　) when replacing the fusant.

  A. different B. the same C. updated

137. During the installation of electrical lines, the connection process between conducting wires or between conducting wire and the electrical bolt that is most likely to cause fire is (　).

  A. copper wire and aluminum wire twisted joint

  B. aluminum wire and aluminum wire twisted joint

  C. copper aluminum transition crimping joint

138. In the case of electrical fire, cut off the power supply before putting out the fire. However, if the position of the switch is not clear or unknown, (　).

  A. several wires should be cut quickly at the same time

  B. different phase wires should be cut at different positions

  C. cut the wires one by one at the same location

139. In a uniform magnetic field, when the magnetic flux passing through a plane is the maximum, the plane is (　) to the magnetic field line.

  A. paralleled B. vertical C. oblique

140. The starting of using (　) to reduce the voltage applied to the three-phase stator windings is called auto coupling step-down starting.

  A. autotransformer B. frequency sensitive transformer

  C. resistor

141. When the heat resistance grade of insulating material is Class E, its limit operating temperature is (　) °C

  A. 90 B. 105 C. 120

142. When the capacitor is found to be damaged or defective, it should be (　).

  A. self repaired B. returned for repair

  C. discarded

143. Common step-down starting of cage asynchronous motor includes (　) starting, autotransformer step-down starting and Y-Δ step-down starting.

  A. rotor with resistance in series

  B. step-down with resistance in series

  C. rotor with frequency sensitivity transformer in series

144. The circuit breaker is connected by manual or electric operating mechanism, and

---

Solutions: 136.B;　137.A;　138.B;　139.B;　140.A;　141.C;　142.B;　143.B;　144.C trips automatically by ( ) device to achieve fault protection.

    A. automatic   B. movable device   C. tripping device

145. The insulation inspection of each winding of the motor should be conducted according to the following requirements: insulation resistance should be ( ) for every 1kV working voltage of the motor.

    A. less than $0.5M\Omega$ B. greater than or equal to $1M\Omega$

    C. equal to $0.5M\Omega$

146. Electric injury is caused to human body by the ( ) effect of current.

    A. thermal   B. chemical   C. thermochemistry and Mechanical

147. When the leakage protection circuit breaker works normally, the switch remains closed when the phasor sum of circuit current ( ).

    A. is positive   B. is negative   C. is zero

148. When the current relay is used, its suction coil is ( ) directly or through the current mutual inductor in the controlled circuit.

    A. connected in parallel   B. connected in series

    C. connected in parallel or in series

149. 150mm of long nose pliers refers to ( ).

    A. its insulating handle is 150mm long

    B. its total length is 150mm

    C. its opening is 150mm

150. The long-term allowable current of the conducting wire connecting the capacitor should not be less than ( )% of the rated current of the capacitor.

    A. 110   B. 120   C. 130

151. The color of protective wire (grounding or neutral wire) should be ( ) according to the standard.

    A. blue   B. red   C. yellow and green.

152. The special operation personnel has been engaged in this type of work continuously for more than 10 years within the validity period of the operation certificate, without any illegal activities, and the review time of the operation certificate could be extended to ( ) years with the approval of the assessment and certificate issuing authority.

    A. 4   B. 6   C. 10

153. Dry powder extinguisher could be used for live extinguishing of lines below ( ) kV.

    A. 10   B. 35   C. 50

Solutions: 145.B; 146.C; 147.C; 148.B; 149.B; 150.C; 151.C; 152.B; 153.C

154. Low-voltage fuses are widely used in low-voltage power supply and distribution systems and control systems, mainly for (　) protection, and sometimes for overload protection.

　　A. quick disconnecting　　　　　　B. short circuit
　　C. overcurrent

155. The signboard of "No switching on, someone is working" should be made as (　).
　　A. red letter on a white background
　　B. white letter on a red background
　　C. green letter on a white background

156. Poor contact of wire joint and controller contact is an important cause of electrical fire. The essential reason for the so-called "poor contact" is (　).
　　A. overvoltage caused by resistance change of joints and contact points
　　B. the resistance of joints and contact points decreases
　　C. increased power consumption due to increased resistance of joints and contact points

157. The total capacity of each lighting (including fan) branch is generally not greater than (　) kW.
　　A. 2　　　　　　B. 3　　　　　　C. 4

158. The sound of the motor during normal operation is smooth, light, (　) and rhythmic.
　　A. screaming　　B. uniform　　C. frictional

159. When measuring the grounding resistance, the potential probe should be connected at the place (　) m away from the grounding terminal.
　　A. 5　　　　　　B. 20　　　　　　C. 40

160. (　) apparatus is composed of fixed permanent magnet, rotatable coil and shaft, hairspring, pointer, mechanical zero adjustment mechanism, etc.
　　A. Magnetoelectric　　　　　　B. Electromagnetic
　　C. Inductive

161. The phase wire should be connected to the (　) of the screw lamp cap.
　　A. center terminal　　　　　　B. screw terminal
　　C. shell

162. The discharge load of low-voltage capacitor usually uses (　).
　　A. lamp　　　B. coil　　　C. mutual inductor

163. The capacitor bank is prohibited for (　).
　　A. live connecting　　　　　　B. charged connecting

Solutions: 154.B; 155.A; 156.C; 157.B; 158.B; 159.B; 160.A; 161.A; 162.A; 163.B

C. power off connecting

164. When selecting a voltmeter, it's better for the internal resistance to be ( ) the resistance of the load being measured.

    A. far less than   B. much greater than   C. equal to

165. ( ) instrument could be directly used for AC and DC measurement, but its accuracy is low.

    A. Magnetoelectric   B. Electromagnetic

    C. Electric

166. The harm caused by the thermal effect of current on human body is ( ).

    A. electric burn   B. electrobranding   C. skin metallization

167. TN-S is commonly known as ( ).

    A. three-phase-four-wire   B. three-phase-five-wire

    C. three-phase-three-wire

168. When installing the grounding wire, the maintenance equipment should be grounded and ( ) short circuited immediately after it is verified that there is no voltage.

    A. single phase   B. two phases   C. three phases

169. The higher the working voltage of the charged body, the air distance between them is required to be ( ).

    A. the same   B. greater   C. smaller

170. The role of adding steel core to aluminum strands is ( ).

    A. improving conductivity

    B. increasing conducting wire sectional area

    C. improving mechanical strength

171. According to its action mode, low-voltage apparatus could be divided into automatic switching apparatus and ( ) apparatus.

    A. non automatic switching   B. non electric

    C. non mechanical

172. ( ) belongs to the control apparatus.

    A. Contactor   B. Fuse   C. Knife switch

173. The multimeter consists of three main parts: meter head, ( ) and change-over switch.

    A. measuring circuit   B. coil

    C. pointer

174. The two basic principles for correct selection of electrical appliances are safety and ( ).

Solutions: 164.B; 165.B; 166.A; 167.B; 168.C; 169.B; 170.C; 171.A; 172.A; 173.A; 174.B

A. performance   B. economy   C. function

175. In the circuit, the switch should control ( ).

A. neutral wire   B. phase wire   C. grounding wire

176. For the selection of the circuit breaker, the ( ) of the circuit breaker should be determined first, and then the specific parameters should be determined.

A. type   B. rated current   C. rated voltage

177. When the iron case switch is used to control the start and stop of the motor, the rated current should be greater than or equal to ( ) times the rated current of the motor.

A. one   B. two   C. three

178. When measuring the voltage, the voltmeter should be connected with the circuit under test ( ).

A. in parallel   B. in series   C. to positive

179. When the human body contacts the two-phase conductor of the live equipment or line at the same time, the current flows from one phase to the other through the human body. This electric shock phenomenon is called ( ) electric shock.

A. single phase   B. two phases   C. inductive

180. The motor ( ) must be checked for insulation resistance of each winding to be qualified before being energized.

A. that has been in use for less than one day

B. that is not commonly used but has just stopped for less than a day

C. that is new or unused motor

181. The electrical fire of newly installed or unused motor is caused by the existence of dangerous temperature, among which short circuit, equipment failure, abnormal operation of equipment and ( ) might be the factors that cause dangerous temperature.

A. improper selection of conductor section

B. voltage fluctuation

C. long running time of equipment

182. Clamp ammeter is made by using the principle of ( ).

A. current mutual inductor   B. voltage mutual inductor

C. transformer

183. In the insulation inspection of each winding of 380V motor, if the insulation resistance ( ) is found, it could be preliminarily determined that the motor is damp, and the motor should be dried.

A. less than 10MΩ B. greater than 0.5MΩ

Solutions: 175.B; 176.A; 177.B; 178.A; 179.B; 180.C; 181.A; 182.A; 183.C

C. less than 0.5MΩ

184. The shunt capacitor should be configured in ( ).

A. Δ   B. Y   C. rectangle

185. Class Ⅱ hand electric tools are equipment with ( ) insulation.

A. basic   B. protective   C. double

186. The mechanical life of AC contactor refers to the number of operations without load, generally up to ( ).

A. less than 100000 times   B. 6 to 10 million times

C. more than 100 million times

187. ( ) could not be used as wires.

A. Copper stranded wire   B. Steel stranded wire

C. Aluminum stranded wire

188. Capacitors belong to ( ) equipment.

A. hazardous   B. movable   C. static

189. Generally, three-phase asynchronous motor could be directly started with power below ( ) kW.

A. 7   B. 10   C. 16

190. The power incoming wire of screw fuse should be connected at ( ).

A. the upper end   B. the lower end

C. the front end

191. The relationship between the voltages at both ends of each resistor in a series circuit is ( ).

A. the voltage at both ends of each resistor is equal

B. the lower the resistance, the higher the voltage at both ends

C. the higher the resistance, the higher the voltage at both ends

192. ( ) belongs to power distribution appliances.

A. Contactor   B. Fuse   C. Resistor

193. The wire joint is not tightly connected, which would cause the joint ( ).

A. to heat up   B. to insufficient insulation

C. to be non-conductive

194. The capacitor could be measured with a multimeter at the ( ) grade.

A. voltage   B. current   C. resistance

195. Capacitor banks above 1kV should use ( ) connected into Δ as the discharge

Solutions: 184.A; 185.C; 186.B; 187.B; 188.C; 189.A; 190.B; 191.C; 192.B; 193.A; 194.C; 195.C

device.

    A. incandescent lamp   B. current mutual inductor

    C. voltage mutual inductor

196. In general lighting lines, (　) is used to determine whether there is no electricity.

    A. megger measurement   B. test pencil measurement

    C. voltmeter measurement

197. When the combined switch is used for the reversible control of the motor, (　) for reverse connection.

    A. it is not necessary for the motor stops completely

    B. it is ok after the motor stops

    C. only after the motor stops completely, it is ok

198. When disconnecting the knife switch, if there is an arc, it should (　).

    A. disconnect quickly   B. connect immediately

    C. disconnect slowly

199. The rated working voltage of AC contactor refers to the (　) voltage that could ensure the normal operation of electrical appliances under specified conditions.

    A. minimum   B. maximum   C. average

200. The rated current of fuse should be (　) starting current of motor.

    A. greater than   B. equal to   C. less than

201. When the rubber case knife switch is connected, the power wire is connected to (　).

    A. upper end (static contact)   B. lower end (moving contact)

    C. both ends could be used

202. The highest power factor of the following lamps is (　).

    A. incandescent lamp   B. energy saving light

    C. fluorescent lamp

203. The anti-static grounding resistance should not be greater than (　) Ω.

    A. 10   B. 40   C. 100

204. The electrical diagram of the steel case switch is (　), and the character symbol is QS.

    A.   B.   C.

205. Pointer multimeter could generally measure AC/DC voltage, (　) current and resistance.

    A. AC/DC   B. AC   C. DC

Solutions: 196.B; 197.C; 198.A; 199.B; 200.C; 201.A; 202.A; 203.C; 204.B; 205.C

206. The ceramic bushing we usually call is called (　) in electrical engineering.
   A. insulating bushing
   B. isolator
   C. insulating element

207. (　) is unipolar semiconductor device.
   A. Diode
   B. Bipolar diode
   C. MOSFET

208. When low-voltage electrical fire occurs, the first thing to do is (　).
   A. leave the site quickly and reporting to the management
   B. try to cut off the power quickly
   C. try to cut off the power quickly and use dry powder or carbon dioxide extinguisher to extinguish the fire quickly

209. In case of electrical fire, the power supply should be cut off first and then put out the fire. However, when the power supply could not be cut off, only live fire fighting could be carried out. The fire extinguisher available for 500V low-voltage distribution cabinet is (　).
   A. carbon dioxide extinguisher
   B. foam fire extinguisher
   C. water based fire extinguisher

210. The three elements for determining the sine quantity are (　).
   A. phase, initial phase, phase difference
   B. maximum value, frequency, initial phase angle
   C. period, frequency, angular frequency

211. Single phase short circuit of the line refers to (　).
   A. too much power
   B. too much current
   C. direct connection of neutral & live wires

212. The (　) instrument could be directly used for AC and DC measurement with high accuracy.
   A. magnetoelectric
   B. electromagnetic
   C. electric

213. In three-phase AC circuit, phase A is marked with (　) color.
   A. red
   B. yellow
   C. green

214. The power of the capacitor is (　).
   A. active power
   B. reactive power
   C. apparent power

215. According to the line voltage level and user, power lines could be divided into distribution lines and (　) lines.
   A. power
   B. lighting
   C. power transmission

Solutions: 206.C; 207.C; 208.B; 209.A; 210.B; 211.C; 212.C; 213.B; 214.B; 215.C

216. In the electrical component symbols in the following figure, the electrical symbol belonging to capacitor is (   ).

A. ⊥ B. ⊤ C. ⌒

217. The thermal relay has a certain (   ) automatic adjustment and compensation function.

    A. time   B. frequency   C. temperature

218. When the intermediate joint of the conductor is twisted connnected, first twist (   ) turns in the middle.

    A. 1   B. 2   C. 3

219. In order for inspection, the power could be cut off for a short time, the capacitor must (   ) before touching.

    A. be fully discharged   B. be long power cut off
    C. be cooled

220. (GB/T 3805-2008) Limits of Extra Low Voltage (ELV) stipulates that under normal conditions, the limit of effective value of power frequency voltage under normal operation is (   ) V.

    A. 33   B. 70   C. 55

221. The specification of the screwdriver is expressed by the length and (   ) of the shank outside.

    A. radius   B. thickness   C. diameter

222. Carbon exists in nature in two forms: diamond and graphite, of which graphite is (   ).

    A. insulator   B. conductor   C. semiconductor

223. Components with inverse time A/S characteristics have short-circuit protection and (   ) protection capabilities.

    A. temperature   B. mechanical   C. overload

224. (   ) is the organizational measures to ensure the safety of electrical operation.

    A. Work permit system   B. Power cut-off
    C. Install grounding wire

225. When the motor runs under the rated working condition, the (   ) applied to the stator circuit is called the rated voltage.

    A. wire voltage   B. phase voltage   C. rated voltage

Solutions: 216.B; 217.C; 218.C; 219.A; 220.A; 221.C; 222.B; 223.C; 224.A; 225.A

226. The unit of resistance measured with a megger is ( ).

    A. Ohms   B. Kiloohm   C. Megaohm

227. When measuring the resistance with a multimeter, the black probe is connected to ( ) of the power supply in the meter.

    A. two polarities   B. negative polarity   C. positive polarity

228. For low-voltage distribution network, when the distribution capacity is below 100kW, the grounding resistance should not exceed ( ) Ω.

    A. 10   B. 6   C. 4

229. Insulating gloves is ( ) safety tools.

    A. direct   B. auxiliary   C. basic

230. The fuse in the general circuit has ( ) protection.

    A. overload   B. short circuit   C. overload and short circuit

231. According to international and national standards, ( ) wire could only be used as protective grounding or protective neutral wire.

    A. black   B. blue   C. yellow and green

232. When a fuse protects a lamp, the fuse should be connected in series ( ) the switch.

    A. in front of   B. behind   C. in

233. In general, the resistance of human body under the action of 220V power frequency voltage is ( ) Ω.

    A. 500~1000   B. 800~1600   C. 1000~2000

234. Alarm type leakage protector should be installed without automatically cutting off the power supply in ( ).

    A. guest house socket circuit   B. electrical equipment for production

    C. fire elevator

235. The mechanical strength of the conducting wire joint should not be less than ( )% of the mechanical strength of the original conductor.

    A. 80   B. 90   C. 95

236. The allowable voltage loss of lines in general lighting places is ( ) of the rated voltage.

    A. ±5%   B. ±10%   C. ±15%

237. The emergency lighting generally adopts ( ).

    A. fluorescent lamp   B. incandescent lamp

    C. high pressure mercury lamp

238. On each single-phase circuit of the lighting system, the number of lamps and

Solutions: 226.C; 227.C; 228.A; 229.B; 230.C; 231.C; 232.B; 233.C; 234.C; 235.A; 236.A; 237.B; 238.B

sockets should not be more than (　).

　　A. 20　　　　B. 25　　　　　C. 30

239. The clamp ammeter could measure the current under the condition of (　) circuit.

　　A. open　　　B. short circuited　　C. closed

240. The color of zero wire in low-voltage line is (　).

　　A. navy blue　　B. light blue　　C. yellow and green

241. When checking the grounding fault point where there may be high step voltage, the distance to the fault point should not be less than (　) m indoor.

　　A. 2　　　　　B. 3　　　　　C. 4

242. The symbol of the fuse is (　).

　　A. ─▭─　　B. ─▯─　　C. ─▷│─

243. The function of connecting power capacitor in parallel is (　).

　　A. reduce power factor　　　B. increase power factor

　　C. maintain current

244. The protection feature of fuse is also called (　).

　　A. arc extinguishing characteristic　　B. ampere second characteristic

　　C. time characteristic

245. Y-Δ step-down startup refers to the configuration of stator three-phase windings (　) when it starts.

　　A. in Δ　　　B. in Y　　　C. in extended triangle

246. Concealed switches and sockets should have (　).

　　A. clear mark　　B. cover　　C. warning sign

247. When using bamboo ladders, the included angle between the ladder and the ground should be (　).

　　A. 50°　　　B. 60°　　　C. 70°

248. The non-automatic switching device works by (　) for operation.

　　A. external force (such as manual control)

　　B. electric drive

　　C. induction

249. In the distribution system of civil buildings, (　) circuit breakers are generally used.

　　A. frame type　　B. electric　　C. leakage protection

250. When several lines are erected on the same pole, it must be ensured that the high-voltage line is (　) the low-voltage line.

Solutions: 239.C; 240.B; 241.C; 242.A; 243.B; 244.B; 245.B; 246.B; 247.B; 248.A; 249.C; 250.C

A. on the left of B. on the right of   C. above

251. When the capacitor is measured with a multimeter, the pointer should (　) after swinging.

  A. remain still   B. swing back gradually

  C. swing back and forth

252. AC 10kV bus voltage refers to AC three-phase-three-wire system (　).

  A. wire voltage   B. phase voltage   C. line voltage

253. The fuse plays a/an (　) protective role in the circuit of the motor.

  A. overload   B. short circuit   C. overload and short circuit

254. The replacement of fusant or fusant pipe must be carried out under the condition of (　).

  A. being charged   B. being discharged

  C. being loaded

255. According to the counting method, electrical instruments are mainly divided into pointer type apparatus and (　) type apparatus.

  A. electric   B. comparison   C. digital

256. For transformers and high-voltage switch cabinet, the main measure to prevent damage caused by lightning intrusion is (　).

  A. installing lightning arrester   B. installing lightning conductor

  C. installing lightning protection net

257. (　) apparatus is composed of fixed coil, rotatable iron core, rotating shaft, hairspring, pointer, mechanical zero adjustment mechanism, etc.

  A. Magnetoelectric   B. Electromagnetic

  C. Electric

258. Brain cells are most sensitive to hypoxia. Generally, hypoxia for more than (　) min would cause irreversible damage and lead to brain death.

  A. 5   B. 8   C. 12

259. The luminous voltage of the high-voltage electroscope should not be higher than (　)% of the rated voltage.

  A. 25   B. 50   C. 75

260. Clamp ammeter consists of current mutual inductor and magnetoelectric meter head with (　).

  A. measuring circuit   B. rectifier

  C. pointer

Solutions: 251.B; 252.A; 253.B; 254.B; 255.C; 256.A; 257.B; 258.B; 259.A; 260.B

261. The electric energy meter is an instrument for measuring ( ).
    A. current   B. voltage   C. electric energy

262. The resistance in human body is about ( ) Ω.
    A. 200   B. 300   C. 500

263. Artificial respiration should be carried out for the adult injured with electric shock. The volume of air blown into the injured each time should reach ( ) ml to ensure sufficient oxygen.
    A. 500~700   B. 800~1200   C. 1200~1400

264. When the power is supplied by a special transformer, if the motor capacity is less than ( ) time of the transformer capacity, direct starting is allowed.
    A. 0.6   B. 0.4   C. 0.2

265. If an employee needs to suspend his/her work for treatment due to an accident or occupational disease, he/she may stop work with salary, but the period of suspension with salary generally does not exceed ( ) months.
    A. 6   B. 10   C. 12

266. ( ) tools should be used when working in narrow places such as boilers, metal containers and pipes.
    A. Class I   B. Class II   C. Class III

267. Low-voltage electrician operation refers to the installation, commissioning, operation, etc., of electrical equipment below ( ) V.
    A. 250   B. 500   C. 1000

268. ( ) V is preferred for general lighting.
    A. 220   B. 380   C. 36

269. Among the following phenomena, ( ) could be determined that the contact is poor.
    A. fluorescent lamp is difficult to start   B. the bulb flickers
    C. the bulb doesn't work

270. Special operation personnel must be at least ( ) years old.
    A. 18   B. 19   C. 20

271. The minimum insulation resistance of Class II tools should be ( ) MΩ.
    A. 5   B. 7   C. 9

272. The general temperature of bare conducting wire in operation should not exceed ( ).
    A. 40 ℃   B. 70 ℃   C. 100 ℃

Solutions: 261.C; 262.C; 263.B; 264.C; 265.C; 266.C; 267.C; 268.A; 269.B; 270.A; 271.B; 272.B

273. If the heart beat stops and breathing exists, (　) should be carried out immediately.
　　A. artificial respiration
　　B. extrathoracic cardiac compression
　　C. artificial respiration and extrathoracic cardiac compression

274. If special operation personnel fail to receive special training on safe operation and obtain corresponding qualifications as required, (　) should be ordered to work by the production and business operation entity.
　　A. making corrections within a time limit
　　B. imposing a fine
　　C. suspending production and business for rectification

275. In the distribution line, when the fuse is used for overload protection, the rated current of the fusant should not be greater than (　) times the allowable ampacity of the conducting wire.
　　A. 1.25　　　　B. 1.1　　　　C. 0.8

276. The lighting used in flammable and explosive places should be (　) lamps.
　　A. explosion proof type　　　　B. moistureproof type
　　C. ordinary type

277. The rated voltage of the fuse is the maximum working voltage of the specified circuit from the perspective of (　).
　　A. overload　　　　B. arc extinguishing.
　　C. temperature

278. Earplugs is (　).
　　A. noise protection article　　　　B. hearing aids article
　　C. hearing protection article

279. The insulation resistance of Class I electric tools should not be lower than (　).
　　A. 1MΩ　　　　B. 2MΩ　　　　C. 3MΩ

280. For the red signal lights installed on the mechanical equipment under construction that affect the passage of aircraft or vehicles at night, the power supply is set (　) the main switch.
　　A. in front of　　B. behind　　C. on the left side of

281. It could be seen from the actual accidents that more than 70% of the accidents are related to (　).
　　A. the technical level　　　　B. people's emotions
　　C. human errors

282. In a closed circuit, the current intensity is proportional to the power supply

---

Solutions: 273.B; 274.A; 275.C; 276.A; 277.B; 278.A; 279.B; 280.A; 281.C; 282.A

electromotive force and inversely proportional to the sum of the internal resistance and external resistance in the circuit, which is called ( ).

    A. Ohm's law of full circuit   B. Current law of full circuit
    C. Ohm's law of partial circuits

283. The action current of the leakage alarm device used to prevent leakage fire is about ( ) mA.

    A. 6~10   B. 15~30   C. 30~50   D. 100~200

284. When a fuse protects a lamp, the fuse should be connected in series ( ) the switch.

    A. in front of   B. behind   C. in

285. The withstand voltage of wire pliers with plastic handle used by electricians is below ( ) V.

    A. 380   B. 500   C. 1000

286. ( ) gloves should be worn when handling corrosive products.

    A. Canvas   B. Rubber   C. Cotton

287. The unit of battery capacity is ( ).

    A. W   B. A·h   C. kW·h

288. Hand-held electric tools with 回 symbol are ( ) tools.

    A. class I   B. class II   C. class III

289. The conductivity of the diode is ( ) conductive.

    A. unidirectional   B. bidirectional
    C. three-way

290. As shown in the figure, connect a "220V 40W" bulb L_0 at the fuse. If S is closed, both $L0$ and L are dark red, it could be determined that the branch is normal; When S is closed, L_0 would normally emit light, and L would not emit light, so it could be determined that L lamp cap is ( ).

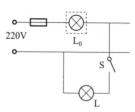

    A. short circuited   B. open circuited
    C. electricity leaked

291. Electric injury is caused to human body by the ( ) effect of current.

Solutions: 283.D; 284.B; 285.B; 286.B; 287.B; 288.B; 289.A; 290.A; 291.C

A. thermal B. chemical
C. thermal, chemical and mechanical

292. Sometimes, before measuring the current with a clamp meter, open and close the jaw several times for the purpose of ( ).
A. residual current elimination B. remanence elimination
C. residual stress relief

293. In the production process, static electricity is harmful to human body, equipment and products. To eliminate or reduce static electricity, spray humidifier could be used for the purpose of ( ).
A. leaking static charge through air
B. making the static charge radiate and leak around
C. making static electricity leak along the surface of insulator

294. ( ) is one of the technical measures to ensure the safety of electrical operation.
A. The work card system B. Electricity inspection
C. The work permit system

295. In order to prevent the step voltage from harming people, the minimum distance between the lightning protection grounding device and the building entrance and sidewalk should not be less than ( ) m.
A. 2.5 B. 3 C. 4

296. When replacing and overhauling electrical equipment, the best safety measure is ( ).
A. cutting off the power B. operating on a stool
C. operating with rubber gloves

297. Clean the dirt and dust inside the motor with ( ).
A. wiping with wet cloth
B. wiping by cloth with gasoline, kerosene, etc
C. blowing with compressed air or wiping with dry cloth

298. In the neutral grounded power supply system, the metal shells of all electrical equipment are reliably connected with the zero wire of the system, it is ( ) to use protective grounding instead of protective neutral connection.
A. allowed B. prohibited C. able

299. The connecting capacity of AC contactor refers to the capacity that would not cause ( ) when the switch is closed to make current.
A. contact melting B. arc
C. voltage drop

Solutions: 292.B; 293.C; 294.B; 295.B; 296.A; 297.C; 298.B; 299.A

300. Respiratory protective articles that are inhaled and filtered through a canister (box) to remove toxins in the air to prevent toxins from being inhaled into the respiratory tract are: ( ).
   A. self contained air breathing apparatus
   B. long tube mask
   C. filter type gas mask

301. Judging from the path of current flowing through the human body during electric shock, the most dangerous electric shock path is ( ).
   A. anterior chest to left hand   B. anterior chest to right hand
   C. hands to feet

302. ( ) instrument has high sensitivity and accuracy, and is mostly used to make portable voltmeter and ammeter.
   A. Magnetoelectric   B. Electromagnetic
   C. Electric

303. The work cards used for the written basis of electrical operation should be in ( ) copies.
   A. 2   B. 3   C. 4

304. High strength copper core rubber sheath soft insulation ( ) should be adopted for power supply of mobile electrical equipment.
   A. conducting wire   B. cable   C. stranded wire

305. The skin protection articles that could remove oil, dust, poison and other dirt on the skin and protect the skin from damage are called ( ).
   A. waterproof skin care agent   B. oil proof skin care agent
   C. cleansing skin care agent

306. Dry powder fire extinguisher could be used for live extinguishing of lines below ( ) kV.
   A. 10   B. 35   C. 50

307. When working on electrical equipment of 10kV and below, the safe distance between the normal activity range and live equipment is ( ) m.
   A. 0.2   B. 0.35   C. 0.5

308. The general temperature of plastic insulated conducting wire in operation should not be higher than ( ).
   A. 40 ℃   B. 70 ℃   C. 100 ℃

309. In the process of underground production in the coal mine, if fracture occurs on somebody, other personnel should adopt the first aid principle of ( ).
   A. waiting for the ambulance to arrive   B. sending to the hospital immediately
   C. fixing before handling

Solutions: 300.C; 301.A; 302.A; 303.A; 304.B; 305.C; 306.C; 307.B; 308.A; 309.C

310. When selecting the knife switch, the rated voltage of the knife switch should be greater than or equal to the actual ( ) voltage of the line.

　　A. rated　　　　B. maximum　　　C. fault

311. In the case of measuring the insulation resistance of stator windings and housing phase by phase with a megger. When the handle is rotated, the pointer points to zero, indicating the winding ().

　　A. connects the housing　　　　B. is short circuited

　　C. is open circuited

312. The lightning rod is a common lightning protection device. During installation, the lightning rod should be equipped with an independent grounding device. If it is in a non high resistivity area, its grounding resistance should not be higher than ( ) Ω.

　　A. 2　　　　B. 4　　　　C. 10

313. Silicosis is a serious occupational disease caused by ( ).

　　A. productive dust　　　　B. productive poison

　　C. benzene

314. It is specified that the terminal voltage of fixed battery is ( ) when discharging at discharge rate for 10 hours (25 ℃).

　　A. 1.8V/one　　B. 2V/one　　C. 2.2V/one

315. When the injured person suddenly stops breathing in underground operation well, ( ) should be done.

　　A. timely carrying out artificial respiration and then transfer to hospital

　　B. sending him to the ground hospital immediately

　　C. reporting to the mine dispatching room immediately

316. Thunder that could produce electric effect, thermal effect and mechanical effect on buildings, electrical equipment and structures and has greater destructive effect is ( ).

　　A. spherical thunder　　　　B. inductive thunder

　　C. direct strike lightning

317. When the protective grounding of low-voltage power supply lines and the lightning protection grounding grid of buildings need to be shared, the grounding grid resistance requirements are ( ).

　　A. ⩽ 2.5 Ω　　B. ⩽ 1 Ω　　C. ⩽ 10 Ω

318. In a three-phase-four-wire system with a phase voltage of 220V, the working grounding resistance R_N is 2.8 Ω. The electrical equipment in the system adopts the grounding protection mode, and the grounding resistance is R_A=3.6 Ω. If the electrical equipment leaks,

Solutions: 310.B; 311.A; 312.C; 313.A; 314.A; 315.A; 316.C; 317.B; 318.B

the grounding voltage of the leakage equipment before troubleshooting is ( ) V.

    A. 34.375   B. 123.75   C. 96.25

319. For high-voltage power supply users with a receiving capacity of more than 160kV · A, the monthly average power factor standard is ( ).

    A. 0.8   B. 0.85   C. 0.9

320. The distance between the grounding device of the lightning rod and the road entrance should not be less than ( ) m.

    A. 1   B. 2   C. 3   D. 5

321. The bimetallic sheet of thermal relay would ( ) deform after being heated.

    A. expand and   B. stretch and   C. be damaged and   D. be bent and

322. The daily average maximum temperature of the environment where the power capacitor is located should not be higher than ( ) ℃.

    A. 35   B. 40   C. 45   D. 50

323. *The Labor Law of the People's Republic of China* has been formally implemented since ( ) in 1995.

    A. January 1st B. May 1st C. June 1st D. October 1st

324. When the multi-layer power lines erected on the same pole/tower are connected with grounding wires, the installation sequence should be ( ). The removal sequence is reverse.

    A. hang the low-voltage first, then the high-voltage; Hang the lower layer first, then the upper layer; Hang the far side first, then the near side

    B. hang the high-voltage first, then the low-voltage; Hang the lower layer first, then the upper layer; Hang the near side first, then the far side

    C. hang the low-voltage first, then the high-voltage; Hang the lower layer first, then the upper layer; Hang the near side first, then the far side

325. *The Law on the Prevention and Control of Occupational Diseases of the People's Republic of China* has come into force since ( ) 2002.

    A. January 1st B. October 1st C. May 1st D. December 1st

326. The fire resistance rating of capacitor room should not be lower than grade ( ).

    A. one   B. two   C. three   D. four

327. The ( ) fuses have the largest breaking capacity.

    A. quartz sand filled cartridge   B. fiber tube

    C. ceramic plug-in   D. open

328. The megger has three terminals, L, E and G. Where, the function of the terminal G is ( ) when measuring the cable.

Solutions: 319.C; 320.C; 321.D; 322.B; 323.A; 324.C; 325.C; 326.B; 327.A; 328.C

A. to make mechanical zero adjustment
B. used for grounding protection
C. to connect the short circuit ring
D. to connect the conductor to be measured

329. Mobile electrical apparatus should regularly ( ).
A. test the running voltage   B. replace the power wire
C. check the load current   D. measure the insulation resistance

330. Sudden fainting or convulsion during insulation resistance measurement is ( ) heatstroke symptom.
A. premonitory   B. mild   C. severe

331. The occupational diseases specified in *The Catalogue of Occupational Diseases* (No. 108, 2002) issued by the Ministry of Health are classified in ( ) catagories.
A. 8   B. 9   C. 10

332. ( ) are inorganic dust.
A. Cement, emery, glass fiber   B. Cotton, hemp, flour, wood
C. Animal hair, cutin, hair

333. The occupational disease diagnosis institution should issue an *Occupational Disease Diagnosis Certificate* to the employee who has been diagnosed as suffering from occupational disease, and enjoy the ( ) stipulated by the state.
A. medical insurance benefits
B. industrial injury insurance benefits
C. endowment insurance benefits

334. The two resistors marked with "100 Ω 4W" and "100 Ω 36W" are connected in series, and the maximum allowable voltage is ( ) V.
A. 20   B. 40   C. 60

335. When overhead lines are adjacent to explosive gas environment, the distance between them should not be less than ( ) times of pole/tower height.
A. 3   B. 2.5   C. 1.5

336. During low-voltage live working, ( ).
A. both wearing insulating gloves and being supervised are necessary
B. wear insulating gloves is necessary, but supervision is not
C. it is unnecessary to wear insulating gloves when someone is monitoring

337. Occupational disease hazardous factors include various harmful ( ) factors existing in occupational activities and other occupational harmful factors generated in the

---

Solutions: 329.D; 330.C; 331.C; 332.A; 333.B; 334.B; 335.C; 336.A; 337.A

process of operation.

    A. chemical, physical, biological   B. chemical, physical, physiological

    C. physiological, physical, biological

338. The grounding wire should be multi strands soft bare copper wire, and its sectional area should not be less than ( ) mm².

    A. 6   B. 10   C. 25

339. When a miner rescues himself, if he feels the dry and hot air he inhales after wearing a self rescuer, he should ( ).

    A. not take it off, keep using it, and leave the disaster area

    B. take it off, walk evenly, and leave the disaster area

    C. take it off, wait for rescue, and leave the disaster area

340. Hand electric tools are divided into ( ) categories according to electric shock protection methods.

    A. 2   B. 3   C. 4

341. Productive poisons could cause occupational poisoning. If small amount of poison enters the body continuously for a long time, it might cause ( ).

    A. acute poisoning   B. chronic poisoning

    C. subacute poisoning

342. In accordance with the provisions of *The Production Safety Law of the People's Republic of China*, the production and business operation entity ( ) conclude an agreement with the employees to exempt or mitigate its legal liability for the casualties of employees due to production safety accidents.

    A. could

    B. getting the approval of the relevant departments to

    C. should not

    D. is generally not allowed to

343. The purpose of formulating *The Production Safety Law of the People's Republic of China* is to prevent and reduce production safety accidents and ensure the ( ) safety of the people.

    A. lives   B. properties

    C. lives and properties   D. lives and health

344. The effective value limit of power frequency safety voltage is ( ) V according to national standards.

    A. 220   B. 50   C. 36   D. 6

Solutions: 338.C; 339.A; 340.B; 341.B; 342.C; 343.C; 344.B

345. The explosion-proof certificate number is marked with the symbol "( )", indicating that the equipment has a component certificate.
　　A. W　　　　B. U　　　　　　C. X　　　　　　　D. F

346. For general work at heights, ( ) should be worn.
　　A. hard soled shoes　　　　　B. antiskid shoes with soft soles
　　C. ordinary rubber shoes　　　D. canvas shoe

347. For capacitor banks in operation, patrol inspection in summer should be arranged ( ).
　　A. at the highest room temperature　　B. when the system voltage is the highest
　　C. at 2 o'clock every afternoon　　　　D. at any time

348. The grade marked AC on the multimeter is ( ).
　　A. AC current　　B. AC voltage　　C. DC current　　D. DC voltage

349. Fire hazard zone 21 is a fire hazard environment with combustible ( ).
　　A. liquid　　　B. powder　　　C. fiber　　　D. solid

350. The work that underage worker could do is ( ).
　　A. engaging in underground work
　　B. work with Grade IV physical labor intensity stipulated by the state
　　C. toxic and harmful work
　　D. work with Grade III physical labor intensity stipulated by the state

351. During the operation of power capacitor, the current should not be greater than ( ) times the rated current of capacitor for long time operation.
　　A. 1.1　　　B. 1.2　　　C. 1.3　　　D. 1.5

352. Ammonium phosphate dry powder extinguisher could not be used to extinguish ( ).
　　A. the initial fire of solid substances
　　B. metal burning fire
　　C. the initial fire of flammable and combustible liquid and gas
　　D. the initial fire of charged equipment

353. When extinguishing a fire with a dry powder extinguisher, the extinguisher could be carried to the fire site by hands or shoulders, and put the extinguisher down about ( ) from the burning place.
　　A. 3m　　　B. 5m　　　C. 7m　　　D. 9m

354. The minimum frequency of compressing the heart is ( ).
　　A. 10 times/min　B. 50 times/min　　C. 100 times/min　　D. 200 times/min

355. According to Decree 260 of the provincial government, if a production safety

---

Solutions: 345.B; 346.B; 347.A; 348.A; 349.A; 350.D; 351.C; 352.B; 353.B; 354.C; 355.B

accident occurs due to the contract letting or leasing to an entity or individual that does not have the conditions for safe production or the corresponding qualifications, or the failure to sign an agreement on safe production management with the contractor or lessee, what responsibilities should the production and business operation unit and the contractor and lessee bear? ( )

    A. all and joint liability for compensation
    B. the main and joint liability for compensation
    C. the main and full liability for compensation
    D. all and full liability for compensation

356. ( ) refers to tools mainly used to further strengthen the insulation strength of basic safety appliances.
    A. Insulation safety apparatus   B. General protective safety equipment
    C. Basic safety apparatus   D. Auxiliary safety equipment

357. ( ) is the most dangerous current path.
    A. From right hand to foot   B. From left hand to right hand
    C. From left hand to chest   D. From left hand to foot

358. When the electrified body has a grounding fault, ( ) could be used as a basic safety apparatus to protect the step voltage.
    A. low-voltage test pencil   B. insulated shoes
    C. signboards   D. temporary barrier

359. The working current of single-phase socket in indoor non-production environment could generally be ( ) A.
    A. 1   B. 1.5   C. 2   D. 2.5

360. If the power cut-off line operation also involves the lines that other entities cooperate with the power cut-off, the person in charge of the work should not start the work until the designated contact person of the operation and management unit of the cooperation cut-off equipment notifies that these lines have been cut-off and grounded, and performs ( ).
    A. work permit
    B. telephone release of work permit
    C. face to face notification of work permit
    D. written procedures for work permit

361. When the handle of the megger has not been turned, the pointer of the megger that is horizontally placed in good condition should point to ( ).
    A. the leftmost end of the dial   B. the rightmost end of the dial

---

Solutions: 356.D; 357.C; 358.B; 359.D; 360.D; 361.D

C. the center of the dial　　　　D. the random position

362. Before the installation of arc welding machine, check whether the insulation resistance is qualified, and the primary insulation resistance should not be lower than (　) MΩ.

A. 1　　　　B. 2　　　　C. 3　　　　D. 4

363. When the load current reaches the rated current of the fuse fusant, the fusant would (　).

A. melt immediately　　　　B. melt after a long time delay
C. melt after a short time delay　　　　D. not melt

364. When removing welding slag or excess metal, what should be done to reduce the danger? (　)

A. The direction of removal must be close to the body.
B. Wear personal protective articles such as goggles and gloves
C. The fan must be turned on to strengthen the air circulation and reduce the inhalation of metal mist
D. Without any protective articles.

365. Electrostatic phenomenon is a very common electrical phenomenon, (　) is its greatest harm.

A. human body discharge　　　　B. killing people directly
C. high-voltage breakdown insulation　　D. easy to cause fire

366. Tools above (　) V must be of double insulation structure.

A. 36　　　　B. 70　　　　C. 220　　　　D. 380

367. The following mandatory clauses of the labor contract are (　).

A. remuneration　　　　B. probation period
C. keeping business secrets　　　　D. fringe benefits

368. Except for mine methane, explosive gas, steam and mist are classified into (　) classes according to the minimum ignition current ratio.

A. 3　　　　B. 4　　　　C. 5　　　　D. 6

369. (　) is not the cause of the fusant fusing of the primary fuse of the voltage mutual inductor.

A. Phase to phase short circuit of primary winding
B. Inter turn short circuit of primary winding
C. Short circuit of primary winding terminal
D. Phase to phase short circuit of secondary winding

370. When using dry powder extinguisher containers, if the wall temperature has been higher than the spontaneous ignition point of the flammable liquid, it is very easy to

Solutions: 362.A; 363.D; 364.B; 365.C; 366.C; 367.A; 368.A; 369.B; 370.A

cause the phenomenon of re-ignition after fire extinguishing. If it is used with ( ), the fire extinguishing effect is better.

  A. foam fire extinguisher   B. CO_2 extinguisher
  C. acid alkali fire extinguisher   D. gas cylinder fire extinguisher

371. The extrathoracic cardiac compression method is applied about ( ) times per minute.

  A. 3~5   B. 10~20   C. 30~40   D. 60~80

372. The voltage of portable lamps used in metal containers should be ( ) V.

  A. 220   B. 110   C. 50   D. 12

373. On August 31_{st}, 2014, the 10th Meeting of the Standing Committee of the Twelfth National People's Congress of the People's Republic of China deliberated and adopted the Decision of the Standing Committee of the National People's Congress on Amending the *Production Safety Law of the People's Republic of China*, and the revised *Production Safety Law of the People's Republic of China* was officially implemented from ( ).

  A. October 1st, 2014   B. December 1st, 2014
  C. January 1st, 2015   D. January 3rd, 2015

374. The voltmeter and relay coil in the secondary circuit of the voltage mutual inductor are connected in ( ) with the secondary coil of the mutual inductor.

  A. parallel B. series   C. Y   D. Δ

375. The safety belt should be ( ).

  A. hung flat during use   B. hung low and used high
  C. hung high and used low   D. hung securely during use

376. The lines behind the leakage protector are only allowed to connect ( ).

  A. to the working zero wire of the electrical equipment
  B. to PE
  C. to PEN
  D. to grounding

377. When measuring the grounding resistance between the grounding terminal in the socket and the equipment frame or between the equipment frame and the control panel, the grounding resistance should generally be less than ( ).

  A. 0.1 Ω   B. 0.2 Ω   C. 0.3 Ω   D. 0.4 Ω

378. The distance between the crane and the nearest point of the line below 1kV should not be less than ( ) m.

  A. 1.5   B. 2   C. 2.5   D. 3

Solutions: 371.D; 372.D; 373.B; 374.A; 375.C; 376.A; 377.A; 378.A

379. The three elements for determining the sine quantity are ( ).
   A. phase, initial phase, phase difference
   B. maximum value, frequency, initial phase angle
   C. period, frequency, angular frequency

380. When the conducting wire of 6~10kV overhead line passes through the residential area, the minimum distance between the wire and the ground is ( ) m.
   A. 4   B. 5   C. 6.5

381. The period of validity of the special operation certificate is ( ) years.
   A. 12   B. 8   C. 6

382. Y-Δ step-down starting refers to the ( ) configuration of three-phase stator windings during starting.
   A. Δ   B. Y   C. extended triangle

383. To improve the power factor, 40W lamp should be equipped with ( ) μF capacitance.
   A. 2.5   B. 3.5   C. 4.75

384. There are many methods of artificial respiration, but ( ) artificial respiration is the most convenient and effective.
   A. mouth to mouth blowing
   B. lying down and compressing the back
   C. supine chest compression method

385. The climbing board and rope should be able to withstand ( ) N tensile test.
   A. 1000   B. 1500   C. 2206

386. Enclose the sound source with certain materials, structures and devices to achieve the purpose of noise transmission. Such technical measure to control noise propagation is ( ).
   A. noise absorption   B. noise elimination
   C. noise insulation

387. The signboard "( )" should be hung on the iron frame and ladder used by the staff to get up and down, and "No Climbing, High Voltage Danger!" should be hung on other nearby live frames that may be boarded by mistake.
   A. Work here!   B. No climbing, high-voltage danger!
   C. Upwards and downwards here!

388. When electric shock is found in the underground human body, rescue should be organized so that the person who gets electric shock ( ), and quickly remove phlegm,

---

Solutions: 379.B; 380.C; 381.C; 382.B; 383.C; 384.A; 385.C; 386.C; 387.C; 388.C

mucus and other secretions in the mouth or nose.

    A. lie on his side B. sit half down   C. lie flat

389. Insulating gloves and insulating shoes should be tested regularly, and the test cycle is generally ( ) months.

    A. 1   B. 3   C. 6   D. 12

390. Provincial Government Decree No. 260 stipulates that ( ) a safety director should be appointed.

    A. high-risk production and operation entities with more than 100 employees
    B. high-risk production and operation entities with more than 300 employees
    C. high risk production and business units with more than 100 employees and other production and business units with more than 500 employees
    D. that high risk production and business units with more than 300 employees and other production and business units with more than 1000 employees

391. If Class II equipment is used in narrow and special dangerous places such as boilers, metal containers and pipelines, ( ) protection must be installed.

    A. short circuit   B. overload
    C. voltage loss   D. leakage

392. According to the provisions of the *The Production Safety Law of the People's Republic of China*, the employees of production and business operation entities have the right to know the risk factors, preventive measures and ( ) in their workplaces and posts.

    A. employment conditions   B. safety technical measures
    C. safety investment   D. accident emergency measures

393. According to the provisions of the *Emergency Response Law of the People's Republic of China*, according to the emergency degree, development trend and possible harm degree of the emergency, the accident early warning is divided into four levels, of which the highest level is ( ) early warning.

    A. red   B. yellow   C. blue   D. orange

394. When working with face shield goggles, replace the protective sheet at least once for ( ) hours.

    A. 5   B. 6   C. 7   D. 8

395. Serious external bleeding caused by accident should be ( ).

    A. wrapped after cleaning the wound
    B. directly wrapped with cloth to prevent bleeding

Solutions: 389.C; 390.B; 391.D; 392.D; 393.A; 394.D; 395.A

C. absorbed with cotton wool

D. allowed to bleed, and the blood platelets in the blood coagulate the blood flowing out

396. The key measure for positive pressure electrical equipment is that the pressure of the protective gas (fresh air or inert gas) inside the equipment is at least (   ) higher than the ambient pressure.

  A. 40Pa   B. 50Pa   C. 60Pa   D. 70Pa

397. *The Provisions on the Special Protection of underaged Workers* was promulgated by the former Ministry of Labor in (   ).

  A. 1994   B. 1995   C. 1996   D. 1997

398. In case of a fire in a single heading roadway, it is required to (   ) while maintaining the normal ventilation of the local ventilator.

  A. evacuate safely   B. extinguish actively

  C. rescue quickly   D. close quickly

399. *The Production Safety Law of the People's Republic of China* stipulates that the production and storage entities of hazardous substances, as well as the mining and metal smelting entities, should appoint (   ) engaged in the management of production safety.

  A. part-time safety production management personnel

  B. full time or part-time safety production management personnel

  C. technician

  D. certified safety engineer

400. When the shocked person is disconnected from the power supply, such as deep coma, respiratory and cardiac arrest, the first thing to do is (   ).

  A. finding an ambulance and waiting for it to arrive

  B. emergency transport to hospital

  C. carrying out mouth to mouth (nose) artificial respiration and extrathoracic cardiac compression on the spot

  D. letting the shocked lie down

401. An enterprise has revised the Safety Production Responsibility System in the process of safety production standardization construction, which should be implemented after being signed and issued by the enterprise's (   ).

  A. person in charge of safety   B. chief engineer

  C. person in charge of production   D. principal responsible person

402. Construction activities should ensure (   ).

Solutions: 396.B; 397.A; 398.B; 399.D; 400.C; 401.D; 402.B

A. economy

B. the quality and safety of construction projects

C. technology advanced

D. conducive to promoting local economic development

403. The three-phase-four-wire line and three-phase lighting line shared by power and lighting must use ( ) pole protector.

  A. one   B. two   C. three   D. four

404. ( ) belong to the permitted signboards.

  A. No smoking   B. No switching on, someone is working!

  C. Work here!   D. Danger:High Voltage!

405. In a dry environment, the human body resistance is about ( ) within the range of 100~220V contact voltage.

  A. 1000~3000 Ω B. 10~30k Ω   C. 100~300k Ω   D. 100~300 Ω

406. During emergency rescue, it is advisable to press the adult's chest to make the sternum sink ( ) cm.

  A. 1~2   B. 2~4   C. 3~5   D. 5~7

407. The correct wiring of single-phase three hole sockets from facing the three sockets, which are a 品 shape is ( ).

  A. connecting L on the left, N on the right, and PE on the top

  B. connecting L on the left, PE on the right, and N on the top

  C. connecting N on the left, L on the right, and PE on the top

  D. connecting PE on the left, L on the right, and N on the top

408. For carbon dioxide extinguishers without jet hoses, pull the nozzle upward ( ).

  A. 10°~30°   B. 30°~70°   C. 50°~70°   D. 70°~90°

409. The sectional area of copper core flexible wire of indoor pendant lamp in civil buildings should not be less than ( ) mm^2.

  A. 0.5   B. 0.4   C. 0.3   D. 0.2

410. ( ) refers to the activity of restoring the electrical equipment that has been used or stored for a period of time but is not necessarily faulty to a fully usable state.

  A. Repair   B. Overhauling   C. Maintainance   D. Restoring

411. The minimum depth of low-voltage power cable when directly buried is ( ) m.

  A. 0.6   B. 0.7   C. 1   D. 1.05

412. The valid period of plastic safety helmet is ( ) from the date of completion of product manufacturing

---

Solutions: 403.D; 404.C; 405.A; 406.C; 407.C; 408.D; 409.B; 410.B; 411.B; 412.B

A. 3 years   B. 2 and half years   C. 2 years   D. 1 and half years

413. The Power supply voltage of the portable lamp should not exceed ( ) V.

A. 40   B. 36   C. 24   D. 12

414. The national emergency leading agency for production safety accidents and disasters is the Safety Committee of the State Council, ( ) is specifically responsible for the emergency management of production safety accidents and disasters.

A. Office of the Security Committee of the State Council
B. National Production Safety Emergency Rescue Command Center
C. State Administration of Production Safety
D. State Administration of Production

415. The power capacitor should not be closed with residual charge. The capacitor should be discharged for at least ( ) min before re-connecting.

A. 1   B. 2   C. 3   D. 10

416. Partial lighting lamps and hand lighting lamps used in the environment with electric shock hazard should use a safe voltage not exceeding ( ) V.

A. 12   B. 24   C. 36   D. 220

417. The protective shoes that could eliminate the accumulation of static electricity in human body in time and prevent electric shock from power supply below ( ) are anti-static shoes.

A. 36V   B. 100V   C. 250V   D. 1000V

418. In accordance with the provisions of *The Production Safety Law of the People's Republic of China*, the trade union should ( ) the production safety work according to law.

A. supervise   B. monitor
C. evaluate and compare   D. control

419. If the electricity is tested on a wooden pole, wooden ladder or wooden frame, and no indication could be given without grounding, the grounding wire could be connected to the end of the electroscope insulating rod, but it should be approved by the person in charge of operation on duty or ( ).

A. the special supervisor   B. the person issuing the work card
C. the person in charge of the work   D. the work permit person

420. The noise voltage of the telephone scale of the high-frequency switching power supply system is ( ).

A. $\leqslant$ 2mV   B. $\leqslant$ 100mV   C. $\leqslant$ 40mV   D. $\leqslant$ 50mV

421. The mark of positive voltage electrical equipment is ( ).

A. d   B. p   C. o   D. s

Solutions: 413.B; 414.B; 415.C; 416.C; 417.C; 418.B; 419.C; 420.A; 421.B

422. The function of thermal release of low-voltage circuit breaker is ( ).
   A. short circuit protection
   B. overload protection
   C. leakage protection
   D. phase failure protection

423. The wheeled carbon dioxide fire extinguisher is generally operated by two people. When using, two people together push or pull the extinguisher to the burning place about ( ) away from the combustor.
   A. 4m   B. 6m   C. 8m   D. 10m

424. *The Production Safety Law of the People's Republic of China* stipulates that the relevant production and business operation entities should, in accordance with the provisions, withdraw and use the budget of production safety for ( ).
   A. technology improvement
   B. process improvement
   C. production safety conditions improvement
   D. strengthening safety training for employees

425. According to Decree 260 of the provincial government, the standard of compensation for death in production safety accidents is 20 times of ( ).
   A. the reference value that not less than the per capita disposable income of urban residents in the previous year at the place where the accident occurred
   B. the reference value not less than the per capita disposable income of urban residents in this city in the previous year
   C. the reference value not less than the per capita disposable income of urban residents of the province in the previous year
   D. the reference value not less than the per capita disposable income of urban residents in the previous year

426. The installation height of low-voltage switch is generally ( ) m.
   A. 0.8~1.0   B. 1.1~1.2   C. 1.3~1.5   D. 1.6~2.0

427. The height of the incoming line crossing the main traffic road from the ground should not be less than ( ).
   A. 6m   B. 5m   C. 4.5m   D. 3.5m

428. The action time of thermal relay ( ) with the increase of current.
   A. increases sharply
   B. increases slowly
   C. decreases
   D. remains unchanged

429. The minimum horizontal distance between the low-voltage overhead line conducting wire and the building is ( ) m.

Solutions: 422.B; 423.D; 424.C; 425.C; 426.C; 427.C; 428.C; 429.C

A. 0.6　　　　B. 0.8　　　　C. 1　　　　D. 1.2

430. Explosive gases, vapors and mists are classified by ( ).

　　A. explosion limit　　　　B. flash point
　　C. ignition point　　　　D. ignition temperature

431. Common strong alkali chemical burns include potassium hydroxide, sodium hydroxide and ( ) burns.

　　A. hydrated lime　　　　B. calcium hydroxide
　　C. slaked lime　　　　D. quick lime

432. The distance between high temperature lamps such as spotlights, iodine tungsten lamps and combustibles should not be less than ( ) m. Otherwise, thermal insulation and heat dissipation measures should be taken.

　　A. 0.1　　　　B. 0.2　　　　C. 0.3　　　　D. 0.5

433. The validity of the first and second types of work cards is limited to the ( ) day for maintenance.

　　A. applied　　B. scheduled　　C. approved　　D. agreed

434. When flat steel is used as downlead of lightning protection device, its sectional area should not be less than ( ) mm^2.

　　A. 24　　　　B. 48　　　　C. 75　　　　D. 100

435. In case of electrical fire in the 10kV high-voltage control system, if the power supply could not be cut off and it is necessary to extinguish the fire with electricity, the optional fire extinguisher is ( ).

　　A. dry powder fire extinguisher
　　B. the extinguishers that the distance between nozzle and body and electrified body shall not be less than 0.4m
　　C. atomizing water gun, wear insulating gloves and insulating boots, and the water gun head is grounded, and the water gun head is more than 4.5m away from the charged body
　　D. the extinguishers that the distance between nozzle and electrified body is not less than 0.6m

436. If the labor dispatching unit is unable or evades to pay for the work-related injuries and occupational diseases of the labor dispatching personnel, ( ).

　　A. the expend should be from financial special items at all levels
　　B. it should be advance payment by dispatched workers
　　C. it should be paid by the employer in advance
　　D. it should be paid by medical unit in advance

Solutions: 430.D; 431.D; 432.D; 433.C; 434.D; 435.A; 436.C

437. For leakage protection devices with rated leakage action current of 30mA and below, when the leakage current is the rated leakage action current, the action time should not exceed ( ) s.

  A. 0.5   B. 0.2   C. 0.1   D. 0.04

438. *The Production Safety Law of the People's Republic of China* stipulates that the main person in charge of the production and business operation entities and the management personnel of production safety must have ( ) corresponding to the production and business operation activities of the entities.

  A. production safety training

  B. production safety management capability

  C. production safety knowledge

  D. production safety knowledge and management ability

439. The critical fusing current of the lighting circuit fusant should not be greater than ( ) times the allowable current of the line conducting wire.

  A. 1.1   B. 1.25   C. 1.45   D. 2

440. The state protects the ( ) of trade unions from infringement.

  A. rights and interests   B. functions and interests

  C. legitimate rights   D. legitimate interests

441. If the work of the type I work card has not been completed by the scheduled time, ( ) should handle the extension formalities.

  A. the issuer of the work card   B. the special supervisor

  C. work permitter   D. the person in charge of the work

442. The buried depth of electric pole should not be less than ( ) m.

  A. 1   B. 1.2   C. 1.5   D. 1.8

443. When pressing the chest, press quickly with the weight of the upper body to make the sternum sink ( ), and the heart receives compression, then relax, press and relax rhythmically, 60~70 times per minute.

  A. 1~2cm   B. 3~4cm   C. 4~5cm   D. 5~6cm

444. When the foam extinguisher is about ( ) away from the ignition point, turn the cylinder upside down. Hold the bail with one hand, and hold the bottom ring of the cylinder with the other hand to aim the jet at the combustor.

  A. 6m   B. 8m   C. 10m   D. 12m

445. The screw of the screw lamp holder should be connected with ( ).

  A. working zero wire   B. neutral wire

  C. protective grounding wire   D. phase wire

Solutions: 437.B; 438.D; 439.C; 440.C; 441.D; 442.C; 443.B; 444.C; 445.A

446. The clamp ammeter should be used with a larger range first, and then the range should be changed depending on the magnitude of the measured current. When switching the measuring range, ( ) should be done.

A. rotation range switch direct
B. stepping away from the conducting wire first
C. shifting the range switch again
D. shifting the range switch while stepping to the conducting wire

447. The period of validity of the production safety license is ( ) years.
A. 1	B. 3	C. 4	D. 5

448. The breaking current of low-voltage circuit breaker should be ( ).
A. greater than the minimum short circuit current at the installation site
B. less than the minimum short circuit current at the installation site
C. greater than the maximum short circuit current at the installation site
D. less than the maximum short circuit current at the installation site

449. The standard position of the upper arm tourniquet is ( ).
A. the upper 1/3 of upper arm	B. the upper 1/4 of upper arm
C. the upper 1/2 of upper arm	D. the upper 1/5 of upper arm

450. The "four phase" protection of female workers in the process of female physiological function change, namely ( ) labor protection.
A. menstruation, pregnancy, delivery and menopause
B. menstruation, pregnancy, delivery and lactation
C. menstruation, pregnancy, lactation, menopause
D. pregnancy, childbirth, lactation, menopause

451. After the diesel engine is started, it is allowed to enter full load operation after the engine oil temperature is higher than ( ).
A. 45 ℃	B. 65 ℃	C. 75 ℃	D. 55 ℃

Solutions: 446.B; 447.B; 448.C; 449.A; 450.B; 451.A

Judgment question bank

Low Voltage Electrician Certification Training Course

1. To measure the insulation resistance to ground and between phases of the motor, a megger is often used instead of a multimeter. (√)

2. The person in charge of the work (supervisor) should be approved by the leader in charge of production in the workshop (sub field) or work area (office). (×)

Analysis: The person in charge of the work (supervisor) should generally have the corresponding qualification of the person in charge of the work card, having more than 5 years of experience in overhaul and maintenance operations, and be a person who has passed the examination of the enterprise's "full-time safety supervisor of the work card".

3. After the leakage switch trips, it is allowed to check the circuit by means of shunt power cut-off and power transmission. (√)

4. It is not necessary to cut-off the power supply of the tested equipment before using the megger. (×)

Analysis: The power supply of the equipment to be tested must be cut-off before using the megger for measurement. It is prohibited to measure the insulation resistance of live equipment.

5. The bimetallic sheet of thermal relay is rolled from a metal material with different thermal expansion coefficient. (×)

Analysis: The bimetallic sheet of thermal relay is rolled from a composite metal material with different thermal expansion coefficients composed of two or more metals or other materials with appropriate properties.

6. It is strictly forbidden to loosen the power line by suddenly cutting the conducting wire. (√)

7. The values measured by AC ammeter and voltmeter are effective values. (√)

8. The time required for each AC cycle is called period T. (√)

9. Insulating rod is used for connecting or disconnecting high-voltage isolating switch and drop fuse, assembling and disassembling portable grounding wire, and for auxiliary measurement and test. (√)

10. The bending speed of bimetallic sheet of thermal relay is related to the current. The greater the current is, the faster the speed is. This characteristic is called proportional time characteristic. (×)

Analysis: This characteristic is called inverse time characteristic, not proportional time characteristic.

11. Special operation personnel must be at least 20 years old and not exceed the national statutory retirement age. (×)

Analysis: Special operation personnel must be at least 18 years old and not exceed the national statutory retirement age.

12. For multi overhead lines erected on the same pole, the maintenance work when one circuit is powered off should be regarded as live working. (√)

13. After the ground wire is removed, the line should be considered live and no one is allowed to climb the pole again for work. (√)

14. If the number of employees is less than 100, there is no need to have full-time or part-time management personnel for production safety. (×)

Analysis: It should be assigned according to the regulations, and the specific number of personnel could be determined according to the business situation and actual needs of the enterprise.

15. If the test could not be carried out on the electrical equipment, the power frequency high-voltage generator could be used to verify that the electrical equipment is in good condition. (√)

16. The production and business operation entities must provide the employees with labor protection articles that meet the national standards or their own standards. (×)

Analysis: According to Article 42 of *The Production Safety Law of the People's Republic of China*, the production and business operation entities must provide the employees with labor protection articles that meet the national or industrial standards, and supervise and educate the employees to wear and use them according to the rules of use.

17. The text symbol of the button is SB. (√)

18. Electricity inspection is one of the technical measures to ensure the safety of electrical operation. (√)

19. The color of normal exhaust smoke of diesel engine is blue. (×)

Analysis: After the diesel engine fuel is completely burned, the normal color is generally light gray, and it is dark gray when working under load.

20. The whole industry must carry out post emergency knowledge education, self-rescue and mutual rescue, and escape skills training for employees, and organize regular assessment. (√)

21. Safety isolation transformer is a low-voltage transformer with double insulation structure and reinforced insulation structure. (√)

22. The protection characteristics of the thermal relay should be close to the overload characteristics of the motor when protecting the motor. (√)

23. Check the insulation of each winding of the motor. If the measured insulation resistance is unqualified, it is not allowed to operate with power on. (√)

24. The insulation resistance between live parts and accessible conductors of Class I equipment should not be less than 2M Ω. (√)

25. The action value and release value of the intermediate relay could be adjusted. (×)

Analysis: It has been determined in the manufacturing process and could not be adjusted in the use process.

26. The positioning structure of universal change-over switch generally adopts the radial structure of roller clamp shaft. (×)

Analysis: Generally, the radial structure of roller clip ratchet wheel is adopted.

27. In the three-phase-four-wire distribution system, the lighting load should be evenly distributed in the three-phase as far as possible. (√)

28. According to the magnitude of the current passing through the human body and the different reaction states of the human body, the current could be divided into sensing current, escape current and ventricular fibrillation current. (√)

29. In a work card, the issuer of the work card and the work permit should not concurrently serve as the person in charge of the work. (√)

30. Escape from the fire quickly, and the faster the action, the better. When escaping, pay attention to closing the doors and windows on the passage to prevent and delay the smoke from flowing to the escape passage. (√)

31. The iron case switch could be used to infrequently start three-phase asynchronous motors below 28kW. (√)

32. The way to discharge a capacitor is to connect its two ends with wires. (×)

Analysis: It should be discharged through load (resistance or bulb, etc.). Because the power compensation capacitor of the power system usually stores some electric energy after it is out of service, and the voltage is relatively high, direct discharge with the wire would endanger personal safety.

33. Lightning could be divided into direct strike lightning and inductive lightning according to its transmission mode. (×)

Analysis: lightning could be divided into direct lightning, inductive lightning and spherical lightning according to its transmission mode.

34. The smaller the distance between the joint and the insulating layer after the wire is connected, the better. (√)

35. When someone is found to have an electric shock, others should immediately pull him/her away from the power supply with bare hands. (×)

Analysis: See Chapter 3 of this book for details. There are many methods to make the person who gets a low-voltage electric shock separate from the power supply, such as pulling, shearing, picking, padding and pulling. To ensure the safety of the rescuer, the method of pulling with bare hands is absolutely not allowed.

36. In the case of using a multimeter to measure the resistance, ohmic zero adjustment should be conducted every time when the ohmic switch is changed. (√)

37. If the test is conducted in an explosive hazardous area where some production devices have been put into operation, the concentration of explosive gas in the environment around the equipment must be monitored, and the power on test could only be conducted after confirming that the relevant hot work conditions are met. (√)

38. The discharge load of capacitor should not be equipped with fuse or switch. (√)

39. The wiring of hand electric tools could be lengthened at will. (×)

Analysis: The Management of Hand Electric Tools stipulates that the flexible cables or cords of tools should not be arbitrarily extended or replaced.

40. Fire brought into the underground would detonate gas and ignite combustibles to cause fire. (√)

41. "No climbing, high voltage danger!" belongs to prohibited signboard. (×)

Analysis: It belongs to warning signs.

42. The automatic air switch has overload, short circuit and undervoltage protection. (√)

43. When making business decisions involving production safety, the production and business operation entities should listen to the opinions of the administration of production safety and the personnel in charge of production safety. (√)

44. The fuse with rated voltage of 380V could be used in 220V lines. (√)

45. When wearing the helmet, the safety helmet belt must be fastened to ensure that it would not fall off under various conditions. (√)

46. When the main switch is switched down, the line is considered dead. (×)

Analysis: After switching down the main switch, it is also necessary to conduct an electrical test to confirm whether there is no electricity.

47. When rotating electrical equipment catches fire, it is not suitable to use dry powder extinguisher to extinguish the fire. (√)

48. Safety measures for power cut-off operation could be divided into predictive measures and protective measures according to the security role and safety measures. (√)

49. Regenerative power generation braking is only used when the motor speed is higher than the synchronous speed. (√)

50. The right hand rule is used to determine the direction of induced current generated when a straight conductor moves to cut the magnetic line of force. (√)

51. The electroscope must be in good condition before use. (√)

52. When employees find an emergency that endangers their personal safety, they should stop working and leave the workplace immediately. (×)

Analysis: Article 47 of *The Production Safety Law of the People's Republic of China* stipulates that when employees find an emergency that directly endangers their personal safety, they have the right to stop operations or leave the workplace after taking possible emergency measures. However, this right does not apply to certain employees engaged in special occupations, such as pilots, ship drivers, vehicle drivers, etc., who could not leave their workplaces or posts in the event of an emergency that endangers their personal safety.

53. Emergency lighting is not allowed to share the same line with other lighting. (√)

54. The method of energy consumption braking is to convert the kinetic energy of the rotor into electrical energy and consume it on the resistance of the rotor circuit. (√)

55. During the use of anti-static shoes and conductive shoes, the resistance should be re-measured every half a year, and they could be used again if they meet the requirements. (√)

56. When using the resistance grade of the pointer multimeter, the red probe is connected to the positive pole of the internal power supply of the multimeter. (×)

Analysis: The red probe is connected to the negative pole of the internal power supply of the multimeter.

57. The correct pressing point of external cardiac compression is at the upper left of the cardiac fossa. (×)

Analysis: The correct compression point for external cardiac compression is the junction of the middle and lower 1/3 of the sternum, where the sternum has a large range of motion and is not easy to fracture during compression.

58. Insulating boots, gloves and other rubber products could contact with petroleum grease. (×)

Analysis: Rubber products such as insulating boots, gloves and pads are easy to react with petroleum grease, which would corrode and reduce the insulation of insulating materials, and cause premature aging and failure.

59. The high-voltage electroscope is generally tested once every six months. (√)

60. After the injury, the injured person only has headaches and dizziness, indicating that it is a minor injury. In addition, there are dilated pupils, hemiplegia or convulsion, which is at least a moderate brain injury. (√)

61. Solder transistor and other weak current components should use 100W electric soldering iron. (×)

Analysis: Weak current components such as transistors are generally soldered with an electric soldering iron of about 20W. Otherwise, the temperature of the electric soldering iron is too high, and if it stays too long, it is easy to burn the electronic components.

62. When giving first aid to an electric shock, first of all, the person who gets an electric shock should be quickly disconnected from the power supply. It is better to stand on a dry board or stool with one hand, or wear insulating shoes, and pay attention not to touch other

grounding bodies. (✓)

63. The special operation certificate should be reviewed by the assessment and certificate issuing department once a year. (✕)

Analysis: The special operation certificate should be reviewed once every three years.

64. The repeated grounding device of electrical equipment could be connected with the grounding device of independent lightning rod. (✕)

Analysis: The grounding device of electrical equipment belongs to equipotential grounding of equipment, while the grounding device of independent lightning rod is lightning protection grounding. They belong to different types, so they should be set separately and cannot be shared.

65. In the case of electrical fire, the power supply should be cut off immediately. If it is impossible to cut off the power supply, non-conductive fire extinguishers such as dry powder and carbon dioxide should be quickly selected to extinguish the fire. (✓)

66. The working voltage of the conducting wire should be greater than its rated voltage. (✕)

Analysis: The rated voltage, also known as the nominal voltage, refers to the standard voltage for long-term stable operation of electrical equipment (including wires). When the operating voltage of electrical equipment (wires) is higher than the rated voltage, it is easy to damage the equipment, while when it is lower than the rated voltage, it would not work normally.

67. Under the same conditions, AC is more harmful than DC. (✓)

68. The dust generated in the production process is called productive dust, which is harmful to the health of operators. (✓)

69. Static electricity is a very common electrical phenomenon, and its harm is not small. Solid static electricity could reach more than 200kV, and human static electricity could also reach more than 10kV. (✓)

70. The current is measured with an ammeter, which is connected in parallel in the circuit. (✕)

Analysis: When measuring DC current, the ammeter should be connected in series with the load in the circuit, and pay attention to the polarity and range of the instrument. If the ammeter is connected in parallel in the circuit for measurement, the ammeter may be burnt due to overload.

71. When measuring the capacitor, the pointer of the multimeter stops after swinging, indicating that the capacitor is short circuited. (✓)

72. The electrical angle between the voltage and the current would decrease after the capacitor is connected in parallel with an inductive load. (✓)

73. The maximum value of 220V AC voltage is 380V. (✕)

Analysis: 220V is the effective value of AC voltage, and the effective value of sinusoidal AC=maximum value/$\sqrt{2}$. It could be calculated that the maximum 220V AC voltage is 311V, not 380V.

74. The withstand voltage level of low-voltage insulating materials is generally 500V. (√)

75. The conductor joint should be located at the fixed position of the insulator as far as possible to facilitate unified wire binding. (×)

Analysis: The joint position of the conducting wire should not be at the fixing position of the insulator, and the distance between the joint position and the fixing position of the conducting wire should be more than 0.5m, so as not to interfere with the wire binding.

76. The leakage protection device could prevent single-phase electric shock and two-phase electric shock. (×)

Analysis: The leakage protector is mainly used to protect against electric shock in case of electric leakage fault of equipment and fatal danger. It has overload and short circuit protection functions and could effectively protect against direct and indirect electric shock in low-voltage power grid.

77. Electrical control system diagram includes electrical schematic diagram and electrical installation diagram. (√)

78. When measuring idling current of motor with clamp meter, it could be directly measured without grade change. (×)

Analysis: Before measurement, determine the rated current of the motor first, switch the clamp ammeter to the appropriate current grade, and then conduct the measurement.

79. The tensile strength of the conducting wire joint must be the same as that of the original conductor. (×)

Analysis: Different. The mechanical strength of the connecting joint of the conducting wire should not be less than 90% of the mechanical strength of the conducting wire.

80. Quickly transfer the carbon monoxide poisoning victim to a ventilated and warm place to lie flat, and unfasten the collar and belt to facilitate smooth breathing. (√)

81. According to international practice and China's legislation, the industrial injury insurance compensation implements the principle of "liability compensation", that is, no fault compensation. (×)

Analysis: According to international practices and China's legislation, industrial injury insurance compensation implements the principle of "no liability compensation", that is, no fault compensation, which is based on the theory of occupational risk. Proceeding from the concept of protecting the rights and interests of employees to the maximum extent, compensation should be made as long as injuries are caused on business.

82. Undervoltage protection refers to automatic power off when the voltage is lower than

a certain proportion of the rated value. (√)

83. Grounding or neutral connection measures should be taken for Class II equipment and Class III equipment. (×)

Analysis: Class II equipment does not need to be grounded or connected to neutral, while Class III equipment needs to be grounded or connected to neutral.

84. The accident investigation report should be published to the public in a timely manner according to law. (√)

85. The main person in charge of the production and business operation entity should organize the formulation of the rules and regulations on safe production and operating procedures of the entity; Employees should organize themselves to formulate and implement their own safety production education and training plans. (×)

Analysis: Article 18 of *The Production Safety Law of the People's Republic of China* stipulates that the main person in charge of the production and business operation entity should organize the formulation of its own rules and regulations on production safety and operating procedures.

86. The lightning rod could be welded with galvanized steel pipe, its length should be more than 1m, the diameter of steel pipe should not be less than 20mm, and the thickness of pipe wall should not be less than 2.75mm. (×)

Analysis: The lightning rod could be welded with galvanized steel pipe. When the length of the rod is less than 1m, the diameter of round steel should not be less than 12mm, and the diameter of steel pipe should not be less than 20mm. When the rod is 1~2m long, the diameter of round steel should not be less than 16mm; The diameter of steel pipe should not be less than 25mm. The diameter of round steel on the top of independent chimney should not be less than 20mm; The diameter of steel pipe should not be less than 40mm, and its thickness should not be less than 2mm.

87. The entity using labor protection articles should provide labor protection articles that conform to the state regulations for workers free of charge, and should not replace labor protection articles with currency or other articles. (√)

88. Eliminating toxic substances and productive dust in the production process is the fundamental measure for dust prevention and poison prevention. (×)

Analysis: In the production process, the fundamental measure for dust prevention and poison prevention is to reform the process, and those processes that do not produce dust and poison substances in the production process or eliminate them in the production process should be preferred. For example, wet operation is adopted to prevent dust from flying; Strengthen the ventilation of the workplace and remove the dust from the site; Strengthen personal protection, and protect workers with dust masks and air supply helmets that meet national standards.

89. The barrier should be made of dry insulating materials, not metal materials. (√)

90. Before the reform and opening up, China emphasized that aluminum should be used as the conducting wire instead of copper to reduce the weight of the conducting wire. (×)

Analysis: The conductivity of copper is better than that of aluminum. The same cross section is 1.5 times that of aluminum. The physical strength of copper is higher than that of aluminum, and its chemical stability is also better than aluminum. Before the reform and opening up, due to the low output of copper, which is very valuable, and the conductivity of aluminum is second only to copper, aluminum wires were used as distribution lines. After the reform and opening up, copper mines could be imported from abroad, which improved the situation.

91. As the power on time increases, the resistance of the human body increases due to sweating, resulting in the decrease of the current passing through the human body. (×)

Analysis: The longer the power on time is, the lower the resistance of the human body due to sweating and other reasons, resulting in an increase in the current passing through the human body and an increase in the risk of electric shock.

92. The protective neutral connection is applicable to the distribution system with the neutral point directly grounded. (√)

93. When using electrical equipment, because the wire section is too small, when the current is large, it would also cause fire due to excessive heating. (√)

94. If the magnetic induction intensity of each point in the magnetic field is the same, the magnetic field is uniform. (×)

Analysis: Uniform magnetic field means that the magnetic induction intensity of each point in the magnetic field is equal in size and direction.

95. Major accident potential and major hazard source are the source of major accidents, so the concepts of the two are the same. (×)

Analysis: The hazard source is objective. For example, the natural gas used at home is a source of danger, and the electrical equipment is also a source of danger, but it could not be called a hidden danger. Only when the protection is reduced or even invalid could it be called a hidden danger.

96. Safety barrier is a kind of protective component, commonly used are voltage stabilizing transistor type safety barrier and triode type safety barrier, which are voltage and current limiting devices set between intrinsically safe circuit and non intrinsically safe circuit to prevent the impact of energy of non intrinsically safe circuit on intrinsically safe circuit. (√)

97. When the motor runs according to the nameplate value, the rated working system for short-term operation is indicated by S2. (√)

98. The time delay of electric time relay is not affected by the fluctuation of power supply voltage and ambient temperature. (√)

99. According to the specification requirements, the sectional area of copper core wire used for insulated conducting wire through pipe should not be less than 1mm². (√)

100. Productive noise only damages people's hearing. (×)

Analysis: Noise not only affects the auditory system, but also affects non auditory systems such as nervous system, cardiovascular system, endocrine system, reproductive system and digestive system.

101. The frequency of sinusoidal alternating current in China is 50Hz. (√)

102. The flameproof effect of the flameproof shell is to realize flameproof by using the flange clearance of the shell. The larger the flange gap is, the more bioenergy would be generated by the explosion passing through the gap, the stronger the explosion transmission and the better the explosion-proof performance. (×)

Analysis: The flange clearance of the flameproof electrical equipment shell is the key to whether the flameproof electrical equipment could be flameproof, provided that the strength of the flameproof shell is appropriate, that is, even if the inside is explosive, the shell could not be broken, cracked, or otherwise damaged.

103. For safety and reliability, all switches should control phase wire and neutral wire at the same time. (×)

Analysis: The function of the switch is to disconnect or connect the circuit at a certain point of the closed circuit. Generally, the switch has only two terminals (one in and one out), and both of them control the phase wire. If the switch controls the zero wire, there is still voltage on the electrical appliance after the circuit is disconnected, which would bring inconvenience and danger to the use and maintenance.

104. For current mutual inductor that is not used temporarily in operation, the two ends of their secondary windings could not be short circuited. (×)

Analysis: The secondary circuit of the current mutual inductor must not be open circuit during operation, because the current is zero when the secondary side is open circuit, so it could not generate magnetic flux to offset the effect of the primary side magnetic flux, while the secondary side could induce a voltage of about 1000V, which endangers the safety of people and equipment, and the iron core of the current mutual inductor itself would also be seriously heated. Therefore, for current mutual inductors that are not used temporarily in operation, the secondary coils of current mutual inductors must be short circuited first when disassembling the instrument.

105. On the way to transfer the injured, the critically ill should pay close attention to their breathing, pulse, and unblock the respiratory tract. (√)

106. The main function of low-voltage knife switch is to isolate electrical equipment from power supply during maintenance. (√)

107. During the measurement, the change-over switch of the multimeter should not be turned, and the switch grade must be shifted after exiting from the wire. (√)

108. 30~40Hz current is the most dangerous. (×)

Analysis: When the human body touches the alternating current, the human body is actually a capacitive impedance. Different frequency currents have different effects on human body. When the frequency is high, the capacitive reactance is small and the harm to human body is large. Theoretically, 50~60Hz AC is the most dangerous; The danger of alternating current is greater than that of direct current.

109. In case of fire on the high-voltage line, evacuate the site quickly and call 119 for alarm. (×)

Analysis: In case of fire on high-voltage lines, insulation tools with corresponding insulation grades should be used to quickly open the isolating switch and cut off the power supply, or relevant departments should be notified to cut off the power quickly. At the same time, call 119 to the fire agency.

110. Power safety tools and apparatus must strictly perform the acceptance procedures. The safety supervision department is responsible for organizing the acceptance and signing on the acceptance form for confirmation. (×)

Analysis: In this question, the content that the safety supervision department is responsible for organizing the acceptance should be instead by: the procurement department is responsible for organizing the acceptance, and the safety supervision department sends personnel to participate.

111. The power supply of movable electrical equipment is generally protected by erecting or through pipes. (√)

112. Under no circumstances should the fusant of fuse be replaced under load. (√)

113. When measuring the idling current of the motor with a clamp meter, it could be directly measured with a small current range. (×)

Analysis: Generally speaking, the range of clamp ammeter should be selected according to the rated current of the motor. The value of the so-called small current range in the question is unknown, so it is impossible to measure the idling current of the motor at one time.

114. The IT system is the protective neutral connection system. (×)

Analysis: IT system refers to a system in which the neutral point of the power supply is not grounded and the exposed conductive part of the electrical equipment is directly grounded. IT system is a protective grounding system, not a protective neutral system.

115. When measuring the active electric energy of an AC circuit, the voltage coil, current coil and two ends of each coil could be connected to the line at will because they are AC. (×)

Analysis: When measuring the active electric energy of an AC circuit, the current coil should be connected in the circuit to measure the current; The voltage coil should be connected in the circuit in parallel to measure the voltage. Meanwhile, if the electric energy meter is connected through the mutual inductor, it should also be noted that the head and tail ends of the current mutual inductor wire should not be wrongly connected, and the head and tail ends of the voltage mutual inductor should not be wrongly connected.

116. If the person who gets an electric shock is unconscious and has a heartbeat but stops breathing, mouth to mouth artificial respiration should be carried out immediately. (√)

117. For safety, it is better to disconnect the load when replacing the fuse. (×)

Analysis: When replacing the fuse, the power supply must be disconnected instead of the load.

118. There are many methods of artificial respiration, but the supine chest compression method is the most convenient and effective. (×)

Analysis: There are many methods of artificial respiration, such as mouth to mouth air blowing, prone back pressing and supine chest pressing, but mouth to mouth air blowing is the most convenient and effective.

119. The single pole switch of lighting should not be connected to the ground wire, but must be connected to the neutral wire or phase wire. (×)

Analysis: The single pole switch of the lighting must be connected to the phase wire. If the switch is connected to the neutral wire (ground wire), although the electrical appliance does not work when the switch is disconnected, there is still 220V voltage between the electrical appliance circuit and the "ground", which is inconvenient for maintenance and unsafe. In case of the switch is connected to the phase wire, when the switch is disconnected, the electrical circuit is connected to the "ground", which is convenient for maintenance and not easy to get electric shock.

120. Three-phase-four-wire system and single-phase-three-wire system should be adopted for power supply in explosive hazardous areas. (×)

Analysis: According to the requirements of the specification, single-phase-three-wire (live wire L, neutral wire N and grounding wire PE), three-phase-five-wire system should be used for power supply in flammable and explosive places.

121. The selection of RCD must consider the influence of normal leakage current of electrical equipment and circuits. (√)

122. Low-voltage distribution panel assembles relevant low-voltage primary and secondary equipment according to a certain wiring program, and each main circuit program

corresponds to one or more auxiliary programs, thus simplifying the engineering design. (√)

123. The burn of DC arc is more serious than that of AC arc. (√)

124. When measuring the current with a clamp meter, try to place the wire in the middle of the iron core of the jaw to reduce the measurement error. (√)

125. The electric energy meter is a device specially used to measure the power of equipment. (×)

Analysis: Electric energy refers to the ability to use electricity to do work (i.e., generate energy) in various forms. The electric energy meter is an electrical instrument for measuring electric energy. Its role in the circuit is to measure the electricity consumption for a period of time.

126. Special electrical operators should have high school education or above. (×)

Analysis: Electrical special operation personnel should have junior high school education or above.

127. When dressing the wound, it should be wound from the bottom up, usually from left to right, from bottom to top. (√)

128. As long as the diode works in the reverse breakdown area, it would be broken down. (×)

Analysis: Not always. Only when the reverse voltage is greater than the withstand voltage of the diode (reverse breakdown voltage) could breakdown occur.

129. The lightning protection device should be laid along the external wall of the building and grounded in the shortest way. If there are special requirements, it could be concealed. (√)

130. For the maintenance of associated equipment or intrinsically safe circuits, it is allowed to conduct only after the circuit connection with the hazardous area is disconnected. (√)

131. Those who have obtained the senior electrician certificate could engage in electrical work. (×)

Analysis: Senior electrician certificate is the qualification certificate of electrician, not the work license. He/she must obtain the electrician's special operation certificate before taking the post, or he/she would be treated as taking the post without certificate in case of any accident. If he/she works without certificates, he/she may be ordered to stop the special equipment once it is found, and he/she may be ordered to train and obtain certificates, and a fine of 10,000 to 30,000 yuan may be imposed as well.

132. When constructing in the area where underground cables and power cables intersect or are buried in parallel, the position must be checked repeatedly, and the operation could only be carried out after confirmation. (√)

133. For asynchronous motors, the national standard stipulates that motors below 3kW

are configured in △. (×)

Analysis: According to the national standard, asynchronous motors above 4kW are configured in △ and below 4kW are configured in Y. Generally, three-phase motors below 3kW are Y-configured and directly started.

134. Fence is a kind of shield for equipment installed to prevent workers from accidentally touching live equipment, which is divided into temporary fence and permanent fence. (√)

135. If the special operation personnel has not received special safety operation training, have not obtained the corresponding qualification, and work on their posts, resulting in accidents, the relevant personnel of the production and business entity should be held accountable. (√)

136. When the lamp could not reach the minimum height, the voltage below 24V should be adopted. (×)

Analysis: If the lamp could not reach the minimum height, 36V and below voltage should be used for power supply.

137. To check the lubrication of the motor bearing, turn the motor shaft with power to see if it could rotate flexibly and listen for abnormal noise. (×)

Analysis: The most direct method is to check the running sound of the bearing with a sound listening stick when the motor is running, and judge whether there is mechanical friction sound and impact sound. If it is heard "grunt, grunt" noise, it means that the bearing is short of oil.

138. The expenses for occupational health check-up should be paid by the employees themselves. (×)

Analysis: According to Article 36 of the Law of the People's Republic of China on the Prevention and Control of Occupational Diseases, the cost of occupational health check-up should be paid by the employer.

139. All wires should be laid open indoors, and protective sleeves must be provided when crossing walls and floors. (√)

140. Lightning could discharge through other charged bodies or directly to human body, causing great damage to human body until death. (√)

141. When checking electricity with a test pencil, stand barefoot to ensure good contact with the ground. (×)

Analysis: When checking the electrical circuit with an test pencil, be sure to wear shoes with rubber soles. If necessary, wear protective gloves and make all preparations to prevent electric shock.

142. Mutual inductor transforms high-voltage and large current on the line into low-

voltage and small current in a certain proportion, which could keep measuring instruments and relay protection devices away from high-voltage and is conducive to safety. (√)

143. When the bulb is turned on, there would be current on the zero wire, and the human body could no longer touch the zero wire. (×)

Analysis: Under normal circumstances, the zero wire has current but no voltage, and the voltage with the same potential as the ground is zero. Once people touch it, there is no danger, but generally they should not touch it. If the zero wire of the appliance is broken in use, the contact end between the zero wire and the appliance has no current, but the voltage to the ground rises to 220V. Meanwhile, it must not be touched, which would endanger life.

144. The current carrying conductor must be affected by the magnetic force in the magnetic field. (×)

Analysis: According to the calculation formula of ampere force, $F=IBL\sin\theta$, Only when there is a certain angle between the current carrying conductor and the magnetic field could there be magnetic force. If the angle between conductor and magnetic field is $0°$ or $180°$, there is no magnetic force.

145. The total voltage of the parallel circuit is equal to the sum of the voltage of each branch. (×)

Analysis: The total voltage of the series circuit is equal to the sum of the voltage of each branch. In a parallel circuit, the voltage of each branch is equal, and the voltage of each single branch in parallel on the power supply is equal to the power supply voltage.

146. The simple work of installing and disassembling grounding wires could be done by one person without supervision. (×)

Analysis: According to the regulations, two people must be assigned to install and remove the grounding wire. One operator and one supervisor.

147. When the communication line works near the power line, it is not necessary to contact the power department in advance to stop the power transmission, but it could work safely according to the situation. (×)

Analysis: In case of the special situation that the power lines cross over the top of the communication pole with small spacing, the power department must be contacted to operate after power cut-off, and the tools and materials used must not be close to the power line and its ancillary facilities, and the head of the operator must not exceed the pole top.

148. In the case of live fire extinguishing, if the spray water gun is used, the nozzle of the water gun should be grounded, and insulated boots and gloves should be worn before fire extinguishing. (√)

149. The production and business operation entities should cooperate with the supervisors and inspectors of the departments responsible for the supervision and

administration of production safety in performing their duties of supervision and inspection according to law, and should not refuse or obstruct them. (√)

150. When selecting the circuit breaker, the rated connecting and disconnecting capacity of the circuit breaker should be greater than or equal to the maximum load current that may occur in the protected circuit. (×)

Analysis: The connecting and disconnecting capacity of the open circuit should be greater than or equal to the maximum fault current that may occur in the protected line.

151. Labor protection articles could be converted into cash and distributed to individuals. (×)

Analysis: As far as the protective performance and special functions of labor protection articles are concerned, they should be regarded as production tools, belonging to the category of means of production, and are indispensable. Converting labor protection articles into cash and distributing them to individuals in the form of welfare violates the relevant national policies on labor protection and occupational safety and health.

152. The alternator generates electricity by applying the principle of electromagnetic induction. (√)

153. The motors' speed is regulated by changing the number of pole pairs are generally wound rotor motors. (×)

Analysis: The motor for speed regulation by changing the number of magnetic pole pairs is not a wound type motor, but a two speed or three speed cage type asynchronous motor.

154. The function of hearing protection articles is to avoid excessive stimulation of hearing by noise and protect hearing. (√)

155. For safety, high-voltage lines usually use insulated conducting wires. (×)

Analysis: High-voltage power transmission is generally transmitted underground by cables with insulating layers in cities, and in the field, it is often transmitted by bare conducting wires through overhead lines carried by iron towers.

156. When using the foot buckles to climb the pole, each step of up and down must make the foot buckles buckle the pole completely and reliably to move the body, otherwise it would cause an accident. (√)

157. The text symbol of the fuse is FU. (√)

158. Open plastic pipe wiring should not be used in places prone to mechanical damage. (√)

159. The power supply entering the explosion hazardous area should be separated from the neutral wire and ground wire, that is, three-phase-five-wire system. If the three-phase-four-wire system is adopted, it should be converted to three-phase-five-wire system in safe places, and the grounding resistance of the protective ground wire should meet the provisions

of relevant standards. (√)

160. To ensure the safety of the neutral wire, the three-phase-four-wire neutral wire must be equipped with a fuse. (×)

Analysis: It is not allowed to install fuses on the zero main wire of three-phase-four-wire system lines, because once the fuse is melted, it is equal to the open circuit of the zero wire, and the voltage of the one phase circuit with heavy load in the three-phase line passing through the zero wire is higher, which makes the voltage of other single-phase lines rise sharply, up to 260~300V, and the electrical appliances may be burnt.

161. The working zero wire in explosion hazardous areas should be used together with the protective zero wire. (×)

Analysis: The working zero wire is also called the neutral wire, and the code is "N"; The protective zero wire is also called the ground wire, and the code is "PE". There is also a kind of protection neutral wire and working neutral wire "combined into one" to form a common protection line, which is called protection neutral wire, and is coded as PEN. Except that the PEN wire allows the PE wire to share with the N wire, the protective zero wire and the working zero wire could not be electrically connected at any other branch. In explosion hazardous areas, the working neutral wire (N) and the protective neutral wire (PE) are set separately and should be mutually insulated, and the equipment shell is only connected with the PE wire.

162. The greater the internal resistance of the voltmeter, the better. (√)

163. The knife switch is extinguished by lengthening the arc. (√)

164. Electric shock accident is caused by electric energy acting on human body in the form of current. (√)

165. Implement the national safety production laws and regulations, and implement the safety production policy of "safety first, prevention first, and comprehensive treatment". (√)

166. The $NaHCO_2$ dry powder extinguisher is applicable to the initial fire of flammable and combustible liquid, gas and charged equipment. (√)

167. The legislative purpose of the Law of the People's Republic of China on the Prevention and Control of Occupational Diseases is to prevent, control and eliminate occupational disease hazards. (√)

168. The installation height of sockets in kindergartens, primary schools and other children's activity places should not be less than 1.8m. (√)

169. In any case, the triode has the function of current amplification. (×)

Analysis: The triode has three working states. When it is in the amplification state, it has the current amplification function, but when it is in the cut-off state and saturated conduction state, it would lose the current amplification function.

170. Where there are more than 100 employees, an administrative organization for production safety should be established or full-time personnel for the administration of production safety should be assigned. (√)

171. There are only three ways for productive toxicants to invade human body: respiratory tract, skin and digestive tract. (√)

172. The socket of safety voltage should be equipped with a zero or grounding plug or slot. (×)

Analysis: The safe voltage refers to the voltage below 36V, which has no danger of electric shock to people. Moreover, no strong current would be generated in human body, which belongs to harmless voltage, so grounding protection is unnecessary.

173. For obvious difference, switches of the same model installed in parallel should be of different heights and arranged in an orderly manner. (×)

Analysis: The switches of the same model installed in parallel should be arranged in parallel with the same height, neat and beautiful.

174. The generator set could be shut down with load. (×)

Analysis: Before the diesel engine shuts down, the load should be removed, and the speed should be gradually reduced, and the engine should run for several minutes without load. The machine should not be shut down immediately with load or after the load is suddenly removed.

175. Employees of enterprises and public institutions who engage in special operations without special operation certificates are in violation of regulations. (√)

176. RCD with residual action current less than or equal to 0.3A belongs to high sensitivity RCD. (×)

Analysis: RCD with residual action current $\leq$ 0.03A belongs to high sensitivity RCD, RCD with residual action current greater than 0.3 $\leq$ 1A belongs to medium sensitivity RCD, and RCD with residual action current $>$ 1A belongs to low sensitivity RCD.

177. When the knife switch is used as the disconnecting switch, the rated current of the knife switch should be greater than or equal to the actual fault current of the line. (×)

Analysis: The knife switch is mainly used for power isolation of various equipment and power supply lines, and could also be used to convert circuits. The rated current should be greater than the actual operating current of the circuit (generally about 1.5 times).

178. The method of phase wire touching ground wire could be used to check whether the ground wire is well grounded. (×)

Analysis: This is very dangerous and belongs to illegal operation. Check whether the ground wire is well grounded. The correct operation method is to use a grounding resistance meter for measurement.

179. High-voltage shock wave is generated by overhead lines after lightning, which is called direct lightning strike. (✗)

Analysis: When lightning discharges, it hits power transmission and distribution lines, poles and towers or their buildings. A large amount of lightning current passes through the object to be struck and is grounded through the impedance of the object to produce a voltage drop on the impedance, resulting in a very high potential at the point to be struck. This high potential is called direct lightning over-voltage or lightning over-voltage.

180. People with Meniere's disease should not be engaged in electrical work. (✓)

181. For the installation of sockets, sockets with different voltage levels should be obviously different, so that they could not be inserted incorrectly; The height of concealed socket to the ground should not be less than 0.3m. (✓)

182. All elements in the electrical schematic diagram are drawn according to the state when they are not powered on or no external force acts. (✓)

183. Labor protection articles are different from general commodities and directly involve the life safety and health of workers, so they must comply with national standards. (✓)

184. According to the nature of electricity, power lines could be divided into power lines and distribution lines. (✗)

Analysis: According to the nature of electricity, power lines could be divided into power lines and lighting lines.

185. The clamp ammeter could be made to measure both AC current and DC current. (✓)

186. The size of the explosion-proof gap is the key to whether the explosion-proof enclosure could be flameproof. (✓)

187. When putting out a fire, just lift or carry the fire extinguisher to the fire site, put down the fire extinguisher about 1m away from the combustor, pull out the safety pin, hold the handle at the bottom of the nozzle with one hand, and hold the handle of the on-off valve with the other hand. (✗)

Analysis: When putting out a fire with a CO_2 extinguisher, just lift or carry the extinguisher to the fire site, put down the extinguisher about 5m away from the combustor, pull out the safety pin, and hold the handle at the root of the nozzle with one hand and the handle of the on-off valve with the other hand. For CO_2 extinguishers without jet hoses, the nozzle should be pulled up 70°~90°.

188. The power factor of electronic ballast is higher than that of inductive ballast. (✓)

189. Anti-noise earplugs should be used in production noise environment. (✓)

190. The enterprise must carry out post emergency knowledge education, self-rescue and mutual rescue, and escape skills training for employees, and organize regular assessment. (✓)

191. In order to ensure that the major property of the enterprise would not be lost, the person in charge of the enterprise may organize employees to take risks. (✕)

Analysis: According to *The Production Safety Law of the People's Republic of China* (Amendment, 2020), the main technical responsible person of the production and business entities should have the decision-making and command power of production safety technology within the scope of authorization of the main responsible person. In case of an accident, the person in charge of the enterprise should organize rescue in an orderly manner according to the accident emergency plan, but should not organize employees to take risks.

192. The line investment must be considered when selecting the conducting wire, but the sectional area of the conducting wire should not be too small. (✓)

193. Communication power supply could be divided into AC grounding, protective grounding and lightning protection grounding according to the purpose of grounding system. (✕)

Analysis: The grounding system of communication power supply could be divided into AC grounding system and DC grounding system according to the electrification nature. It is divided into working grounding system, protective grounding system and lightning protection grounding system according to the purpose.

194. The method of changing rotor resistance is only applicable to wound asynchronous motor. (✓)

195. The labor protection articles must be provided with the Certificate of Conformity, manufacturing date and product instructions when leaving the factory. (✕)

Analysis: When the labor protection articles leave the factory, they must obtain the Product Inspection Certificate and Product Qualification Certificate, and have the manufacturing date and product instructions.

196. When the pendant lamp is installed above the table, the vertical distance from the table should not be less than 1.5m. (✕)

Analysis: Generally, when the home mounted pendant lamp is installed above the table, the vertical distance between the pendant lamp and the table should not be less than 0.5m. The distance between the pendant lamp and the table is generally 1.5~1.8m, which is the most appropriate distance. The vertical distance between the pendant lamp and the ground is generally about 2.5~2.8m.

197. There should be good natural lighting in the capacitor room. (✕)

Analysis: The capacitor room should be equipped with good ventilation, which is correct. Because the capacitor bank would generate heat during operation, it is necessary to add fan or ventilator, mainly to ensure the air flow between the capacitor room and the outside. There is no special regulation on whether there is natural lighting in the capacitor room.

198. During maintenance, "Stop! High-voltage danger!" should be hung on the operating handles of circuit breakers and isolation switch that could be powered to the work site once they are switched on. (×)

Analysis: "No switching on, someone is working" should be hung in the substation; The signboard of "No switching on, someone working on the line" should be hung on the transmission line.

199. The total resistance after several resistors are connected in parallel is equal to the sum of the reciprocal of the parallel resistors. (×)

Analysis: It is incorrectly stated above. It should be said that in a resistance parallel circuit, the reciprocal of the total resistance is equal to the sum of the reciprocal of the resistance of each branch.

200. After the fluorescent lamp is turned on, the ballast could reduce voltage and limit current. (√)

201. If an enterprise or institution uses persons without corresponding qualifications to engage in special operations, and a heavy casualty accident occurs, it should be sentenced to fixed-term imprisonment of not more than three years or criminal detention. (√)

202. It is strictly prohibited to install bedside switches in residential buildings. (√)

203. Overload means that the current in the line is greater than the calculated current or allowable ampacity of the line. (√)

204. The object whose conductivity is between conductor and insulator is called semiconductor. (√)

205. The insulating oil of oil filled equipment vaporizes and decomposes under the action of electric arc, which could form explosive mixed gas. (√)

206. The background of prohibited signboards should be white and the characters should be red. (√)

207. Escape current is the maximum current that a person could get rid of charged body independently, and the power frequency escape current is about 10A. (×)

Analysis: For different people, escape current value is also different. Generally speaking, people's escape current is about 10mA.

208. The rated acting current of RCD refers to the maximum current that could make RCD act. (×)

Analysis: The rated acting current of RCD refers to the minimum current that could make RCD act. If it is less than the rated acting current, the leakage protector would not act.

209. Mobile electric welding machine is not allowed to be used in flammable, explosive

or volatile places. (√)

210. No joint is allowed for the wire in the pipe. If a joint is required, it should also be connected in the junction box and insulated. (√)

211. When Y-Δ step-down starting is used, the starting current is 1/2 of the starting current when Δ-configuration is directly used. (×)

Analysis: The motor adopts Y-Δ step-down starting. The starting current of Y-configuration is 1/3 of that of Δ-configuration. The calculation basis is that the wire current of Y-configuration is equal to the phase current.

212. When measuring the resistance of a multimeter, the pointer is the most accurate in the middle of the dial. (√)

213. Automatic frequency regulation and compensation is a function of thermal relay. (×)

Analysis: The thermal relay is mainly used to prevent the control circuit from being disconnected when the current is too large, which has nothing to do with the frequency. The thermal relay does not have the function of automatic frequency regulation.

214. When filling in the type I work card valid for several days, at the end of each day, if the grounding wire installed at the work site is removed and the work resumes the next day, it is not necessary to re-check the grounding wire. (×)

Analysis: If the work is resumed the next day, the electricity should be checked again and the grounding wire should be installed.

215. Each circuit of street lamp should be protected, and each lamp should be equipped with a separate fuse. (√)

216. The entities engaged in the production, marketing and storage of hazardous substances should set up an organization for the administration of production safety or be staffed with part-time personnel for the administration of production safety. (×)

Analysis: According to *The Regulations on the Safety Management of Dangerous Chemicals*, the production, operation and storage units of dangerous goods should have safety management organizations and full-time safety management personnel.

217. In the magnetic circuit, when the magnetoresistance is constant, the magnetic flux is inversely proportional to the magnetomotive force. (×)

Analysis: According to Ohm's law of magnetic circuit, the magnetic flux passing through the magnetic circuit is proportional to the magnetomotive force and inversely proportional to the magnetoresistance.

218. The grounding wire is an important tool to ensure the staff when the voltage is unexpectedly appeared on the equipment and lines that have been powered off. As required, the grounding wire must be made of bare copper flexible wire with a sectional area of more than 25mm^2. (√)

219. The table lamp with screw lamp cap should adopt three-hole socket. (√)

220. The short circuit test of the motor is to apply a voltage of about 35V to the motor. (×)

Analysis: The short-circuit voltage of the motor with rated voltage of 380V is generally 75~90V.

221. The coil resistance of the transformer could be measured by using the multimeter resistance range. (×)

Analysis: Generally, the coil resistance of transformer is relatively small, which could be measured by multimeter, but the measured data is inaccurate and could only be used as a reference value. If it is necessary to accurately measure its resistance value, a DC bridge could be used.

222. In DC circuits, brown is often used to indicate positive. (√)

223. During the operation of the equipment, the cause of fire is the current heat is the indirect cause, while the spark or arc is the direct cause. (×)

Analysis: Current and spark or arc are the direct causes.

224. The grounding resistance meter is mainly composed of hand generator, current mutual inductor, potentiometer and galvanometer. (√)

225. The insulation resistance of valve arrester operating below 10kV should be measured once a year. (×)

Analysis: Once every six months. As the case may be, if the operating conditions are complex, it could be done every 3 months.

226. The capacity of lead acid battery should not decrease with the increase of discharge rate. (×)

Analysis: The capacity of lead acid battery decreases with the increase of discharge rate. The larger the discharge current of lead acid battery is, the smaller the output capacity is.

227. Ohm's law points out that in a closed circuit, when the conductor temperature is constant, the current passing through the conductor is inversely proportional to the voltage applied on both ends of the conductor, and is proportional to its resistance. (×)

Analysis: Ohm's law refers to that in the same circuit, the current passing through a section of conductor is proportional to the voltage at both ends of the conductor and inversely proportional to the resistance of the conductor.

228. In the case of electrical fire, if the power supply could not be cut off, the fire could only be extinguished with electricity, and dry powder or CO_2 extinguishers should be selected, and water-based extinguishers should be used as little as possible. (×)

Analysis: "Power off before putting out the fire" is the basic principle of putting out electrical fires. CO_2 extinguisher should be used as far as possible in case of fire in electrical equipment, and other extinguishers may damage electrical equipment. CO_2 extinguisher is applicable to extinguish the initial fire of flammable liquid and gas, as well as the fire of live equipment.

229. In case of heavy work task and tight time, the special supervisor could also do other work. (✕)

Analysis: In order to concentrate on the supervision work, the Safety Regulations stipulate that the dedicated supervisor is not allowed to do other work concurrently.

230. The grounding of the safety barrier should be disconnected only before disconnecting the electrical circuit in the hazardous area. (✓)

231. Incandescent lamps should be used for lighting in places with frequent switching. (✓)

232. When measuring the voltage, the voltmeter should be connected in parallel with the circuit under test. The internal resistance of the voltmeter is far greater than the resistance of the load being measured. (✓)

233. When using hand electric tools, check whether the power switch is out of order, damaged, firm and loose. (✓)

234. As the density of carbon monoxide (CO) is slightly lighter than that of air, it floats on the upper layer. When rescuers enter and leave the site, it would be safer to crawl. (✓)

235. The rotor connected with frequency sensitive rheostat in series has large starting torque and is suitable for heavy load starting. (✕)

Analysis: Since the starting current and starting torque would decrease at the same time after the rotor connected with frequency sensitive rheostat in series, it is only suitable for no-load and light load starting.

236. *The Production Safety Law of the People's Republic of China* is not only applicable to production and business entities, but also applicable to the management of national security and social security. (✕)

Analysis: According to Article 2 of *The Production Safety Law of the People's Republic of China*, this law is applicable to the production safety of entities engaged in production and business activities within the territory of the People's Republic of China. National security and social security do not belong to production and business activities, so they are not applicable to *The Production Safety Law of the People's Republic of China*.

237. When measuring the current, connect the ammeter in series with the circuit under test. (✓)

238. In case of lightning, it is forbidden for outdoor maintenance and test at heights, and indoor electricity inspection. (✓)

239. The insulating operating rod for measurement should be subject to insulation test once every six months. (✓)

240. Before the electroscope is used to check the electricity, the line should be considered

electrified. (√)

241. The safety extra low-voltage (SELV) is only used as electric shock protection for grounding system. (×)

Analysis: SELV is only used for the safety extra low-voltage of ungrounded system.

242. From the perspective of overload, the rated voltage of fuse is specified. (×)

Analysis: A fuse is an electrical appliance that the fusant melts and breaks the circuit with the heat generated by itself when the current exceeds the specified value. The object of overload is relative to the power supply. Therefore, from the perspective of overload prevention, "rated current of fuse" should be specified, not voltage.

243. In a series circuit, the current is equal everywhere. (√)

244. The linear twisted connection method could be used when the single strand conducts wires with small sectional area. (√)

245. When filling in the type I work card that is valid for several days. At the end of each day, if the grounding wire installed at the work site is removed, the grounding wire should be installed again before resuming work the next day. (×)

Analysis: Before resuming work the next day, the electricity should be checked again, and then the grounding wire should be installed.

246. The user entities must establish a system for regular inspection and invalidation & scrapping of labor protection articles. (√)

247. Although the induced voltage and residual charge have potential, they would not cause harm to human body. (×)

Analysis: In the power system, due to the electromagnetic induction and electrostatic induction of live equipment, certain potential could be induced on the nearby power failure equipment, and electric shock accidents caused by induced voltage occur frequently, even causing casualties. Both the phase to phase insulation and insulation to ground of electrical equipment have capacitive effects, which would retain a certain amount of residual charge on the power cut equipment that has just disconnected the power supply. If it is not fully discharged, the human body would be shocked by the residual charge when touching it.

248. For easy access, hand tools should be placed at the edge of the workbench. (×)

Analysis: Hand tools should not be placed beside the workbench. It is easy to bump or fall to the ground.

249. When repairing the line with power, stand on the insulating mat. (√)

250. Symmetrical three-phase power supply is a power supply system composed of 3 sinusoidal power supplies with the same amplitude and 120° phase difference in sequence. (×)

Analysis: The symmetrical three-phase AC power supply is a sinusoidal AC power supply with the same amplitude and 120° phase difference in sequence generated by the three-phase alternator. It is not composed of three independent sinusoidal AC power supplies "connected".

251. If the temperature is too high when the capacitor is running, the ventilation should be strengthened. (×)

Analysis: Ventilation is an environmental condition, which could be strengthened if allowed. If the capacitor temperature is too high due to voltage rise or other reasons, strengthening ventilation to cure the symptoms rather than the root cause.

252. When measuring, the range of the voltmeter should be greater than or equal to the voltage of the measured line. (√)

253. Parallel compensation capacitors are mainly used in DC circuits. (×)

Analysis: Parallel compensation is used in AC circuit. It is a technology to realize reactive compensation by connecting devices with capacitive power load (several capacitors) and inductive power load in the same circuit.

254. For the burned part, the wounded could wipe toothpaste, ointment and other oily substances to the burn wound to reduce the chance of wound pollution and ease the difficulty of treatment when going to the hospital. (×)

Analysis: Wiping toothpaste and other oily substances on the burn site would cause the hot air of the wound to be covered by the toothpaste and could not be dispersed, so it has to spread to the deep subcutaneous tissue, resulting in a deeper scald.

255. "Stop! High-voltage Danger!" belongs to warning signs. (√)

256. AC clamp ammeter could measure AC and DC current. (×)

Analysis: Although the clamp ammeter could be used to measure both AC current and DC current, some clamp ammeters could only measure AC covvld not. Therefore, carefully read the instruction manual before using the clamp ammeter to find out whether it is an AC or AC/DC clamp ammeter.

257. The low-voltage circuit breaker could only effectively connect and disconnect the load current, but the fuse must disconnect the short circuit current. (×)

Analysis: Low-voltage circuit breaker has overload and short circuit protection functions. It could automatically trip and cut off the power supply in case of short circuit or overload of the line, so as to effectively protect the equipment from damage and minimize the accident.

258. All plastic protective wiring should be used for lighting wiring in corrosive places, and all joints should be sealed. (√)

259. The safety helmet is calculated from the date when the product is manufactured. The validity period of glass fiber reinforced plastic and plastic safety helmet is 3 years. The safety helmet beyond the validity period should be scrapped. (×)

Analysis: The validity period of safety helmets varies according to the materials used in their manufacture. The validity period is calculated from the date when the product is manufactured. For example, the validity period of FRP helmet is 2.5 years and that of rubber helmet is 3 years.

260. In daily life, attention should be paid when contacting with flammable and explosive substances. Some media are easy to generate static electricity and even cause fire and explosion. For example, it is not allowed to contain oil in metal drums at gas stations. (×)

Analysis: Adding gasoline or diesel directly into the plastic bucket at the gas station is easy to generate static electricity, which is very easy to cause fire or explosion. If a plastic bucket is used to contain gasoline, first add the oil into the iron bucket and move it into the plastic bucket at a safe place.

261. The buttons could be of open type, waterproof type, anti-corrosion type, protective type, etc., according to the use occasion. (√)

262. Insulating materials refer to materials that are absolutely non-conductive. (×)

Analysis: The insulating material is non-conductive under the allowable voltage, but not absolutely non-conductive. Under the action of a certain external electric field strength, the processes of conduction, polarization, loss, breakdown, etc., would also occur, and aging would occur after long-term use. Its resistivity is very high, usually in the range of $10 \sim 10\ \Omega \cdot m$.

263. Before the electricity inspection, the test should be carried out on the electrical equipment to confirm that the electroscope is in good condition; If the test could not be carried out on the electrical equipment, the power frequency high-voltage generator could be used to verify that the electrical equipment is in good condition. (√)

264. When working on the live line tower, if the distance from the live body is sufficient, there is no need to assign a special person to monitor it. (×)

Analysis: special personnel should be assigned to monitor the work on live line pole/towers. The personnel on the pole should prevent objects from falling, and the tools and materials used should be transmitted with ropes, and should not be thrown randomly; Pedestrians should be prevented from staying under the pole/towers.

265. In a series circuit, the total voltage of the circuit is equal to the sum of the partial voltages of each resistor.(√)

266. When checking the capacitor, just check whether the voltage meets the requirements. (×)

Analysis: There are many inspection items for the capacitor, such as electrical capacity, voltage, insulation, whether the capacitor has expansion, oil injection, oil leakage, whether the ceramic part is clean, whether there is discharge trace, and whether the grounding wire is firm.

267. When the low-voltage electrical equipment is powered off for maintenance, in

order to prevent the maintenance personnel from going to the wrong position, entering the live space by mistake and approaching the live part too closely, the barrier is generally used for protection. (×)

Analysis: The following technical measures must be taken for power cut-off maintenance of low-voltage electrical equipment: power cut-off, power inspection, discharge of equipment with capacitors, installation of grounding wires, installation of barriers and hanging of signs.

268. The relevant production and business operation entities should, in accordance with their own or enterprise's financial situation, appropriately adjust and use the production safety expenses, which should be used exclusively to improve the production safety conditions. (×)

Analysis: *The Production Safety Law of the People's Republic of China* stipulates that relevant production and business entities should withdraw and use production safety expenses in accordance with the provisions, and improve production safety conditions.

269. The lighting switch of the warehouse should be installed outside the warehouse to ensure safety. (√)

270. DC ammeter could be used for AC circuit measurement. (×)

Analysis: DC ammeters could not measure AC, and AC ammeters could not measure DC, because the working principles of the two ammeters are different and their purposes are different.

271. When the overhead line is not high enough to the ground, the method of tightening the line could be used to increase the height of the conductor. (×)

Analysis: The safe distance of the line refers to the minimum allowable distance between the conducting wire and the ground (water surface), tower components, and crossing objects (including power lines and weak current lines). When the overhead line is not high enough to the ground, insulation isolation protection measures could be taken to solve the problem. If the sag of the line is too small due to tightening, it is easy to exceed the strength of the overhead line under the maximum stress meteorological conditions, which would lead to a line breaking accident, and it is difficult to ensure the safe operation of the line.

272. If someone is found to have an electric shock, the finder should first call a doctor, and then start the first aid immediately after the doctor arrives. (×)

Analysis: When someone gets an electric shock at low voltage, the finder should try to turn off the power supply in time as soon as possible. Get the wounded away from the power supply, the finder should immediately give first aid at the site. Meanwhile, the finder should call 120 for a doctor.

273. Various short circuits of windings, short circuit of iron core, bearing damage or lack of oil, severe overload or frequent starting might cause excessive temperature rise of the

motor. (√)

274. The sockets with different voltages should be obviously different. (√)

275. The dull sound when the motor is running is the sound when the motor is running normally. (×)

Analysis: When the motor operates normally, the sound is smooth, light and uniform; If there are shrieks, dullness, friction, vibration and other harsh noises, the motor is faulty.

276. Electrician operation is divided into high-voltage electricians and low-voltage electricians. (×)

Analysis: Electrician operation is divided into high-voltage electrician operation, low-voltage electrician operation and explosion-proof electrical operation.

277. China's safety production policy is "safety first, prevention first, and comprehensive governance". (√)

278. According to Article 71 of *The Production Safety Law of the People's Republic of China*, any entity or individual has the right to report to the department responsible for the supervision and administration of production safety on potential accidents or violations of production safety. (√)

279. When the capacitor exploded, check it immediately. (×)

Analysis: When the capacitor exploded, the connection between the capacitor and the power grid should be cut off first, and the inspection could only be carried out after manual discharge.

280. The AC contactor could cut off the short circuit current. (×)

Analysis: A separate AC contactor would not actively cut off the short circuit current. In practices, electromagnetic starters are often combined with appropriate thermal relays or electronic protective devices to protect circuits that may have overload or phase failure.

281. All bridges are used to measure DC resistance. (×)

Analysis: In addition to measuring DC resistance, the bridge could also be used to measure the parameters of capacitors, inductors and other components.

282. Red is used in the safety color code to indicate prohibition, stop or firefighting. (√)

283. When the conductor temperature is constant, the current passing through the conductor is proportional to the voltage at both ends of the conductor and inversely proportional to its resistance. (√)

284. When the wounded is disconnected from the power supply, the general condition of the wounded should be checked immediately, especially the respiration and heartbeat. If the respiration and heartbeat stop, the funeral should be prepared immediately. (×)

Analysis: In case of respiratory and cardiac arrest, the wounded should be immediately rescued at the site, and 120 should be contacted urgently to send the wounded to the nearest hospital for further treatment; On the way to the hospital, the rescue work cannot be interrupted.

285. People with cardiac arrest should take artificial respiration first. (✗)

Analysis: If the injured person's heartbeat stops and breathing is normal, external cardiac compression should be carried out immediately for first aid.

286. The sign board of "Stop, High-Voltage Danger" is white with red edge and red arrow. (✓)

287. Anti-corrosion measures must be taken when connecting wires. (✓)

288. The function of the travel switch is to send out the electrical signal of the length of mechanical travel. (✗)

Analysis: The function of the travel switch is to send out the electrical signal at the limit position of the mechanical travel. It is the limited position, not the length.

289. In three-phase-four-wire distribution system, PEN line represents the common neutral wire for working neutral and protection neutral. (✓)

290. Electric shock injury is divided into electric shock and electric injury. (✓)

291. After the multimeter is used, the range switch could be placed at any position. (✗)

Analysis: After the multimeter is used, place the range switch at the highest or neutral position of AC voltage. This is mainly for the safety of the meter. It is not easy to burn the meter when the meter is opened for measurement next time.

292. The cross sectional area of overhead insulated copper conducting wire crossing railways, highways, etc., should not be less than 16mm^2. (✓)

293. The handle of No. 1 electrician knife is longer than that of No. 2 electrician knife. (✓)

294. The production and business operation entities should document the major hazards, conduct regular testing, evaluation and monitoring, formulate emergency plans, and inform the employees and relevant personnel of the emergency measures that should be taken in case of emergency. (✓)

295. Use a tourniquet to stop bleeding. The tourniquet usually lasts no more than 1 hour. If it is too long, it may cause necrosis of the body. (✓)

296. The prenatal examination of pregnant female workers should be counted as working time. (✓)

297. In electrical power safety operation regulations, safety technical measures include work card system, work permit system, work monitoring system, work interruption transfer and termination system. (✗)

Analysis: The measures described above are measures for electrical safety. Electrical safety technical measures include power on, power inspection, grounding wire installation, etc.

298. After the leakage switch is installed, the equipment shell does not need to be grounded or connected to zero. (×)

Analysis: When the leakage switch is installed, it is not necessary to connect to neutral, but the grounding protection must be done. For example, the grounding protection wire in the three hole plugs at home is directly connected to the equipment shell of the electrical appliance. If it is connected to zero, it would cause the leakage circuit breaker to trip and cannot be used.

299. For distribution lines equipped with overload protection, the allowable current carrying capacity of the insulated conductor should not be less than 1.25 times the rated current of the fuse. (√)

300. The employment injury insurance fund should pay the disability allowance on a monthly basis, and the payment standard for Level Ⅲ disability should be 70% of my salary. (×)

Analysis: Grade Ⅰ disability is 24 months' salary, Grade Ⅱ disability is 22 months' salary, Grade Ⅲ disability is 20 months' salary, and Grade Ⅳ disability is 18 months' salary. If an employee is identified as Class V or Class VI disabled due to work, the labor relationship with the employer should be maintained, and the employer should arrange appropriate work. If it is difficult to arrange work, the employer should pay a disability allowance on a monthly basis. The standard is: 70% of his/her salary for Grade 5 disability, and 60% of my salary for Grade 6 disability.

301. The symbol "A" indicates AC power supply. (×)

Analysis: A represents current in amperes. AC power supply is expressed in AC.

302. Leakage protector or leakage switch should not be installed near the high current bus. (√)

303. Protective articles could eliminate accidents, so it is safe enough to use protective articles at work. (×)

Analysis: Protective article is a necessary defensive equipment to protect the personal safety and health of workers in the production process, and plays a very important role in reducing occupational hazards. Labor protection articles must be used correctly when working on the production site.

304. The use of lightning rods and lightning strips is the main measure to prevent lightning from damaging power equipment. (×)

Analysis: The main measure to prevent lightning from damaging power equipment is not lightning rod or lightning strip, but lightning arrester.

305. In China, ultrahigh voltage (UHV) transmission lines are basically laid overhead. (√)

306. The voltage when the insulator is broken down is called breakdown voltage. (√)

307. The residual current acting protection device is mainly used for low-voltage systems

below 1000V. (√)

308. Class I equipment should be provided with good neutral or grounding measures, and the protective conductor should be separated from the working neutral wire. (√)

309. After the motor is overhauled, the no-load test and short circuit test could be carried out on the motor after all inspections are qualified. (√)

310. The distance between the ceiling lamp and the ground in the hazardous area should not be less than 3m. (×)

Analysis: 2.5m indoor and 3m outdoor. In dangerous and special dangerous places, when the height of the lamp from the ground is less than 2.4m, lighting lamps with rated voltage of 36V and below should be used or special protective measures should be taken.

311. The underground distance between the grounding body and the grounding body of the independent lightning rod should not be less than 3m. (√)

312. Lighting installed in damp places should be moisture-proof to ensure safety. (√)

313. For three-phase asynchronous motor with full voltage starting, the fuse on the power line mainly plays the role of short circuit protection; The thermal relay on the power line mainly plays the role of overload protection. (√)

314. The grounding resistance meter is an instrument to measure the insulation resistance of the circuit. (×)

Analysis: The grounding resistance is relatively small, while the insulation resistance is relatively large. The grounding resistance meter could not test the insulation resistance. The insulation resistance should be measured with an insulation resistance meter (megger).

315. Because the insulation resistance depends on the insulation structure and its state, the insulation resistance could be measured after power cut-off or without power cut-off. (×)

Analysis: The insulation resistance could only be measured in case of power cut-off, and the measurement should be suspended in case of high humidity.

316. Electricians should monitor the operation of electricity users in special places. (√)

317. For places easy to generate static electricity, keep the ground moist or lay a floor with good conductivity. (√)

318. The automatic switching device works automatically depending on the change of its own parameters or external signals. (√)

319. There are 6 types of short-time operation durations specified in China when the motor operates under short-time rating condition. (×)

Analysis: There are four types for the duration of short-time motor operation: 10min, 30min, 60min and 90min.

320. The magnetic force line is a closed curve. (√)

321. Breaking current capacity is one of the main technical parameters of various knife

switches. (√)

322. When the combination switch is selected as the direct control for motor, its rated current should be 2-3 times of the rated current of the motor. (√)

323. The employees should have the right to criticize, report and accuse the problems existing in the work of production safety of their own entities. (√)

324. The specification of multi-purpose screw driver is expressed by its full length (handle plus screw rod). (√)

325. Dust, poisons and noise existing in the operation are harmful factors causing occupational diseases. (√)

326. If the heartbeat stops and breathing exists, artificial respiration should be carried out immediately. (×)

Analysis: External cardiac compression should be performed.

327. In case of cardiac arrest with breathing, external cardiac compression should be carried out immediately. (√)

328. After the line connected to the shunt capacitor is cut off, the capacitor bank must be disconnected. (√)

329. The rescuer stood on the side of the injured person's head, took a deep breath, and blew air into the injured person's mouth (two mouths should be close to each other without air leakage), causing exhalation. (×)

Analysis: When performing artificial respiration, the rescuer stands on the side of the injured person's head, takes a deep breath, and blows air into the injured person's mouth (the two mouths should be close to each other without air leakage), causing inspiration.

330. The leakage switch would only act when someone gets an electric shock. (×)

Analysis: In case of short circuit, overload, leakage, overvoltage, undervoltage, etc., the leakage protection switch would trip to protect the circuit.

331. In order to improve the starting and running performance of the motor, the rotor iron core of the cage asynchronous motor generally adopts a straight slot structure. (×)

Analysis: The rotor iron core of cage asynchronous motor generally adopts the chute structure.

332. The most basic requirement for first-aid on the spot of thermal burn is to quickly get away from the heat source, take off the burning clothes or use water to extinguish the fire on the body. (√)

333. For wound type asynchronous motor, the contact between brush and collector ring and the wearing, pressure and spark of brush should be checked frequently. (√)

334. When the motor is protected by fuse, if the running current reaches the rated current of the fusant, the fusant would fuse immediately. (×)

Analysis: The fuse of motor is selected according to 1.5~2 times of its rated current. Therefore, when the current of motor overload is less than 1.5~2 times of its rated current, the fusant would not melt, and the motor might burn out due to overload heating.

335. The literal symbol of the contactor is KM. (√)

336. High pressure mercury lamp has a high-voltage, so it is called high-voltage mercury lamp. (×)

Analysis: High pressure mercury lamp is an electric light source that uses high pressure (0.2~1MPa) mercury vapor generated during mercury discharge to obtain visible light. High pressure mercury lamp has nothing to do with high-voltage.

337. When measuring the insulation resistance of electrical equipment with large capacitance, after reading and recording, disconnect the L terminal first, then stop swinging, and then discharge. (√)

338. The rated current of AC contactor is the current value under rated working conditions. (√)

339. According to the provisions of *The Production Safety Law of the People's Republic of China*, the person in charge of the production and business operation entity should promptly take effective measures to organize first aid after receiving the report on personnel accidents at the site of the accident. (√)

340. If multiple intrinsically safe circuit systems have electrical connections, comprehensive system intrinsically safe performance evaluation should be conducted. (√)

341. In thunderstorm weather, do not repair electrical circuits, switches, sockets, etc. at home even indoors. If it is necessary to repair, turn off the main power switch at home. (×)

Analysis: If it is necessary to repair, turn off the main switch of the home power supply and ground the line.

342. "No switching on, someone is working!" belongs to prohibited signboard. (√)

343. Kirchhoff's first law is the law of node current, which is used to prove the relationship between currents on a circuit. (√)

344. The competent department responsible for the supervision and administration of production safety should be informed of the appointment and removal of the personnel in charge of production safety in the entities that produce and store dangerous goods and in the mining and metal smelting entities. (√)

345. When putting out flammable and combustible liquid fires, the dry powder fire extinguisher should spray at the waist of the flame. If the liquid fire being put out is flowing, the root of the flame should be sprayed from near to far, from left to right, until all the flames are put out. (√)

346. Thermal relay is a kind of protective appliance that uses bimetallic sheet to push the

contact act when it is heated and bent. It is mainly used for quick break protection of lines. (✗)

　　Analysis: Thermal relay is a protective appliance for heating and deformation through thermocouple, and is used for overload protection of motor or other electrical equipment and electrical circuit.

　　347. Even if the eyelid wound is bleeding, the eyeball could not be compressed for hemostasis, otherwise the contents of the eye would overflow, making the vision irreversible. (✓)

　　348. The height of the toggle switch of the lighting from the ground is 1.2~1.4m. (✓)

　　349. Before using insulating gloves, the appearance should be checked first. If holes and cracks are found on the surface, the gloves should not be used. (✓)

　　350. Dual insulation refers to working insulation (basic insulation) and protective insulation (additional insulation). (✓)

　　351. For the equipment with motor, check the auxiliary equipment, mounting base and grounding of the motor before the motor is powered on, and then power on the equipment after it is normal. (✓)

　　352. In the zero connection system, the working zero wire (N) connection hole and the protective zero wire connection hole of the single-phase-three-hole socket should not be connected together. (✓)

　　353. The positive direction of the electromotive force is specified as from low potential to high potential, so the positive pole of the voltmeter should be connected to the negative pole of the power supply when measuring, and the negative pole of the voltmeter should be connected to the positive pole of the power supply. (✗)

　　Analysis: The electromotive force is the work done by the non-electrostatic force inside the power source to move the positive charge, and the "direction" refers to the process of doing work. The voltmeter mentioned above measures the result of work, which is just the opposite of the process of work, so the connection of the voltmeter is reversed. The correct method is that the voltmeter is connected in parallel with the power supply, the positive pole is connected to the positive pole of the power supply, and the negative pole is connected to the negative pole of the power supply.

　　354. The employees and the production and business operation entities may appropriately adjust the standards of social insurance compensation and civil compensation for employment injuries according to their own or enterprise financial conditions. (✗)

　　Analysis: Neither the employees nor the production and business operation entities should determine the standard for employees to obtain social insurance compensation and civil compensation for employment injuries.

　　355. When the load current reaches the rated current of the fusant of the fuse, the fusant would melt immediately, thus playing the role of overload protection. (✗)

Analysis: When the load current is equal to the rated current of the fusant of the fuse, the fusant would melt after a delay. Since the temperature of the fusant rises sharply after the current reaches its rated current and is maintained at this temperature for a long time, the internal resistance of the fusant would become larger and larger, so the heating would become stronger and stronger until the fusant is melted.

356. The power transformation and distribution equipment should be equipped with complete shielding devices. (√)

357. Electricians should operate in strict accordance with the operating procedures. (√)

358. The voltage mutual inductor in operation should be out of service in case of ceramic bushing fracture, high-voltage coil breakdown discharge, connection point ignition, serious oil leakage and other faults. (√)

359. The employment injury insurance fund should pay the disability allowance on a monthly basis, and the payment standard for level I disability should be 90% of the salary of the injured. (√)

360. Copper wire and aluminum wire could be directly connected when necessary. (×)

Analysis: The electrochemical properties of copper and aluminum are different. Aluminum wire is easy to oxidize in the air, forming a layer of oxide on its surface. In addition, aluminum is less hard than copper, which would greatly increase the contact resistance at the connection between aluminum wire and copper wire. When the current passes through this connection, the contact resistance would heat up. If the current is large, the heating would be very serious, and the connection would be burned. Copper and aluminum transition terminals could be used when connecting copper wire and aluminum wire, and the copper end should be tinned.

361. The function of the cable protective layer is to protect the cable. (√)

362. In power supply and distribution system and equipment automatic system, knife switch is usually used for power isolation. (√)

363. Text symbol of time relay is KT. (√)

364. The production and business operation entity might negotiate with the employees, and conclude an agreement with the employees based on the results of the negotiation, so as to exempt or mitigate its legal liability for the casualties of employees due to production safety accidents. (×)

Analysis: *The Production Safety Law of the People's Republic of China* stipulates that the employees of production and business operation entities have the right to obtain production safety guarantee according to law, and should perform their obligations in production safety according to law. It also stipulates that the state implements a system of accountability for production safety accidents and investigates the legal liability of those responsible for production safety accidents. Therefore, if a production and business operation entity composes an agreement with its employees to exempt or mitigate its legal liability for the casualties of employees due to production safety accidents, the agreement is invalid.

365. The power circuit of firefighting equipment could only be equipped with leakage alarm devices that do not cut off the power supply. (√)

366. Electrician's pliers, electrician's knives and screwdrivers are common basic tools for electricians. (√)

367. The absorption ratio of large capacity equipment is the ratio of the measured insulation resistance (60s) to the measured insulation resistance (15s). (√)

368. Occupational disease refers to the disease caused by workers' exposure to occupational disease inductive factors in their occupational activities. (×)

Analysis: Occupational disease refers to the disease caused by the workers of enterprises, institutions, individual economic organizations and other employers who are exposed to dust, radioactive substances and other toxic and harmful substances in their occupational activities.

369. When using bamboo ladders for operation, it is better to place the ladders at about 50° to the ground. (×)

Analysis: Bamboo ladders are generally placed at about 60° to the ground. When the angle is too large, it is easy to overturn; If the angle is too small, the ladder and the ground are easy to slide. When people climb to the upper end, the upper ladder is relatively easy to break due to the large lateral force.

370. High strength copper core rubber sheathed hard insulated cables should be used for power supply of mobile electrical equipment. (×)

Analysis: When the movable electrical equipment is working, the cable should be dragged and bent at any time, so the soft sheathed cable should be used. Hard wires would hinder operation.

371. 220V hand lighting lamp should be equipped with complete protective net and insulated handle with heat insulation and moisture resistance; This requirement is not applicable to hand lamps with safe voltage. (×)

Analysis: There are five levels of safety voltage: 42V, 36V, 24V, 12V, 6V. According to the national standard Safety Voltage, when the electrical equipment is used for more than 24V, protective measures must be taken to prevent direct contact with live bodies.

372. The telephone weighing noise voltage with high frequency switching power supply should be ⩽ 20mV. (×)

Analysis: The telephone weighing noise voltage with high frequency switching power supply should be ⩽ 2mV.

373. One of the purposes of labor protection for female workers is to ensure the health of the next generation. (√)

374. The working voltage of Class Ⅲ electric tools should not be greater than 50V. (√)

375. Operators working in flammable, explosive, burning and static electricity places are not allowed to issue and use chemical fiber protective articles. (√)

376. According to statistics of some provinces and cities, electric shock accidents in rural areas are less than those in urban areas. (×)

Analysis: Due to the relatively poor electricity conditions in rural areas, the lack of protective devices, more random connecting, the lack of electrical knowledge and other reasons, more electric shock accidents occurred in rural areas than in urban areas.

377. If someone is found to have an electric shock, the hospital should be immediately notified to send an ambulance for rescue. Before the doctor arrives, the on-site personnel should not rescue the person who has an electric shock to avoid secondary injury. (×)

Analysis: If someone is found to have an electric shock, the finder should try to turn off the power supply in time as soon as possible. After the person who gets an electric shock is disconnected from the power supply, the finder should timely judge whether the wounded has heartbeat, respiration and consciousness, give first aid on the spot, and call 120 for a doctor.

378. When filling in the type Ⅱ work card for power lines, it is not necessary to go through the work permit formalities (√)

379. In addition to the independent lightning rod, the lightning protection grounding device could be shared with other grounding devices on the premise that the grounding resistance meets the requirements. (√)

380. The high-voltage electroscope should be tested on the energized equipment to verify that the electroscope is in good condition. (√)

381. The current would be formed in the rotor conductor of three-phase asynchronous motor, and its current direction could be determined by the right-hand rule. (√)

382. The equipment management personnel should be responsible for the daily maintenance of electrical equipment. (×)

Analysis: The maintenance and repair of daily electrical equipment should be the responsibility of the equipment maintenance personnel (maintenance electrician), not the equipment management personnel.

383. TT system is a system in which the neutral point of distribution network is directly grounded, and the shell of electrical equipment is also grounded. (√)

384. Universal relays could be made into various relays by replacing coils of different properties. (√)

385. In places with explosion and fire hazards, portable and movable electrical equipment should be used as little as possible. (√)

386. The handle of the electrician's knife is not insulated and could not cut on the live wire or equipment to avoid electric shock. (√)

387. When measuring the diode with multimeter at range R × 1kΩ, the resistance measured by connecting the red probe to one pin and the black probe to the other pin is about several hundred ohms. If the resistance is large in reverse measurement, the diode is in good condition. (√)

388. Current and magnetic field are inseparable. Magnetic field always exists with current, and current is always surrounded by magnetic field. (√)

389. Patients with respiratory arrest and cardiac arrest should be sent to the hospital immediately. (×)

Analysis: This is wrong. The correct way is to rescue immediately at the site, unblock the respiratory tract, perform artificial respiration and cardiac compression, call for 120, and closely observe the changes of the condition.

390. The anti-static grounding resistance should not be greater than 4Ω in principle. (√)

391. The occupational disease prevention and control work should adhere to the policy of "prevention first, combine prevention with treatment". (√)

392. Wire stripper is a special tool used to exploit the insulation layer on the surface of small wire head. (√)

393. All parts of the same electrical component are drawn in the schematic diagram dispersedly, and text symbols must be marked in order. (×)

Analysis: The mistake is that "text symbols must be marked in order". The correct one is that all components of the same electrical component are drawn in the schematic diagram dispersedly, and text symbols must be marked, and numerical numbers must be marked in order.

394. The automatic switch belongs to the manual electrical appliance. (×)

Analysis: Automatic switch does not belong to manual electrical appliances. In case of severe overload, short circuit, voltage loss and other faults in the circuit, the automatic switch could automatically cut off the fault circuit and effectively protect the electrical equipment connected behind them. Automatic switch is a commonly used breaker with protection in low-voltage circuit.

395. Insulated rods should be stored on special shelves or vertically suspended on special hangers to prevent bending. (√)

396. When the probability is 50%, the average sensing current of adult male is about 1.1mA, the minimum is 0.5mA, and that of adult female is about 0.6mA. (×)

Analysis: The minimum current that causes people to feel is called sensing current. The experimental data show that the sensing current is different for different people, and the average sensing current of adult male is about 1.1mA; The adult female is about 0.7mA.

397. Temporary overhead lines not used in thunderstorm season might not be installed with lightning protection. (✕)

Analysis: Overhead lines are directly exposed in the open field, and are widely distributed, so they are most vulnerable to lightning strikes. Therefore, temporary overhead lines should be adapted to local conditions, and overhead grounding wires should be erected on the ground wire cross arm above the conducting wire or line lightning arresters should be installed according to the actual situation.

398. The capacity of a capacitor is its capacitance. (✕)

Analysis: The statement above is wrong. The capacity of capacitor is reactive power, and the unit is var. The capacitive reactive power of capacitor could be used to compensate the inductive reactive power of motor, so as to improve the power factor of the circuit; Capacitance refers to the charge storage capacity of the capacitor under a given potential difference. Its unit is Farad (F). Capacitance is the nominal parameter of the capacitor.

399. In three-phase AC circuit, when the load is Y-configured, its phase voltage is equal to the wire voltage of three-phase power supply. (✕)

Analysis: For three-phase-four-wire power system, the voltage between any two of the three phase wires is called wire voltage, and the voltage between any one of the phase wires and the zero wire is called phase voltage. The relationship between wire voltage and phase voltage is that the wire voltage is equal to $\sqrt{3}$ times of the phase voltage. The voltage in three-phase-four-wire system is 380/220V, that is, the wire voltage is 380V; The phase voltage varies with the connection mode: if Y-configuration is used, the phase voltage is 220V; For Δ-configuration, the phase voltage is 380V.

400. At present, the rated current of contactors produced in China is generally greater than or equal to 630A. (✕)

Analysis: The rated current of contactors produced in China is generally less than or equal to 630A. According to the production process and structure, the rated current of the contactor is different, and some products could reach 800A. The current difference before and after the AC contactor is very large, up to more than five times; Generally, it is started with high power and maintained with low power. When selecting the contactor, the rated working current of the main contact should be greater than or equal to the current of the load circuit.

401. The dust and smoke generated in the production process are non-toxic and harmless to operators. (✕)

Analysis: According to different chemical properties, the dust produced in the production process could cause fibrosis, poisoning, sensitization and other effects on the human body. Diameter less than 5 μm (aerodynamic diameter) dust is harmful to the body, and it is easy to reach the depth of respiratory organs. The concentration of dust is also related to the degree of harm to people. Therefore, if the operator need to be exposed to dust for a long time in the working environment, he/she must pay attention to regular physical examination. The hazards of welding fume to health are mainly respiratory mucosa irritation, inflammation, welder pneumoconiosis and poisoning.

402. When the discharge spark energy of static electricity is large enough, it could cause fire and explosion accidents. In the production process, static electricity could also hinder production and reduce product quality. (√)

403. In the electrical installation wiring diagram, all parts of the same electrical element must be drawn together. (√)

404. In general, single-phase electric shock of grounded power grid is less dangerous than ungrounded power grid. (×)

Analysis: Single-phase electric shock refers to an electric shock accident in which a part of the human body touches a phase charged body on the ground or other grounding conductors. The danger degree of single-phase electric shock is related to the operation mode of power grid. In general, single-phase electric shock in grounded power grid is more dangerous than that in ungrounded power grid.

405. Ventilation and dust removal is one of the main measures for dust prevention. (√)

406. Class Ⅱ hand electric tools are safer and more reliable than Class I tools. (√)

407. The slip rate of an asynchronous motor is the ratio of the speed difference between the rotating magnetic field and the motor speed to the speed of the rotating magnetic field. (√)

408. The direction of voltage is from high potential to low potential, which is the direction of potential increase. (×)

Analysis: The positive direction of voltage is defined as the direction of high potential to low potential, that is, the direction of potential drop.

409. In case of leakage or electric shock in the protected circuit, the zero sequence current mutual inductor would generate induced current, after the induced current has been amplified, it would result in the circuit breaker to act to cut off the circuit. (√)

410. It is not necessary to install leakage protection devices for the socket circuits in government offices, schools, enterprises, residences and other buildings. (×)

Analysis: No, leakage protector should be installed (residual current protection device). The following are the national regulations: except for the wall mounted split air conditioner power socket, the power socket circuit should be equipped with residual current protection device; Local equipotential bonding should be carried out for toilets with bath equipment; The main power incoming line of each residence should be provided with residual current action protection or residual current action alarm.

411. After the power is cut off, the motor stops running. When the power grid is powered on again, the motor could start automatically. This is called voltage loss protection. (×)

Analysis: The voltage loss and undervoltage protection means that when the power supply is cut off or the power supply voltage drops too much (undervoltage) for some reason, the protection device could automatically cut off the motor from the power supply. Because when the voltage is lost or undervoltage occurs, the current of the contactor coil would disappear or decrease, and the electromagnetic force is lost or insufficient to hold the moving iron core, so the main contact could be disconnected and the power supply could be cut off. It could be seen that after the loss of voltage or undervoltage protection acts, the motor could only be started again to avoid accidents.

412. In order to prevent major safety accidents caused by short circuit of conducting wires, conducting wires of different circuits, different voltages, AC and DC should not be through in the same pipe. (√)

413. Hand electric tools are classified in two ways, namely, by working voltage and by moisture resistance. (×)

Analysis: Hand electric tools are classified according to the working voltage and electric shock protection mode, not according to the degree of moisture resistance. The national standard "Safety Technical Regulations for the Management, Use, Inspection and Maintenance of Hand Electric Tools" (GB/T 3787—2017) could be divided into three categories: Ⅰ, Ⅱ and Ⅲ.

414. The anti-static safety helmet could be used as a safety helmet for electric power. (×)

Analysis: The role of electric safety helmet is to prevent the head of the operator from being injured by falling objects and other specific factors, playing a protective role; Compared with other types of helmets, anti-static helmets have special anti-static functions. They have different functions and could not be replaced.

415. Movable electrical equipment could be used according to relevant requirements of hand electric tools. (√)

416. Incandescent lamp belongs to thermal radiation light source. (√)

417. The frequency on the nameplate of the AC motor is the frequency of the AC power supply used by the motor. (√)

418. Generally, the patients with good conditions could be directly sent to the hospital with appropriate conditions for treatment, since they are estimated not to be in danger of life during the transfer process; If the patient's condition is critical, he or she should be transferred

to the next higher level hospital to avoid losing the chance of rescue. (×)

Analysis: If the patient's condition is critical, he or she should be treated nearby. Never ignore the patient's condition and transfer him or her without permission, so that the patient would lose the chance of rescue.

419. The characteristic of fuse is that the higher the voltage through the fusant, the shorter the melting time. (×)

Analysis: The fuse refers to an electrical appliance that uses the heat generated by itself to melt the fusant and disconnect the circuit when the current exceeds the specified value. Fuse is a kind of current protector that could break the circuit by melting the fusant with its own heat after the current exceeds the specified value for a period of time.

420. Water conducts electricity better than metal. (×)

Analysis: The resistivity of water is related to the amount of salt in the water, the concentration of ions in the water, the charge number of ions and the movement speed of ions. Therefore, the pure water resistivity is large, and the ultra-pure water resistivity is even greater. The purer the water, the greater the resistivity. Silver is the best conductive material at room temperature, followed by copper.

421. The main cause of asphyxiation accident in the well is water shortage. (×)

Analysis: The main reason for the asphyxiation accident in the well is that the gas and other harmful gases exceed the limit.

422. Reducing the electrical spacing could improve the safety of the enhanced safety electrical equipment to a certain extent. (×)

Analysis: Electrical spacing refers to the shortest space distance between two conductive parts. Increasing electrical spacing could improve the safety of increased safety electrical equipment to a certain extent.

423. The treatment method for human body exposed to productive toxicants could be repeatedly washed with clean water. (√)

424. When checking with a test pencil, it means that the circuit must be electrified when the pencil lights up. (×)

Analysis: Generally speaking, electricity means that there is current in the circuit, because the load could work only when there is current. The light emitted by electroscope (test pencil) could indicate that there is voltage (above 60V) to the ground, but it cannot indicate that there must be current.

425. The smaller the internal resistance of the ammeter, the better. (√)

426. The one-time work death subsidy standard is 20 times of the per capita disposable income of urban residents in the previous year. (√)

427. Power frequency current is more likely to cause skin burns than high-frequency

current. (✗)

Analysis: Generally speaking, 50~60Hz power frequency current is the most dangerous to human body. An important characteristic of high-frequency current is skin effect. Compared with power frequency current, high-frequency current has another characteristic that the fatality rate of people after electric shock would be reduced, and the higher the frequency, the lower the fatality rate.

428. High-voltage and dazzling light generated by lightning could damage electrical installations, buildings and other facilities, and damage to power facilities or power lines may lead to large-scale power failures. (✗)

Analysis: The dazzling light would not cause damage to other facilities, and would not damage buildings and electrical equipment. Electrical devices, buildings, power facilities and power lines are equipped with reliable and repeated lightning protection or lightning protection devices, which are inspected regularly and generally would not be struck by lightning.

429. Class I equipment should be provided with good neutral or grounding measures, and the protective conducting wire should be separated from the working neutral wire. (√)

430. The common rated maximum working voltage of AC contactor reaches 6000V. (✗)

Analysis: The rated working voltage of contactor from low to high is 220V, 380V, 660V, 1140V, 3300V, 6000V and 10000V.

431. The installation height of concealed sockets in special places should not be less than 1.5m. (✗)

Analysis: The actual installation height of the socket should be subject to the use environment and requirements. For example, in some laboratories, workshops and other special places, the height of the concealed socket is more than 1.5m, or even about 2.3m. In homes and similar places, the installation height of ordinary sockets is 30cm above the ground, 120cm for washing machines, 140cm for electric water heaters, 150cm for refrigerators, and 180cm for indoor units (wall mounted) of air conditioners.

432. The current on each branch in a parallel circuit is not necessarily equal. (√)

433. The electronic ballast of fluorescent lamp could make the fluorescent lamp obtain high-frequency alternating current. (√)

434. The fuse could provide overload protection in all circuits. (✗)

Analysis: Strictly speaking, it could not. The current action curve of the fuse is steep drop type, similar to the varistor diode in the electronic circuit. Only at the critical point could act quickly. Its action current is inversely proportional to time. The greater the current, the smaller the action time. The rated action time of the fusant melt is ∞ when conducting 1.25 times of the rated current. The rated action time is 3600s when conducting 1.6 times of the rated current. The rated action time is 1200s when conducting 1.8 times of the rated current. That is to say, it takes 1h when conducting 1.6 times of the rated current, and 20min when conducting 1.8 times of the rated current. If the motor conducts 1.8 times of the current and the fusant acts for 20min, it is easy to damage the insulation and cause the coil to be burnt due to short circuit.

435. The electrotechnical symbol of electrolytic capacitor is shown as ⊥. (√)

436. The yellow green conducting wire could only be used for the protective wire. (√)

437. The production and business operation entities must provide the employees with labor protection articles that meet the national standards or their own standards. (×)

Analysis: Article 42 of *The Production Safety Law of the People's Republic of China* stipulates that the production and business operation entities must provide the employees with labor protection articles that meet the national or industrial standards, and supervise and educate the employees to wear and use them according to the rules of use.

438. The larger the capacity of the compensation capacitor, the better. (×)

Analysis: The larger the capacity of the capacitor, the lower the resonant frequency, and the smaller the frequency range that the capacitor could effectively compensate the current. From the perspective of ensuring the capacity of the capacitor to provide high-frequency current, the idea that the larger the capacitor, the better is wrong, and there is a reference value in general circuit design.

439. Outdoor capacitors are generally installed on a workbench. (√)

440. When the motor is running normally, if smell the smell of burning, it means that the motor speed is too fast. (×)

Analysis: If the statement "Smell the smell of burning" is true, it is generally caused by the serious overheating of the motor, not the short circuit of the motor windings, nor the motor speed being too fast.

441. When the PN junction is positively connected, its internal and external electric fields are in the same direction. (×)

Analysis: When the PN junction is connected with a forward voltage, the PN junction is in a forward bias, and the current flows from the P side to the N side. The holes and electrons move towards the interface, narrowing the space charge area, and the current could pass through smoothly. The direction is opposite to the direction of the electric field in the PN junction.

442. In explosive hazardous areas, three-phase-four-wire system and single-phase-three-wire system should be adopted for power supply. (×)

Analysis: Three-phase-five-wire system and single-phase-three-wire system should be adopted for power supply in explosive hazardous areas.

443. Low-voltage circuit breaker is an important control and protection apparatus. All circuit breakers are equipped with arc extinguishing devices, so they could be safely closed and opened with load. (√)

444. The object whose conductivity is between conductor and insulator is called semiconductor. (√)

445. Shunt capacitor could reduce voltage loss. (√)

446. When testing live equipment with a voltage above 50V to ground, the neon bubble type low-voltage electroscope (test pencil) should show that there is electricity. (×)

Analysis: The measuring range of the electroscope (test pencil) is 60~500V. The starting voltage of neon bubble is about 70V. If the peak value of 50V AC voltage exceeds 70V, the neon bubble could be lit; The 50V DC voltage is not enough to light the neon bubble.

447. Three-pole leakage protection device should be selected for electrical equipment powered by single-phase 220V power supply. (×)

Analysis: Three-pole leakage protection device could not be used for electrical equipment powered by single-phase 220V power supply. If multiple single-phase 220V equipment are distributed on three-phase power supply, three-phase-four-pole leakage protection switch should be used, and the zero wire of load should pass through the leakage switch. If only one phase power supply is used, two-pole leakage switch or 1P+N leakage switch should be selected.

448. It is specified that the direction of the north pole of the small magnetic needle is the direction of the magnetic line of force. (√)

449. In the electrical schematic diagram, when the figure of contact is placed vertically, it is drawn according to the principle of "left open and right close". (√)

450. When the motor runs according to the nameplate value, the rated working mode for short-time running is indicated by S2. (√)

451. The secondary circuit of the current mutual inductor in operation should not be open circuited, and the secondary side of the voltage mutual inductor in operation is allowed to short circuit. (×)

Analysis: Open circuit is not allowed on the secondary circuit side of the current mutual inductor in operation, and short circuit is not allowed on the secondary circuit side of the voltage mutual inductor in operation, which is related to the different working principles of the two types of mutual inductor. Because the internal impedance of voltage mutual inductor is very small, if the secondary circuit is short circuited, a large current would appear, which would damage the secondary equipment and even endanger personal safety. When the current mutual inductor works normally, the secondary circuit side is nearly short circuited. If it is suddenly opened, the secondary circuit side winding would induce a very high spike wave when the magnetic field reaches zero, which could result in thousands or even tens of thousands of volts, threatening the safety of workers and the insulation performance of the instrument.

452. The intermediate relay is actually a voltage relay with adjustable action and release values. (×)

Analysis: The structure and principle of the intermediate relay are basically the same as those of the AC contactor. It is usually used to transmit signals and control multiple circuits at the same time. It could also be used to directly control small capacity motors or other electrical actuators. The intermediate relay is essentially a voltage relay, but it has a large number of contacts and a small capacity. It is an electrical apparatus used as a control switch.

453. For patients suspected of having thoracic, lumbar and vertebral fractures, the method of one person raising his/her head and the other person lifting his/her leg could be used when carrying. (×)

Analysis: This practice is obviously wrong, because it would aggravate the injury of patients with thoracic, lumbar and vertebral fractures. Correct method: When the patient is lifted onto the stretcher, two or three people should hold the shoulder & neck, injured spine, pelvis and lower limbs horizontally with their hands, and lift onto the stretcher. After he/she has been properly fixed, he/she should be escorted to the hospital for relevant diagnosis and treatment in the spine department or orthopaedic department.

454. Safety and reliability are the basic requirements for any switching device. (√)

455. When removing the grounding wires, the sequence is the same as that during installation. Remove the grounding terminals first and then the wire terminals. (×)

Analysis: When installing grounding wires, the grounding terminals must be connected first, then the wire terminals, and the contact must be firm. The sequence of removing the grounding wire is opposite to that of installing the grounding wire.

456. As long as the work is properly arranged, there would be no problem with scheduled power off and transmission. (×)

Analysis: Due to the fact that the power supply and cut off on schedule are divorced from the actual production, there is no way to predict the possible changes in the on-site operation mode at any time, and it is more difficult to implement specific preventive measures for each link. The operation safety has great blindness, so it is a form of illegal operation with great harm.

457. Carbon dioxide (CO_2) fire extinguisher is only applicable to the line below 600V. If it is 10kV or 35kV line, only dry powder fire extinguisher could be selected for live fire extinguishing. (√)

458. During the period of business trip, those who are injured due to work or whose

whereabouts are unknown due to accidents could be identified as employment injuries. (√)

459. In order to avoid explosion accidents caused by electrostatic sparks, the equipment should be separately isolated when processing, transporting, storing various flammable liquids and gases. (×)

Analysis: Not separate isolation, but physical isolation and proper grounding. In order to prevent accidents caused by electrostatic sparks, all metal pipes and non-conductive material pipes of equipment used for processing, storage and transportation of various flammable gases, liquids and powders must be reliably grounded.

460. When selecting the circuit breaker, the single phase to ground short circuit current at the end of the line should be greater than or equal to 1.25 times the setting value of the instantaneous release of the circuit breaker. (√)

461. In the multi-level fuses protection, the rated current of the later stage fusant is larger than that of the former stage, with the power supply end as the front end. (×)

Analysis: This statement is wrong. The place mentioned above is reversed. It should be correct that: in the protection of multi-level fuses, the rated current of the later stage fusant is larger than that of the former stage, and the load end is the frontmost end. Because the former level is regarded as the general insurance, the latter level is divided into sub insurance, otherwise, when the line fails, the general fuse would be burnt out directly and the area of power failure would be expanded.

462. PE wire should not pass through the leakage protector, but PEN wire could pass through the leakage protector. (×)

Analysis: The four-pole leakage protector should be three phase wires with one zero wire, not PEN wire. When the PEN wire is connected, it is directly grounded and short circuited. In this way, the switch would never be closed, because it is implementing the protection function.

463. Safety helmets for construction workers and motorcycle helmets could be used universally. (×)

Analysis: The motorcycle helmet is an all-direction enclosure, and the construction worker's safety helmet is mainly used to protect against the impact of falling objects on the head. The roles of the two are different. Safety helmets belong to special labor protection articles, which are required to have "LA" signs, but motorcycle helmets do not have "LA" signs.

464. Low-voltage electroscope (test pencil) could test the voltage below 500V. (×)

Analysis: The common low-voltage electroscope (test pencil) could detect the voltage of 60~500V, and the voltage below 60V could not be detected. Generally, when the potential difference between the charged body and the ground is less than 36V, the neon bubble does not emit light. When the voltage is 60~500V, the neon bubble emits light. The higher the voltage, the brighter the neon bubble. When the test voltage range is between 6~24V, people often use weak current test pencil.

465. If cables and other pipes are laid on the same trench, measures should be taken to prevent other pipes from affecting the electrical lines (heat or corrosion, etc.). Cables and electrical pipes should be laid along the side of the pipe with low risk. When combustible media are in the pipe, such as media heavier than air, electrical lines should be laid below, otherwise, they should be laid above. (×)

Analysis: The cable and other pipelines could not be laid on the same trench. The strong current and weak current should be separated. The cross lines and straight lines should be separated. The control lines and communication lines should also be separated. The minimum allowable clear distance between cables and cables or other facilities should comply with relevant regulations. It is prohibited to lay the cable directly above or below the pipelines in parallel.

466. When the safety of the staff is threatened by thunder, rain, strong wind or any other situation during the work, the person in charge of the work or the dedicated supervisor could temporarily stop the work according to the situation. (√)

467. For the motor with rotor windings, the external resistance is connected in series into the rotor circuit for starting, and the resistance value is gradually reduced and finally cut off as the motor speed increases, which is called rotor series resistance starting. (√)

468. Before using the megger, it is not necessary to check whether the megger is in good condition, and the insulation measurement could be directly carried out on the tested equipment. (×)

Analysis: Before using the megger, the quality of the megger must be checked, including short circuit check and open circuit check. The measurement could only be carried out after the megger is confirmed to be in good condition through open and short circuit tests.

469. The neutral wire behind the RCD (Residual Current Device) could be grounded. (×)

Analysis: The neutral wire behind the residual current device could not be grounded, because it would cause leakage, switch tripping and other consequences.

470. The combination switch could directly start the motor below 5kW. (√)

471. The equipment installed on the grounded metal frame generally does not need to be grounded. (√)

472. Low-voltage operation and maintenance refer to electrical operations such as installation, operation, maintenance and test on electrical equipment with a voltage of 220V or below to the ground. (×)

Analysis: In the power system, the distinction between high-voltage and low-voltage is based on 1000V (AC) to ground; Above 1000V is high-voltage, below 1000V is low voltage.

473. When handling the capacitor bank fault, although the capacitor bank is discharged automatically by the discharge device, supplementary manual discharge must be carried out to ensure safety. (√)

474. If protective measures are not taken for strong arc light generated during electric welding, operators are prone to cataract and other eye diseases. (√)

475. After being disconnected from the power supply, the person who gets an electric shock should be conscious and should be allowed to walk back and forth to strengthen blood circulation. (×)

Analysis: After being disconnected from the power supply, if the person is conscious, he/she should rest on the spot and to be observed closely. If necessary, he/she should go to the hospital for further examination and treatment.

476. Do not directly measure the internal resistance of the microammeter, galvanometer or battery with the ohm range of the multimeter. (√)

477. After the asynchronous motor is powered on, if the motor does not rotate and makes a "buzz" sound, the power supply should be disconnected immediately. (√)

478. When the safety extra low-voltage is used for direct electric shock protection, the safety voltage of 25V and below should be selected. (√)

479. During high-voltage operation, the distance between the person working without barriers or the tools carried by him/her and the electrified body should not be less than 0.7m. (√)

480. The rated current of the fusant should not be greater than the rated current of the circuit breaker. (√)

481. Employees who die of sudden illness during working hours and posts could be regarded as employment injuries. (√)

482. Isolation switches are used to connect and disconnect the current and isolate the circuit from the power supply. (×)

Analysis: The isolation switch could connect and disconnect the power supply and isolate the circuit from the power supply. The original mistake is the word "current", because the isolation switch has no arc extinguishing device and could not be used to connect and disconnect the current, that is to say, it could not be connected and disconnected with load, and the corresponding circuit breaker must be disconnected, so that the power could be connected and disconnected only after the load is disconnected.

483. After reaching the retirement age and going through the retirement procedures, the injured workers would enjoy basic endowment insurance and disability allowance. (×)

Analysis: According to the Regulations on Work-related Injury Insurance, after reaching the retirement age and going through the retirement procedures, the workers with work-related injuries from level one to level four should stop paying disability allowances and enjoy basic endowment insurance benefits.

484. Electric shock accidents generally occur in the process of operating and using electrical equipment, while electric shock accidents generally do not occur during construction, assembly, disassembly and maintenance. (×)

Analysis: Electric shock accidents might occur during the operation and use of electrical equipment, and in the process of construction, assembly, disassembly and maintenance. In the case of incomplete protection facilities, electric shock accidents are very likely to occur.

485. When putting out the fire of electrical equipment, the power supply must be cut off first. After cutting off the power supply, water could be used to put out the fire. (√)

486. When the motor makes abnormal noise and heats up, and the speed drops rapidly, the power supply should be cut off immediately and the machine should be stopped for inspection. (√)

487. When Y/Δ step-down starting is used, the starting torque is 1/3 of the starting torque when Δ-configuration is directly used. (√)

488. The plastic shell switch is not suitable for direct control of AC motors above 5.5kW. (√)

489. Article 27 of ***The Production Safety Law of the People's Republic of China*** stipulates that the special operation personnel of the production and business operation entities must receive special safety operation training in accordance with the relevant provisions of the state and obtain the corresponding qualifications before they could take up their jobs. (√)

490. The risk of two-phase electric shock is smaller than that of single-phase electric shock. (×)

Analysis: When the human body touches a two-phase charged body, it bears 380V wire voltage, while single-phase electric shock is 220V, so the risk is generally greater than single-phase electric shock. Two-phase electric shock could kill people only in 1~2s.

491. The same work card could be filled in and signed by the person in charge of the work. (×)

Analysis: The person in charge of the work could not issue a work card. If the issuer of the work card is also the person in charge of the work, the completed work card would not be reviewed and rechecked by the person in charge of the work.

492. TN-C-S system is a system fully shared by the protection zero wire and working zero wire of the trunk line. (×)

Analysis: TN-C-S system adopts TN-C grounding form in the first half of low-voltage power distribution system, and the protective zero wire and working zero wire of the trunk part are completely shared. Starting from the general distribution cabinet of the building power incoming line, the protective zero wire and working zero wire are completely separated and converted to TN-S system.

493. Circuit breakers could be divided into frame type and molded case type. (√)

494. Safety belt is a tool for preventing falling casualties for workers working at heights, which is composed of belts, ropes and metal accessories. (√)

495. If a person falls down or suffers from other trauma, severe pain, limited activity, deformity, or friction sound are typical signs of fracture. (√)

496. In order to prevent electric sparks, arcs and other explosives from igniting, explosion-proof electrical equipment with explosion-proof electrical grade and temperature group appropriate to the environment should be selected. (√)

497. When the electric welding machine needs to be moved, it could be moved directly without power cut-off. (×)

Analysis: When moving the electric welding machine, cut off the power supply, and do not move the welding machine by dragging the cable.

498. The absorption ratio is measured with a megger. (√)

499. The relationship between the period and the angular frequency of the sinusoidal alternating current is reciprocal. (×)

Analysis: The relationship between the period and the angular frequency of the sinusoidal alternating current is not reciprocal. The time required for alternating current to change for one cycle is the period T, and the angle that the alternating current wire coil passes within a certain time t is called the angular frequency ω. The relationship between periodic T and angular frequency of sinusoidal alternating current ω is: $\omega = 2\pi/T$.

500. The out-of-using residual current protection device should be tested before use. (√)

501. During the operation of communication lines, the electricity should be checked first; If electricity is found, pay attention to safety and continue to work. (×)

Analysis: The error of this statement is in the second half of the sentence. According to the national regulations on management of special operators, line operators are special operators. Before the communication line operation, observe the surrounding environment and verify whether there is a live conductor. If it is live, stop the operation immediately, find the contact point with the power line along the line, and work after treatment, while ensuring a safe distance.

502. Insulation aging is only a chemical change. (×)

Analysis: Insulation aging refers to a phenomenon that the insulation quality of electrical equipment decreases slowly and the structure is damaged gradually during operation due to the long-term effect of temperature, electric field, humidity, mechanical force and the surrounding environment. The change process of insulation aging is also accompanied by changes in shape and color, which include both chemical and physical changes.

503. In case of fire on the high-voltage line, insulation tools with corresponding insulation grade should be used to quickly to open the disconnecting switch and cut off the power supply, and carbon dioxide or dry powder extinguisher should be selected to extinguish the fire. (×)

Analysis: The insolation switch should not be opened first, and the carbon dioxide extinguisher should not be used. When a fire occurs to a high-voltage line, it is possible that the line has been short circuited, and the insolation switch does not have an arc extinguishing device, so it cannot be used to cut off the load current or short-circuit current. Otherwise, under the effect of high-voltage, the disconnection point would produce a strong arc, which is difficult to extinguish itself, and may even cause flashover (phase to phase or interphase short circuit), burn equipment, and endanger personal safety. This is a serious accident called "pulling the insolation switch with load". Carbon dioxide extinguisher is suitable for putting out the initial fire of books, archives, valuable equipment, precision instruments, electrical equipment below 600V and oil, and cannot be used to put out the fire on high-voltage lines.

504. The rotor and stator of three-phase motor could only work when they are energized at the same time. (×)

Analysis: If it is a three-phase synchronous motor, the stator is generally connected with three-phase AC power supply and the rotor is connected with DC excitation power supply. However, if it is a three-phase asynchronous motor, its rotor does not need to be energized, as long as the stator winding is energized, it could work.

505. The case of the iron case switch must be reliably grounded during installation. (√)

506. As long as the action current of the thermal relay is set smaller, it could play the role of short-circuit protection. (×)

Analysis: The thermal relay is generally used together with the contactor for overcurrent heating protection of the motor. The principle of the thermal relay is that the current flowing into the thermal element generates heat, which deforms the bimetallic sheets with different expansion coefficients. When the deformation reaches a certain distance, the connecting rod would be pushed to act, so that the control circuit would be disconnected, so that the contactor would lose power, the main circuit would be disconnected, and the overcurrent protection of the electric motor would be realized. Under normal conditions, the current setting value of the thermal relay is set as the rated current value of the motor, and the motor beyond the service life should be properly lowered.

507. When there is no electroscope (test pencil) of corresponding voltage level, low-voltage electroscope (test pencil) could be used for the electrical inspection of 10kV line. (×)

Analysis: Do not use low-voltage electroscope (test pencil) to test the 10kV line. Low-voltage electroscope is an ordinary low-voltage testing pen that could test the voltage of 60~500V. The principle of the high-voltage electroscope is that there is a step-down transformer inside, which changes the high-voltage into the induced low-voltage to make the indicator light light-up. When measuring low-voltage, because the voltage is low, the induced voltage of the high-voltage is not enough to make the indicator light-up, so the high-voltage electroscope could not be used to test low-voltage.

508. As long as a safe distance is maintained, it's not necessary to disconnect the power supply when measuring resistance. (×)

Analysis: If the resistance is charged, the ohmmeter would be damaged, and the voltage on the resistance would generate current in the resistance of the meter, damaging the measurement state; Furthermore, the resistance in the meter might change the state of the circuit and damage the circuit in serious cases.

509. Class II equipment itself does not need grounding or neutral connection. (√)

510. Wages should be paid monthly to the workers themselves in monetary form, and should not be withheld or delayed without cause. (√)

511. Common insulation safety protection includes insulating gloves, insulating boots, insulating partitions, insulating mats, insulating platforms, etc. (√)

512. Enclosed and explosion-proof lamps and switches should be used for lighting installation in special places such as flammable and explosive sites. (√)

513. When hanging the climbing pedal, the hook mouth should be outward and upward. (√)

514. In a three-phase AC circuit, when the load is configured in Δ, the phase voltage is equal to the line voltage of the three-phase power supply. (√)

515. During the use of hand electric tools and other electrical equipment, in case of temporary power failure, the power supply should be disconnected. (√)

516. When installing the fusant, bend the fusant counterclockwise and press it under the washer. (×)

Analysis: When installing the fuse, bend the fuse clockwise and press it under the washer. If the fuse is bent counterclockwise, the bending of the fusant would spread naturally during the screw tightening process.

517. In the safety color code, green indicates safety, passing, permission and work. (√)

518. After the mains power is cut off, the team leader could notify the power transmission by telephone. (×)

Analysis: The power transmission work should be carried out in strict accordance with the *Electric Power Safety Work Regulations*. For the relevant work that may endanger the safety of operators, the procedures for power cut-off and transmission must be strictly handled. It is strictly prohibited to cut-off or transmit power by telephone or at an appointed time.

519. In the safety operation regulations of electric power industry, safety organization measures include power cut-off, power inspection, installation of grounding wires, hanging of signboards, installation of barriers, etc. (×)

Analysis: Power cut-off, power inspection, installation of grounding wires, hanging of signboards and installation of barriers are technical measures to ensure safety. Organizational measures to ensure safety generally include work card system, work permit system, work monitoring system, work interruption, transfer and termination, and power transmission restoration system.

520. The capacitor room should be well-ventilated. (√)

521. If the motor stops running due to smell of burning smell, the reason must be found out before it could be powered on for use. (√)

522. Explosive gas, steam and mist belong to Class II explosive hazardous substances. (√)

523. The downlead of lightning protection device should meet the requirements of sufficient mechanical strength, corrosion resistance and thermal stability. If steel strand is used, its sectional area should not be less than 35mm^2. (×)

Analysis: Galvanized round steel or galvanized flat steel are generally selected as downlead of lightning protection. The diameter of round steel is not less than 8mm, the sectional area of flat steel is not less than 48mm^2, and the thickness is not less than 4mm. Steel strand is usually used as overhead ground wire.

524. Defects of electrical equipment, unreasonable design and improper installation are all important causes of fire. (√)

525. Speed relay is mainly used for reverse connection braking of motor, so it is also called reverse connection braking relay. (√)

526. The voltage is measured with a voltmeter, which is connected in series in the circuit. (×)

Analysis: The voltmeter must be connected in parallel with the circuit under test. For a voltmeter, the greater the internal resistance, the better. Because such a voltmeter could measure the line voltage with less influence on the original circuit, the internal resistance of an ideal voltmeter is infinite. With the voltmeter connected in parallel at both ends of the load, the actual working voltage at both ends of the load is measured according to the characteristics of equal voltage of the parallel circuit. If the voltmeter is connected in series in the circuit for measurement, due to the large internal resistance and the small current passing through the voltmeter and the load, the load could not work. The measured voltage is almost the open circuit voltage of the original load circuit, not the working voltage of the load circuit.

References

[1] Zhonganhuabang (Beijing) Safety Production Technology Research Institute. Training and assessment materials for low-voltage electrician operation: Question bank docking version. Beijing: Unity Press, 2013.

[2] Qingde Yang, Zurong Yang. Low voltage maintenance electrician. Beijing: Publishing House of Electronic Industry, 2012.

[3] Qingde Yang, Jian Chen. Fundamentals of Electrical Engineering: Micro-course edition. Beijing: Chemical Industry Press, 2019.

[4] Qingde Yang, Shijin Lu, Zhengzhao Zhao. Fundamentals and Skills of Electrical Technics. Chongqing: Chongqing University Press, 2018.

[5] Qingde Yang, Yongping Zhou, Ping Hu. Question Bank of Electrical Technology and Skills Basic. Beijing: Publishing House of Electronic Industry, 2016.